Oxy-fuel Combustion

Oxy-fuel Combustion

Fundamentals, Theory and Practice

Edited by

Chuguang Zheng
Zhaohui Liu

ELSEVIER

ACADEMIC PRESS
An imprint of Elsevier

Academic Press is an imprint of Elsevier
125 London Wall, London EC2Y 5AS, United Kingdom
525 B Street, Suite 1800, San Diego, CA 92101-4495, United States
50 Hampshire Street, 5th Floor, Cambridge, MA 02139, United States
The Boulevard, Langford Lane, Kidlington, Oxford OX5 1GB, United Kingdom

Library of Congress Cataloging-in-Publication Data
A catalog record for this book is available from the Library of Congress

British Library Cataloguing-in-Publication Data
A catalogue record for this book is available from the British Library

ISBN 978-0-12-812145-0

Publisher: Glyn Jones
Acquisition Editor: Maria Convey
Editorial Project Manager: Katie Chan
Production Project Manager: Omer Mukthar
Cover Designer: Victoria Pearson

Typeset by SPi Global, India

Contents

6. Mineral Behavior During Oxy-fuel Combustion

Liqi Zhang, Yongchun Zhao, Hong Yao

7. Mercury Behavior and Retention in Oxy-fuel Combustion

Jianping Yang, Yongchun Zhao, Junying Zhang, Chuguang Zheng

8. Flame Characteristics of Oxy-fuel Combustion and Burner Design

Junjun Guo, Zhaohui Liu

14. MILD Oxy-fuel Combustion

Pengfei Li, Fan Hu, Biao Xiong, Jianpeng Zhang, Jianchun Mi

15. Oxy-Steam Combustion

Lei Cai, Chun Zou

Contributors

Numbers in parentheses indicate the pages on which the authors' contributions begin.

Lei Cai (31, 325), Huazhong University of Science and Technology, Wuhan, China

Sheng Chen (13), Huazhong University of Science and Technology, Wuhan, China

Juan Chen (87), Shandong University, Jinan, China

Junjun Guo (1, 171, 189), Huazhong University of Science and Technology, Wuhan, China

Xin Guo (87, 339), Huazhong University of Science and Technology, Wuhan, China

Fan Hu (289), Huazhong University of Science and Technology, Wuhan, China

Xiaohong Huang (1), Huazhong University of Science and Technology, Wuhan, China

Bo Jin (223, 239), Hunan University, Hunan, China

Pengfei Li (289), Huazhong University of Science and Technology, Wuhan, China

Zhaohui Liu (1, 171, 189, 209, 223, 239), Huazhong University of Science and Technology, Wuhan, China

Jing Liu (87, 263), Huazhong University of Science and Technology, Wuhan, China

Hao Liu (87), Soochow University, Jiangsu, China

Wei Luo (239), Wuhan Second Ship Design and Research Institute, Hubei, China

Cong Luo (263), Huazhong University of Science and Technology, Wuhan, China

Jianchun Mi (289), Peking University, Beijing, China

Baowen Wang (339), North China University of Water Resources and Electric Power, Zhengzhou, China

Biao Xiong (289), Huazhong University of Science and Technology, Wuhan, China

Jianping Yang (151), Huazhong University of Science and Technology, Wuhan, China

Hong Yao (87, 129), Huazhong University of Science and Technology, Wuhan, China

Baojun Yi (63), Huazhong Agricultural University, Wuhan, China

Liqi Zhang (63, 129), Huazhong University of Science and Technology, Wuhan, China

Junying Zhang (151), Huazhong University of Science and Technology, Wuhan, China

Tai Zhang (209), Huazhong University of Science and Technology, Wuhan, China

Jianpeng Zhang (289), Peking University, Beijing, China

Yongchun Zhao (129, 151), Huazhong University of Science and Technology, Wuhan, China
Haibo Zhao (223, 239), Huazhong University of Science and Technology, Wuhan, China
Chuguang Zheng (1, 151), Huazhong University of Science and Technology, Wuhan, China
Chun Zou (31, 325), Huazhong University of Science and Technology, Wuhan, China

Preface

We are facing unprecedented ecological environment deterioration and a historic global climate crisis.

In 1992 the United Nations Conference on Environment & Development passed a resolution in which the conservation of the global climate was requested as part of mankind's common heritage. Environmental protection can be achieved by the prevention of climate change, by protecting the ozone layer, and by seeking ways to alleviate long-range transport of air pollution.

For centuries we have consumed the Earth's carbon in the form of coal and petroleum that had been stored underground for billion years. Today carbon is emitted into the atmosphere at a high rate, much higher than the carbon released in past volcanos and forest fires. People are also generating new greenhouse gases with unprecedented speed. Thus mankind is damaging the Earth's comfortable, yet fragile equilibrium process established over millions of years.

Many approaches and technologies are attempting to address the issues related to climate change, and carbon capture and storage (CCS) is believed to be one of the most promising technologies. The success of the Paris Agreement, made within the United Nations Framework Convention on Climate Change, could represent a major turning point for CCS. CCS is the potential "sleeping giant" [1] that needs to be awakened to respond to the heightened ambitions of the Paris Agreement.

The International Energy Agency (IEA)'s scenario analysis has consistently highlighted that CCS will be important in limiting future temperature increases to just 2°C, and we anticipate that this role for CCS will become increasingly significant if we are to move toward "well below 2°C." Why is this? Because there is no other technology that can significantly reduce emissions from the generation of coal and gas in power plants, which will remain part of the electricity mix for the foreseeable future. No other technology is capable of delivering the deep reduction in emissions needed across key industrial processes such as steel, cement, and chemicals manufacturing, all of which will remain vital building blocks of modern society. In short, deployment of CCS will be seen as more than optional in implementing the Paris Agreement, it will be seen in IEA's report [1].

Preliminary studies and engineering demonstrations of CCUS technology have been carried out by the government, universities, research institutes, and enterprises in China as well as through recent international cooperation.

Technologies and projects based on post-combustion, oxy-fuel combustion, and pre-combustion CO_2 capture are being developed and constructed. Oxy-fuel technology is considered to be the best potential CCUS technology for large-scale promotion and commercialization [2].

The State Key Laboratory of Coal Combustion at Huazhong University of Science and Technology was first to do oxy-fuel combustion research in China. The laboratory's oxy-fuel combustion research is systematically carried out, and the R&D process not only symbolizes the development of this field in China but also formulates unique R&D features. Based on the accumulation of more than 20 years' of work, this book summarizes the progress and achievements in the study of the fundamental theory, the key R&D technologies, and the industrial demonstration and scaling up of oxy-fuel combustion system.

After 30 years of development, oxy-fuel technology has matured and possesses the fundamental characteristics necessary for commercial application. Importantly, it is suitable for existing coal-fired power plants. In order for China's coal power-dominated energy mix to meet the reduction targets for greenhouse gas emissions and to create the greatest likelihood for the commercialization of oxy-fuel, large-scale demonstrations must be launched as soon as possible.

REFERENCES

[1] IEA. 20 years of carbon capture and storage. 2016. November. International Energy Agency, France.

[2] Toftegaard MB, Brix J, Jensen PA, Glarborg P, Jensen AD. Oxy-fuel combustion of solid fuels. Prog Energy Combust Sci 2010;36:581–625.

Chapter 1

Opportunities and Challenges of Oxy-fuel Combustion

Xiaohong Huang, Junjun Guo, Zhaohui Liu, Chuguang Zheng
Huazhong University of Science and Technology, Wuhan, China

1.1 CLIMATE CHANGE AND CARBON CAPTURE STORAGE

1.1.1 Carbon Emission and Climate Change

Greenhouse gases actually guarantee life by preventing heat from escaping the Earths atmosphere. Without this greenhouse effect, the Earth would be very cold. On the other hand, too big an increase in the greenhouse effect causes too much heat to be captured and may result in the Earth being unsuitable for human, plant, and animal life.

CO_2 is the most significant component of greenhouse gases. In addition to the carbon released naturally into the Earths ecosystem, humans emit a significant amount of carbon through activities such as burning fossil fuels, wood, and other biofuels. Moreover, human activities release carbon more effectively and rapidly into the atmosphere than natural sources do, thus increasing the atmospheric CO_2 concentration. In fact, the atmospheric CO_2 concentration is higher now than at any time during the past 500,000 years or longer.

Fig. 1.1 shows the atmospheric CO_2 levels at 400,000 years [1]. This data was based on the comparison of atmospheric samples contained in ice cores from Greenland and Antarctica and more recent direct measurements. It is significant that the atmospheric CO_2 level fluctuated regularly over the past 400,000 years but that the amount of CO_2 dramatically increased to >300 ppm in just the past few decades. This figure provides evidence-based proof that the Industrial Revolution led to a dramatic increase in atmospheric CO_2 levels.

Fig. 1.2 presents the global CO_2 emissions in the past 250 years [2,3] and shows that global CO_2 emissions significantly increased during the past 100 years. The increasing trend almost looks like an exponential curve. To facilitate assessments of long-term trends of atmospheric global temperature, climatologists from NASA's Goddard Institute of Space Studies (GISS) compared the mean for a base period with the annual mean [4]. The temperature anomaly

Oxy-fuel Combustion. http://dx.doi.org/10.1016/B978-0-12-812145-0.00001-3

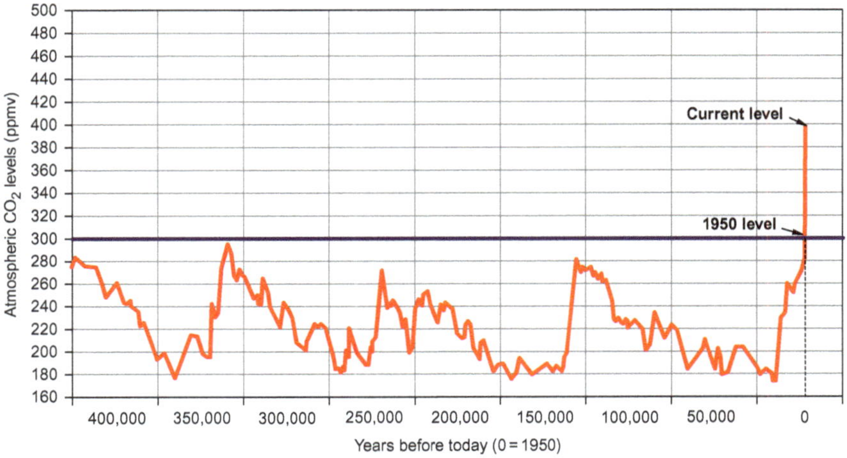

FIG. 1.1 The atmospheric CO_2 levels for 400,000 years.

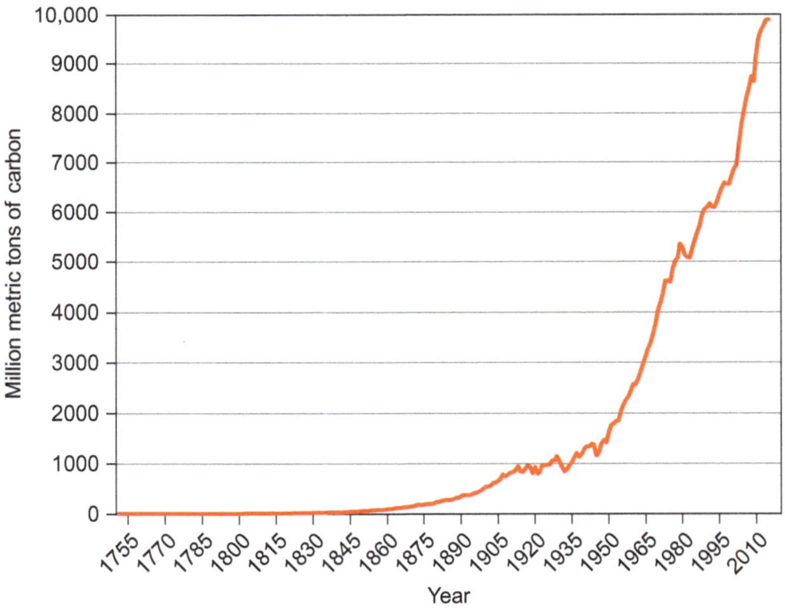

FIG. 1.2 The global carbon emissions in the 1751 to 2015 period.

was used to indicate the change in global temperatures, which is defined as the difference between the baseline mean and the annual mean.

Fig. 1.3 shows the global temperature anomalies between 1880 and 2015 from the Land Meteorological station [5]. In this figure, the period from 1951 to 1980 is used for the baseline period. It is shown that the temperature anomalies were consistently negative during the 1880–1935 period. In contrast, the

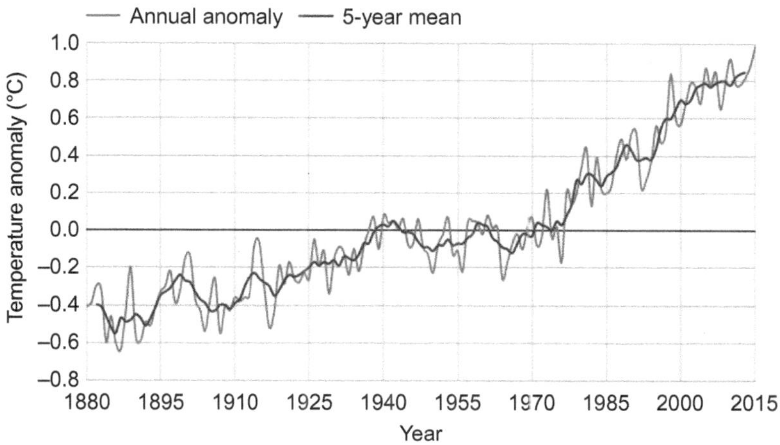

FIG. 1.3 The global annual temperature anomalies in the 1880–2015 period.

anomalies have been consistently positive since 1980. The global temperature has increased continuously. Within the past 120 years, the most recent years show the highest anomalies of +0.8°C.

The climate of the Earth has varied throughout its history, sometimes considerably. Although the warming before 1950 was natural, it has shown that the processes of human-inspired industrialization are the main reasons for recent significant climate warming.

1.1.2 Status of CCS

Like efficiency improvements and low-carbon fuels, carbon capture and storage (CCS) is considered as a potential technology for achieving large-scale reduction of CO_2 emissions from fossil fuel-fired power generation. CCS comprises three stages: CO_2 capture, CO_2 transport, and CO_2 storage.

The aim of CO_2 capture technology is to provide high-concentration, high-pressure CO_2 sources from large point sources such as power plants. Captured CO_2 must be compressed and transported to suitable storage sites. Using pipelines to pump CO_2 is the cheapest transport method, and it is already a well-known and reliable technology. The purpose of CO_2 storage is to store the captured CO_2 and prevent it from entering the atmosphere. There are three options for geological CO_2 storage: saline formations, oil and gas reservoirs, and deep unmineable coal seams [6].

The technologies for each stage of the full CCS chain have different degrees of maturity, each CCS stage is technically available and has been used for many years.

CO_2 capture technologies have long been used in industry to remove CO_2 from gas streams where its not desired or to separate it as a product gas. Some of the earliest CCS projects were in natural gas processing, such as the Sleipner CO_2 storage project in Norway, which captures approximately one million tons of CO_2 per year.

CO_2 transport technology has been used safely for the past 30 years. Thousands of miles of CO_2 pipelines that transport CO_2 deal with naturally occurring CO_2 for enhanced oil recovery (EOR) in the United States. However, challenges exist in the systems that transport CO_2 from industrial power plants, because the amount of impurities in the CO_2 stream captured from the plants will influence the calculations for the hydraulic design of these pipelines.

Geologic CO_2 storage has been carried out for >10 years. Early projects injected CO_2 into different geologic formations, including saline formations and oil and gas reservoirs. Monitoring data has shown that the CO_2 performed as expected after storage. However, more investigation is needed to improve the predictions of CO_2 behavior and to evaluate possible storage sites, because each site will have unique circumstances. The deep saline formations are expected to provide long-term CO_2 storage solutions.

Currently, there are five commercial-scale CCS projects worldwide. The CO_2 streams of the Sleipner, Snøhvit (Norway), and In Salah (Algeria) projects all come from extracted natural gas, which contains a high concentration of CO_2. The CO_2 is separated, captured, and stored in underground geological formations. The Rangely project in North America uses CO_2 from natural gas processing for EOR and storage at the Rangely field in Colorado. At Weyburn-Midale, the CO_2 is captured from a coal-based synfuel plant in North Dakota and piped to an oil field in Canada. In total, five projects store over 5 Mt. of CO_2 per year.

In addition to using CO_2 for EOR, other CCS paths are not financially feasible. CCS reduces efficiency and adds cost. Thus commercial power plants and industrial facilities will not invest in CCS in the current regulatory and fiscal environment. Only when the costs are reduced will CCS be seen as viable in coal combustion [7]. Meanwhile, the driving force behind CCS research and development is mainly government funding.

Validation of the full CCS process is very expensive. However, the only way to solve the problem of integration and scale-up is to build and operate commercial-scale CCS facilities. In the past year, governmental and demonstration activities increased dramatically. Most of the major economies have already invested in large-scale CCS demonstration projects. For example:

- Australia committed AUD 100 million per year for 3 years for the formation of the Global CCS Institute (GCCSI) to foster international collaboration and announced AUD 2 billion for large-scale demonstration.
- Canada announced financial support of CAD 3.3 billion for research and development, mapping, and large-scale CCS project demonstration.
- The European Union (EU) allocated EUR 1.05 billion from an economic recovery energy program for the support of seven CCS demonstrations.
- Norway continues to play a leading role in the development of the Mongstad and Karstø projects.
- The United Arab Emirates is developing three large-scale CCS projects, building on the EOR area.

- The United Kingdom announced funding for up to four CCS projects.
- The United States announced $3.4 billion in new funding for CCS projects.

1.1.3 CO_2 Capture Technologies

It is most effective to capture CO_2 at emission sources, such as large fossil fuel or biomass energy facilities, major CO_2 emissions industries, and so on. The CO_2 concentration in flue gas from coal combustion facilities is approximately 10%–15%, whereas it is approximately 5%–10% in flue gas from natural gas power plants. In other systems, such as coal gasification, the CO_2 concentration can reach approximately 65%. In order to obtain pure CO_2, the CO_2 must be separated from flue gas. CO_2 separation methods are classified as the following three types:

- Chemical absorption method, such as chemical solvents, including amine-based solutions (e.g., MEA and MDEA) and hot potassium carbonate-based processes (e.g., the Benfield process)
- Physical absorption method, such as pressure swing adsorption (PSA) and physical sorbent (e.g., SelexolTM, Rectisol)
- Membrane separation method
- Cryogenic separation processes

However, CO_2 separation consumes too much energy, which makes CCS technology meaningless as a method against global warming. Therefore different technologies with different advantages are competing to become the low-cost solution for the CO_2 capture process. There are three different technologies for CO_2 capture from coal combustion and gasification, including:

- Post-combustion capture: The CO_2 is captured after burning the fossil fuel. This method is applied to most conventional power plants. CO_2 is scrubbed from the flue gas by chemical solvents, solid minerals, and so on.
- Pre-combustion capture: The CO_2 is removed before combustion occurs. This method is used in the production of fertilizer, chemical gaseous fuel, and power. The fossil fuel must first be gasified to form synthetic gas or syngas. Then CO_2 can be separated from this relatively pure exhausted steam. Integrated gasification combined cycle (IGCC) is a typical application of the pre-combustion capture method.
- Oxy-fuel combustion: The pure CO_2 steam is obtained by removing the impurities, such as water vapor. This method can also be applied to most conventional power plants by adding an air separation unit (ASU) and a flue gas recirculation facility.

Impurities in CO_2 streams, such as sulfurs and water, have a significant impact on their phase behavior. There is a significant risk of increased corrosion in pipelines during transport and injection. Generally, the flue gas cleaning device is used initially to remove the impurities in flue gas.

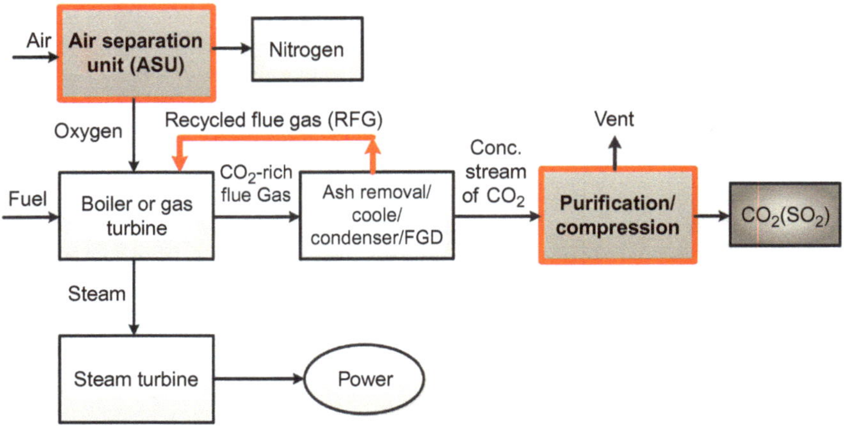

FIG. 1.4 Flowsheet of oxy-fuel combustion technology for power generation with CO_2 capture and storage.

1.1.4 The Concept and Components of Oxy-fuel Combustion Technology

Oxy-fuel combustion technology is one of the most promising technologies being considered for capturing CO_2 from power plants. Fig. 1.4 shows the flowsheet of oxy-fuel combustion technology for power generation with CO_2 capture [8]. The pure oxygen (purity >95%), which comes from the large-scale ASU, is used to replace the air as oxidant. To adjust the temperature and heat transfer characteristics in the furnace, a large portion of the flue gas is recycled into the boiler/gas turbine. As a result, the flue gas consists of a high concentration of CO_2 and water vapor. The fly ash and water in the flue gas can be removed in an ash remover and a condenser, respectively. Finally, the concentrated CO_2 stream is purified and compressed by a CO_2 processing unit (CPU) for transport and storage.

Generally, to capture CO_2 from a power plant successfully, oxy-fuel combustion for power generation plants incorporates the following four major units:

- ASU to provide pure oxygen for combustion
- Boiler or gas turbine to burn the fuel with oxygen and to transfer the combustion heat from the flue gas to the heat exchanger
- Flue gas processing unit to remove the impurities, such as fly ash, SO_x, and H_2O for flue gas cleaning
- CPU to purify and compress the CO_2 stream for transport and storage

1.2 A BRIEF HISTORY OF OXY-FUEL COMBUSTION

1.2.1 R&D History of Oxy-fuel Combustion

The concept of oxy-fuel combustion was first proposed in 1982 by Abraham and coworkers [9] as a way to provide large quantities of CO_2 for EOR. This concept was first tested in a 3 MW pilot-scale facility at the University of North Dakota [10]. Because of the promising prospects for application, the Argonne National Laboratory (ANL) performed research, including techno-economic assessments and pilot-scale studies [11–13].

Oxy-fuel combustion technology has been applied to industrial furnaces for many years [14]. The main purposes of this technology in these industries were to improve productivity, decrease energy consumption, and reduce NO_x emission [15]. However, these industrial applications are much smaller in scale than power generation applications, and also the recycled flue gas (RFG) can't be used. During the early 1990s, the International Flame Research Foundation (IFRF) carried out the first oxy-coal combustion experiment with RFG in a pilot-scale facility in Europe. New Energy and Industrial Technology Development Organization (NEDO) in Japan also conducted research on oxy-coal combustion with RFG with a focus on retrofitting Japanese boilers with this technology. IHI completed the combustion trials [16–19]. Since the late 1990s, because of concerns about greenhouse gas emissions and climate change, interest in oxy-fuel combustion technology with RFG has increased. Babcock and Wilcox (B&W) and Air Liquide led a research alliance to conduct further pilot-scale studies [20]. Several successful projects have been carried out over the past 10 years, such as Vattenfall's Schwarze Pumpe pilot plant (30 MW) [21]. Total's Lacq pilot Project (30 MW) [22], CIUDEN's Technology Development Platform (30 MW oxy-PC and 20 MW oxy-CFB) [23], and HUST's Yingcheng 35 MW pilot plant [24]. Based on these developments, oxy-fuel combustion technology has developed from pilot-scale studies to industrial-scale uses.

The International Energy Agency (IEA) presented the technology roadmap for oxy-fuel combustion in coal-fired power plants using CO_2 capture [25], which clearly indicated that this technology can be commercialized by 2020 and that a full-scale oxy-fuel combustion power plant could be implemented between 2016 and 2020.

1.2.2 Opportunities With Oxy-fuel Combustion

Oxy-fuel combustion plants have significant advantages over conventional air-fired plants. Among these advantages are

- Reduced boiler heat losses

Many theoretical and experimental studies related to oxy-fuel combustion have focused on retrofitting existing PC boilers [26–28]. The geometry of a boiler is determined by the air-fired case, and combustion conditions in oxy-fuel combustion should be similar to those in the air-fired case. Typically, around two-thirds of the flue gas is recycled into the boiler/furnace in oxy-fuel combustion.

In a traditional air-fired boiler/furnace, large amounts of N_2 in the air are heated during the combustion process and then cooled down again as the exhaust gas is released. The temperature of the exhaust flue gas is above the ambient. In a traditional air-fired boiler, the heat loss with the flue gas is up to 10%.

In the oxy-fuel combustion boiler/furnace, bulk nitrogen does not exist in the flue gas, which in turn means that the mass and volume of the flue gas are significantly reduced. Therefore the heat loss with the flue gas can be significantly reduced and the heat efficiency of the boiler/furnace increased.

- Compact boiler design

For retrofitting existing PC boilers, changes in the burner and flue gas system are important. With the deepening understanding of oxy-coal combustion and improvements in simulation tools, it is possible to improve the boiler design for new construction of oxy-fuel combustion boilers. In order to maintain the flame temperature at an acceptable range, the flue gas is circulated internally in the boiler system. The volume of external RFG is reduced. So the boiler geometry and heat dissipation and the electric power requirement for the flue gas recirculating fan are also reduced. Furthermore, the boiler investment cost is significantly reduced because of the reduced size of the boiler.

Unlike a traditional air-fired boiler, an oxy-coal combustion boiler uses pure oxygen as the oxidizer. It is possible to control and optimize the combustion process of pulverized coal by adjusting the concentration and injection position of oxygen in the furnace, which is not possible in air-fired boilers. So the design for oxy-fuel combustion boilers is more flexible than for traditional air-fired boilers, which helps to control combustion conditions, temperature distribution, and emission formation.

- Near zero emission

In oxy-fuel combustion, nitrogen oxide production reduces because of the NO_x reburning chemistry and the absence of nitrogen from oxidants, which decreases the pollutants in the flue gas. This means that the separation is easier and the volume of equipment for flue-gas desulphurization (FGD) and NO_x removal decreases. Therefore the equipment is less expensive than it is for air-fired power plants. For CO_2 sequestration, water and noncondensable gases are removed. During the water condensation process, acid water-soluble pollutants and particles are dissolved and removed. Commercial methods can be used to clean the condensed water. Most of the flue gases are condensable, which makes it possible to compress and separate them.

In summary, with careful design, near zero emission is likely to be achieved in oxy-fuel combustion power plants, not only for CO_2 but also for other pollutants.

1.3 CHALLENGES FOR OXY-FUEL COMBUSTION

After several decades of research and development, the components of oxy-fuel combustion technology are mature. However, the combined operation still needs to be demonstrated in practice. Technical and economic issues are the major challenges for oxy-fuel combustion technology. The technical challenges exist in the boiler design and system operations. The economic challenge is the high-energy cost, in terms of O_2 production and CO_2 separation.

1.3.1 Boiler Design

As mentioned in Section 1.2.2, oxy-fuel boilers have advantages in boiler efficiency improvements and cost reductions. In order to take full advantage of these benefits, a number of challenges associated with oxy-fuel boiler design must be addressed, such as fundamental theory, design, and materials.

Seeking to understand the combustion characteristics and develop the CFD models, many studies have investigated the combustion characteristics of pulverized coal in oxy-fuel conditions [21,29–31]. For both new and retrofitted boilers, more experimental and numerical studies need to be carried out to investigate possible scale-up and optimization of the operating conditions of PC boilers. Unlike N_2, CO_2 and H_2O will play a part in thermal radiation and reaction with char. Thus the coal combustion process and heat transfer characteristics will change in oxy-fuel combustion. Therefore, for the design of boilers, the flame properties under oxy-fuel conditions should be determined, as well as the combustion process, heat transfer, gas phase kinetics, conversion of sulfur and nitrogen, ash behavior, slagging and fouling, and composition of deposits.

Because of the change of flame and ash properties, traditional boiler design theory is not suitable for oxy-fuel combustion boilers. Reducing the external recirculation ratio to reduce the boiler size and increasing thermal efficiency are challenges in boiler design. Another challenge is determining the flue gas recycled point in the flue gas stream. Many studies assumed that the RFG should be extracted before or after the flue gas condenser. Similarly, a strategy for injecting the oxygen into the boiler is also a challenge. Furthermore, the air leakage into the boiler influences CO_2 concentration. Most researchers suggest that the boiler should work with slight positive pressure to minimize air leakage.

The risk of corrosion will increase because of an increase in fouling and the increase of SO_3 in the deposits. Experimental studies in an existing boiler before and after retrofitting to oxy-fuel combustion will contribute to assessing the risk of corrosion. With deeper research on corrosion behavior, the selection of boiler materials can be more accurate.

1.3.2 Oxygen Production

At present, the only choice for air separation in a large-scale facility is the cryogenic air-separation unit (Cryo-ASU). Therefore most pilot- or large-scale oxy-fuel projects employ Cryo-ASUs to produce oxygen. The electric power consumption of a Cryo-ASU may reach 20% of the gross power output for oxy-fuel combustion power plants, which greatly reduces the plants efficiency.

In many studies, the electric power from the grid or internal electric power is used for the compressors motor. However, motor drives for commercial plants must be very large, and the motors are likely driven by steam turbines, which offers a possible area of study on reducing the cost of oxygen production. More efficient heat integration between the power plant and the Cryo-ASU is necessary. Furthermore, there are three non-cryogenic ways to separate oxygen from air: (1) membrane separation, (2) Ceramic Auto-Thermal Recovery, and (3) chemical looping. These new technologies will reduce the cost of oxy-fuel combustion and also benefit other CCS technologies.

1.3.3 CO_2 Purity Requirements and Flue Gas Cleaning

According to the targets for the CO_2 utilization, the requirements of the CO_2 purity may be different. The US Department of Transportation requires a minimum purity of at least 90% for pipeline CO_2 [32]. Moreover, a higher purity standard is required by most CO_2 pipeline operators, usually a minimum of 95% for EOR applications [33]. Therefore the flue gas needs to be compressed and purified at multiple stages in CO_2 capture. This system is used to remove water and other pollutants and to produce high purity CO_2 for transport and storage. The purification process will consume a lot of energy, which also results in a significant reduction in plant efficiency.

1.3.4 Process Integration

All of the previous discussions focus on technical challenges. In general, for power plants, CO_2 capture will reduce the systems efficiency. It is necessary to optimize the energy efficiency to minimize the negative impact of CO_2 capture. The oxy-fuel technology combines the O_2 production, flue gas cleaning, and flue gas recirculation with boiler design and steam recirculation system. There is a significant amount of low-temperature heat that could be used to reduce the heat lost to the environment and thus increase a plants efficiency. In the case of using a Cryo-ASU to produce oxygen, utilizing cold N_2 could also increase the systems efficiency. Therefore additional system-design targets are system integration and cascade utilization of energy.

1.4 ABOUT THIS BOOK

This book is aimed at providing an up-to-date and comprehensive review on various aspects of oxy-fuel combustion technology, including its concept and fundamental theory, pilot practices, a feasibility study of large-scale uses, and related practical issues, such as commissioning and operation of oxy-fuel combustion plants. This book consists of three sessions. Session I discusses the fundamentals and theories of oxy-fuel combustion. Session II focuses on the practice issues of oxy-fuel combustion, and Session III reviews the new concept of oxy-fuel combustion.

In Session I, ignition and combustion characteristics of pulverized coal in oxy-fuel conditions are discussed. This session also presents the conversion behavior of pollutants, such as nitrogen, sulfur, mercury, and other minerals.

Session II discusses the practice issues related to oxy-fuel combustion. Flame and heat characteristics of oxy-fuel combustion are mentioned first because they are very important for burner and boiler designs. Then the experiences of pilot and industrial demonstrations of oxy-fuel combustion are introduced. System integration and optimization for large-scale oxy-fuel combustion systems are discussed in detail. The last chapter of this session presents the control concepts, dynamic behavior, and mode transition strategy for oxy-fuel combustion systems.

Session III introduces the new concept of oxy-fuel combustion, which consists of new oxygen production technologies for oxy-fuel combustion, MILD oxy-fuel combustion, oxy-steam combustion, and chemical looping combustion.

REFERENCES

[1] NASA. Climate change: how do we know? Available from: http://climate.nasa.gov/evidence/.

[2] Varagani R, Chatel-Pélage F, Pranda P, et al. Oxycombustion in pulverized coal-fired boiler: a promising technology for CO_2 capture. In: Proceedings of third annual conference on carbon sequestration. Alexandria, VA, USA; 2004.

[3] Olivier J, Janssens-Maenhout G, Muntean M, et al. Trends in global CO_2 emissions: 2015 report. Maenhout 2015.

[4] NASA. GISS surface temperature analysis. Available from: http://data.giss.nasa.gov/gistemp/graphs_v3/.

[5] Lüthi D. EPICA dome C ice core 800KYr carbon dioxide data. In: IGBP pages/world data center for paleoclimatology data contribution series # 2008-055. Boulder, CO, USA: NOAA/NCDC Paleoclimatology Program; 2008.

[6] International Energy Agency. Technology roadmap: carbon capture and storage. Paris: IEA, http://www.iea.org/publications/freepublications/publication/TechnologyRoadmapCarbonCaptureandStorage.pdf; 2013.

[7] Katzer J, Moniz E, Deutch J, et al. The future of coal: an interdisciplinary MIT study. In: Technical Report. Cambridge, MA: Massachusetts Institute of Technology; 2007.

[8] Wall T, Liu Y, Spero C, et al. An overview on oxy-fuel coal combustion—state of the art research and technology development. Chem Eng Res Des 2009;87(8):1003–16.

[9] Abraham B, Asbury J, Lynch E. Coal-oxygen process provides carbon dioxide for enhanced recovery. Oil Gas J 1982;80(11):68–70.

[10] Wang C, Berry G, Chang K, et al. Combustion of pulverized coal using waste carbon dioxide and oxygen. Combust Flame 1988;72(3):301–10.

[11] Weller A, Rising B, Boiarski A, et al. Experimental evaluation of firing pulverized coal in a CO_2/O_2 atmosphere. IL, USA: Argonne National Laboratory; 1985.

[12] Berry G, Wolsky A. Modeling heat transfer in an experimental coal-fired furnace when CO_2/O_2 mixture replace air. In: ASME 1986 Winter Annual Meeting. Paper 86-WA/HT51; 1986.

[13] Abele A, Kindt G, Clark W, et al. An experimental program to test the feasibility of obtaining normal performance from combustion using oxygen and recycled gas instead of air. IL, USA: Argonne National Laboratory; 1987.

[14] Dugué J, The use of oxygen for industrial combustion. In: Summary of the 17th IFRF topic oriented technical meetings. International Flame Research Foundation Document No. D121/y/6; 2000.

[15] Charon O. Recent developments and future trends in oxy-combustion applications, In International flame research foundation 17th topic oriented technical meeting, Les Vaux de Cernay, France; 2000.

[16] Kiga T, Takano S, Kimura N, et al. Characteristics of pulverized-coal combustion in the system of oxygen/recycled flue gas combustion. Energy Convers Manag 1997;38(96):S129–34.

[17] Kimura N, Omata K, Kiga T, et al. The characteristics of pulverized coal combustion in O_2/CO_2 mixtures for CO_2 recovery. Energy Convers Manag 1995;36(6):805–8.

[18] Nozaki T, Takano S, Kiga T, et al. Analysis of the flame formed during oxidation of pulverized coal by an O_2-CO_2 mixture. Energy 1997;22(2–3):199–205.

[19] Nakayama Y, Kiga T, Miyamae S, et al. Pulverized coal combustion in O_2/CO_2 mixtures on a power plant for CO_2 recovery. Energy Convers Manag 1992;33(5–8):379–86.

[20] Stanger R, Wall T, Spörl R, et al. Oxy-fuel combustion for CO_2 capture in power plants. Int J Greenhouse Gas Control 2015;40:55–125.

[21] Anheden M, Burchhardt U, Ecke H, Faber R, Jidinger O, Giering R, et al. Overview of operational experience and results from test activities in Vattenfall's 30 MWth oxyfuel pilot plant in Schwarze Pumpe. Energy Process 2011;4:941–50.

[22] Fujimoria T, Yamadab T. Realization of oxyfuel combustion for near zero emission power generation. Proc Combust Inst 2013;34:2111–30.

[23] Lupion M, Alvarez I, Otero P, Kuivalainen R, Lantto J, Hotta A, et al. 30 MW_{th} CIUDEN oxy-CFB boiler—first experiences. Energy Process 2013;37:6179–88.

[24] Guo J, Liu Z, Huang X, Zhang T, Luo W, Hu F, et al. Experimental and numerical investigations on oxy-coal combustion in a 35 MW large pilot boiler. Fuel 2017;187(1):315–27.

[25] Santos S. Oxy-coal combustion power plant with CCS current status of development. In: Proceedings of 39th international technical conference on clean coal & fuel systems, Clearwater, FL, USA; 2014.

[26] Singh D, Croiset E, Douglas P, et al. Techno-economic study of CO_2 capture from an existing coal-fired power plant: MEA scrubbing vs. O_2/CO_2 recycle combustion. Energy Convers Manag 2003;44(19):3073–91.

[27] Molburg J, Doctor R, Brockmeier N, et al. CO_2 capture from PC boilers with O_2-firing. In: 18th annual international pittsburgh coal conference. Newcastle, NSW, Australia; 2001.

[28] Fry A, Adams B, Paschedag A, et al. Principles for retrofitting coal burners for oxy-combustion. Int J Greenhouse Gas Control 2011;5:S151–8.

[29] Scheffknecht G, Al-Makhadmeh L, Schnell U, et al. Oxy-fuel coal combustion—a review of the current state-of-the-art. Int J Greenhouse Gas Control 2011;5(12):S16–35.

[30] Chen L, Yong SZ, Ghoniem AF. Oxy-fuel combustion of pulverized coal: characterization, fundamentals, stabilization and CFD modeling. Prog Energy Combust Sci 2012;38(2):156–214.

[31] Guo J, Liu Z, Wang P, et al. Numerical investigation on oxy-combustion characteristics of a 200 MWe tangentially fired boiler. Fuel 2015;140(2):660–8.

[32] WRI (World Resources Institute), CCS guidelines–guidelines for carbon dioxide capture, transport, and storage. Washington, DC; 2008.

[33] Aspelund A, Jordal K. Gas conditioning—the interface between CO_2 capture and transport. Int J Greenhouse Gas Control 2007;1(2):343–54.

Chapter 2

Fundamentals of Oxy-fuel Combustion

Sheng Chen
Huazhong University of Science and Technology, Wuhan, China

2.1 INTRODUCTION

In oxy-fuel combustion, chemical reactions take place in O_2/CO_2 atmosphere where O_2 in the oxidant flow is diluted by CO_2, rather than by N_2 as in a conventional air-firing mode. Because the chemical and physical properties of CO_2 are quite different from those of N_2, the combustion characteristics and reaction pathways in an oxy-fuel combustion condition is altered more significantly than its air-firing counterpart [1]. These differences have a critical impact on the design, operation, and optimization of combustion systems; therefore it is crucial that relevant practitioners understand the chemical and physical effects of CO_2 on combustion features, reaction kinetics, pollution generation, and so on. The purpose of this chapter is to present a brief, up-to-date overview of the progress in these fundamental areas, using oxy-methane combustion as an example, first because methane is the main composition of many sustainable clean fuels (e.g., natural gas and landfill gas) [2] and second because it can help us understand the reaction mechanisms of volatiles released by solid fuels in O_2/CO_2 atmosphere [3].

2.2 EFFECTS OF CO_2 ON COMBUSTION CHARACTERISTICS

The effects of CO_2 on combustion characteristics can be attributed to the following chemical and physical properties: (1) its low thermal diffusivity; (2) its high molar heat capacity; (3) its chemical path alteration, because it is not an inert gas in oxy-fuel combustion; and (4) its modified radiative heat transfer feature. Generally, the influences of CO_2 on combustion characteristics fall into four categories as discussed next.

Oxy-fuel Combustion. http://dx.doi.org/10.1016/B978-0-12-812145-0.00002-5

2.2.1 Burning Velocity in O_2/CO_2 Atmosphere

Burning velocity, the rate of flame propagation relative to unburned gas ahead of the flame front, is a fundamental combustion property of all fuel/oxidant mixtures. A change in the burning velocity of a reactant mixture may cause poor combustion performance and a modified distribution of temperature and species in combustion chambers. Consequently, identifying the variation trend in the burning velocity of methane is very important for the oxy-methane combustion process.

About three decades ago, it was found that substituting N_2 with CO_2 in the oxidizer led to a reduction of the flame speed [4]. Such reductions were observed over a wide range of experimental conditions, for example, with pressures varying from 0.25 atm to 2 atm (atmosphere) and with flame temperatures varying from 1550 to 2250 K [4]. Then Liu et al. gave an explanation for such phenomena from the viewpoint of the chemical properties of CO_2; they believed that the decrease in burning velocity in O_2/CO_2 atmosphere could not be entirely described only by considering the physical properties of CO_2 (the higher molar heat capacity of CO_2 would cause a lower adiabatic flame temperature) [5]. With the aid of the software CHEMKIN, their analysis indicated that the most important reaction associated with the direct chemical participation of CO_2 in these flames is

$$CO + OH \leftrightarrow CO_2 + H \tag{2.1}$$

which reduced the concentration of important radicals, for example, H, O, and OH, in the combustion chamber and thus reduced the burning velocity. In order to provide more detailed experimental support, Heil et al. carried out an experimental study based on a lab-scale furnace at RWTH Aachen University [6]. Their well-designed experimental activities demonstrated that the chemical effects of CO_2 had a significant impact on the production and consumption rates of carbon monoxide in oxy-fuel combustion. CO_2 competed with O_2 for atomic hydrogen and led to the formation of CO through the reverse reaction of reaction (2.1), which could decrease the burning rate of methane in O_2/CO_2 atmosphere. They also found that an increase of O_2 concentration in oxy-fuel combustion might reduce this impact.

Oh and Noh [7] also investigated the burning velocity of methane flame in O_2/CO_2 atmosphere, but with the aid of a lab-scale Bunsen burner. They conducted their experiments in atmospheric conditions (300 K and 1 atm), not only in fuel-lean but also in fuel-rich conditions. However, they observed that the laminar burning velocity of methane flame in O_2/CO_2 atmosphere is faster than its air-firing counterpart. In order to validate their experimental observations, the authors also conducted a numerical verification with the aid of the CHEMKIN software. They claimed that the laminar burning velocity of methane flame in O_2/CO_2 atmosphere depended on the temperature of the reactive mixture, the ambient pressure, the global equivalence ratio, and the mixture fraction of the diluent gases. The maximum burning velocity of methane flame in O_2/CO_2 atmosphere appeared in a fuel-rich condition, in the vicinity of a stoichiometric reaction. It is very interesting that Oh and Noh's major observations are quite opposite those shown in previous literature [4,5]. The discrepancy may result from the differences between their research prototypes. In Refs. [4,5] the adiabatic flame temperature in O_2/CO_2 atmosphere was lower than its air-firing counterpart. On the contrary, in Ref. [7], it was reported that, for their investigated configuration, the adiabatic flame temperature in O_2/CO_2 atmosphere was higher than its air-firing counterpart.

The effects of water vapor on the burning velocity of methane flame in O_2/CO_2 atmosphere was reported in Ref. [8]. It is well known that in the wet cycling of power plants, the concentration of water vapor is so high that its influence cannot be ignored. The authors observed that the laminar burning velocity of methane flame in O_2/CO_2 atmosphere would experience a quasi-linear decrease when the steam molar fraction was increased. Furthermore, it was found that, compared with its air-firing counterpart, such a decrease would be more pronounced in oxy-fuel combustion. Mazas et al. presented an explanation of this phenomena by analyzing the physical and chemical effects of adding steam [9]. They claimed that the effects were mainly attributed to steams high chaperon efficiency in the recombination reaction

$$H + O_2 + M \leftrightarrow HO_2 + M \tag{2.2}$$

which competes with the chain-branching reaction

$$H + O_2 \leftrightarrow OH + O \tag{2.3}$$

They underlined that methane oxidation in the reaction

$$CH_4 + O \leftrightarrow CH_3 + OH \tag{2.4}$$

would be suppressed because steam also favors O radical consumption and OH production through the reaction

$$H_2O + O \leftrightarrow OH + OH \tag{2.5}$$

More recently, Xie et al. [10] investigated, experimentally and numerically, the burning velocity of oxy-methane flame under various pressures. Their results were consistent with those in Refs. [4,5] in that the laminar burning of methane flame in O_2/CO_2 atmosphere decreased with an increase in the CO_2 fraction. Furthermore, the authors found that although the calculated data agreed well with the experimental measures under the atmospheric pressure, the discrepancy became greater with an elevated pressure. They also observed that the discrepancy became greater with a decrease in the CO_2 fraction. The authors explained that the discrepancy was partially caused by radiation reabsorption of the flames, which had a positive effect on the laminar burning velocity when the mixtures CO_2 fraction was high. They emphasized that the GRI-3.0 mechanism, which was used regularly for methane combustion simulation, needed to be revised and validated for the calculation of oxy-methane flames, since it might account for the main part of the discrepancy.

2.2.2 Adiabatic/Maximum Flame Temperature in O_2/CO_2 Atmosphere

Adiabatic flame temperature is a fundamental parameter for theoretical combustion. Maximum flame temperature is another important quality in combustion science and engineering and is more useful for realistic combustion systems. In fact, adiabatic flame temperature is just the maximum temperature of reactants in an adiabatic condition.

Generally, the adiabatic flame temperature decreases with the addition of CO_2 in the mixture for all equivalence ratios. It was reported that the adiabatic flame temperature did not decrease linearly with the increase of the CO_2 fraction because its thermal effects became stronger with a higher concentration [10].

To reveal which effects are the dominant ones for determining the maximum flame temperature of oxy-methane combustion, a detailed comparison analyzing the difference between the physical and chemical effects of CO_2 was presented in Ref. [1]. In their study, the analysis was based on the software CHEMKIN. The research prototype was counterflow diffusion combustion, a benchmark in combustion science. The effects of added steam on the maximum flame temperature was considered; as previously mentioned, the effects of steam on oxy-fuel combustion cannot be omitted in wet cycling of a power plant where the mole fraction of H_2O in the oxidant stream may be as high as 30 vol%. To aid the analysis, artificial species X, X', Y, and Y' were introduced. In their research, X and Y had the same thermal and transport properties of CO_2 and H_2O, respectively, but they were not allowed to be a part of a chemical reaction. Therefore X and Y were regarded strictly as chemically inert species. Moreover, artificial species X' and Y' had the same transport properties as those of CO_2 and H_2O, respectively, while their thermal properties remained the same as N_2, and they were not part of a chemical reaction. Therefore X' and Y' were regarded strictly as thermal and chemical inert species. A total of 11 cases were designed to conduct a comprehensive comparison analysis, as shown in Table 2.1.

Fig. 2.1 illustrates the variation of the maximum flame temperature at various mole fractions of O_2 in the oxidizer stream, with CO_2, X, and X' as additives. The differences in the predictions for the maximum flame temperature between X' and X additive cases are due to the thermal effect. Thus the difference between the additive cases of CO_2 and the artificial species X is due to the chemical effect. The results show that both the thermal and chemical effects of CO_2 decrease the flame temperature. It is clear that the result of the thermal effect is due mainly to the higher heat capacity of CO_2. The main factor for the chemical effect is the thermal dissociation of CO_2, which can be explained by two characteristics: (1) the thermal dissociation of the CO_2 process is endothermic, which is quite noticeable at high mole fractions of CO_2 under oxy-fuel conditions; and

TABLE 2.1 Cases Designed for Comparison Analysis Between Physical and Chemical Effects of CO_2 and H_2O

Case	O_2	CO_2	H_2O	X	X'	Y	Y'
Sim1	20%	–	–	–	80%	–	–
Sim2	25%	–	–	–	70%	–	5%
Sim3	25%	–	–	70%	–	–	5%
Sim4	25%	70%	–	–	–	–	5%
Sim5	25%	70%	–	–	–	5%	–
Sim6	25%	70%	5%	–	–	–	–
Sim7	30%	–	–	–	50%	–	20%
Sim8	30%	–	–	50%	–	–	20%
Sim9	30%	50%	–	–	–	–	20%
Sim10	30%	50%	–	–	–	20%	–
Sim11	30%	50%	20%	–	–	–	–

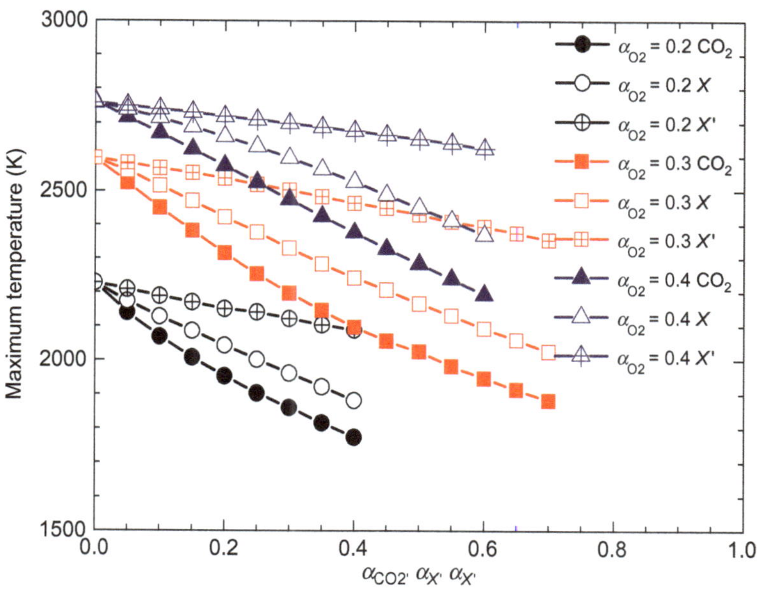

FIG. 2.1 Variation of the maximum flame temperature with the mole fraction of CO_2, X, and X' for various mole fractions of O_2 in the oxidizer stream. *(Reprinted with permission from Wang L, Liu Z, Chen S, Zheng C, Li J. Physical and chemical effects of CO_2 and H_2O additives on counter-flow diffusion flame burning methane. Energy Fuels 2013;27:7602–11. Copyright (2013) American Chemical Society.)*

(2) the thermal dissociation of CO_2 breaks the chemical equilibrium of some reactions related to the production of CO. These factors increase the time for the production and oxidation of CO, thus slowing the release of heat.

In Ref. [1], in addition to the thermal effect of CO_2 on the maximum flame temperature, the influence of the mass diffusion of CO_2 was also covered. In most available literature discussing the physical effects of CO_2 on oxy-fuel combustion, the influence of the latter usually was ignored. In Ref. [1], a type of oxy-methane counterflow combustion was adopted, where 21% O_2 was diluted by 79% CO_2 or CO_2^*. CO_2^* is an artificial species whose diffusion property is the same as N_2. The results indicated that although the lower diffusion coefficient of CO_2 would not change the maximum flame temperature obviously, it would alter the temperature profile significantly. In addition, it was observed that the thermal effects of H_2O decreased the maximum flame temperature because of its higher heat capacity. The difference between the additive cases of H_2O and the artificial species was due to chemical effects. The results indicated that the chemical effects of H_2O slightly increased the flame temperature.

The overall physical and chemical effects of CO_2 and H_2O on the maximum flame temperature of oxy-methane combustion is summarized in Fig. 2.2, where AIR, OCDR, and OCWR denote conventional air-firing, oxy-fuel combustion with dry cycling, and oxy-fuel combustion with wet cycling, respectively. For either OCDR or OCWR, both the thermal and chemical effects of CO_2 induce a decrease in the flame temperature, and the thermal effect shows the greatest importance. However, for OCWR, the proportion of the chemical effect of CO_2

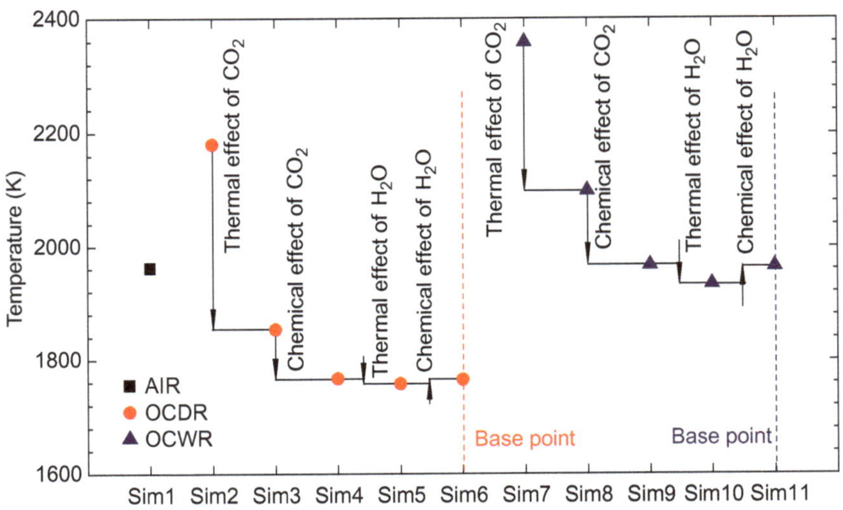

FIG. 2.2 Overall effect for both CO_2 and H_2O on maximum flame temperature of oxy-methane counterflow combustion. *(Reprinted with permission from Wang L, Liu Z, Chen S, Zheng C, Li J. Physical and chemical effects of CO_2 and H_2O additives on counterflow diffusion flame burning methane. Energy Fuels 2013;27:7602–11. Copyright (2013) American Chemical Society.)*

on the flame temperature increases. Moreover, the thermal and chemical effects of H_2O on the flame temperature almost cancel each other out; therefore the existence of H_2O hardly affects the flame temperature.

It is well known that pressure has a significant influence on flame temperature. The effects of various pressures on the adiabatic flame temperature of methane in O_2/CO_2 atmosphere were numerically investigated by Seepana and Jayanti [11]. They found that the adiabatic flame temperature increased monotonically with pressure, ranging from 1 to 30 bar. The authors also made a comparison with its air-firing counterpart. Their results showed that under the same pressure, the adiabatic flame temperature of methane in O_2/CO_2 atmosphere is much lower than its air-firing counterpart because of the higher specific heat of CO_2.

2.2.3 Ignition, Extinction, Flammability, and Flame Instability in O_2/CO_2 Atmosphere

Knowledge about ignition, extinction, and flame stability is crucial for any combustion technology because of performance and safety considerations. Unfortunately, so far the relevant literature on oxy-methane combustion is quite sparse.

A detailed experimental study on the ignition times of oxy-methane combustion [12] was published last year. According to the best knowledge of the present author, before Ref. [12] there was no publication on the influence of high concentration CO_2 on ignition delay times for methane, although there are several studies on the ignition of propane in O_2/CO_2 atmosphere [13]. The experiments were carried out in a shock tube facility, with temperatures ranging from 1577 to 2144 K and pressures varying from 1 to 4 atm. The experimental data in Ref. [13] showed that the ignition of methane was postponed with an increasing CO_2 concentration, although the delay was not significant. An empirical correlation for the ignition delay time was proposed based on the experimental data, which read as

$$\tau = Ae^{E/RT}P^b\varphi^C X_{CO_2}^d \tag{2.6}$$

where τ, A, E, R, T, P, φ, X_{CO_2}, b, c, and d are the ignition delay time, preexponential factor, activation energy, universal gas constant, pressure, equivalent ratio, methane mole fraction, and fitting parameters, respectively. The authors identified the most dominant reaction in their case to be:

$$H + O_2 \leftrightarrow O + OH \tag{2.7}$$

As in oxy-fuel combustion, CO_2 competed for the H radicals through the reaction listed in Eq. (2.1), whereas the reaction presented by Eq. (2.7) was inhibited. The authors claimed that the ignition delay of methane in high concentration CO_2 atmosphere was attributed to the reaction pathway Eq. (2.7). Besides the chemical effects, the authors also thought that the physical effects of CO_2, namely its lower diffusion coefficient and higher heat capacity, contributed to ignition delay. Because there were some uncertainties in the measurements of the ignition delay time, further studies are necessary.

Comparing Refs. [12,13], it seems that the ignition delay times of fuels in O_2/CO_2 atmosphere are extremely sensitive to fuel types. However, as the available open data on this topic are too limited, extensive future investigation on the dependence of ignition delay times on fuel types.

The effects of CO_2 addition on nonpremixed flame extinction was investigated numerically by Seepana and Jayanti [14]. They indicated that the peak flame temperature and the mole fractions of the intermediate radicals H, O, OH, and CH_3 all decreased with the addition of CO_2. They also conducted a comparison between oxy-methane flame extinction and its air-firing counterpart. They found that for their investigated cases, the extinction features of oxy-methane flames were nearly the same with the N_2-diluted air-methane flames.

The effects of radiative heat loss and stretch rate on oxy-methane counterflow flame extinction were reported by Ref. [15]. Their research prototype was a lab-scale counterflow burner. They claimed that the effect of radiative heat loss on stretch extinction limits was not significant in the high oxidizer temperature. Their results also showed that the extinction stretch rate was mainly determined by the total fuel enthalpy flux and specific chemical kinetics. Moreover, the contributions of physical and chemical properties of dilution gases to flame extinction were analyzed numerically. Through a comparison analysis, they claimed that the extinction curves of air-methane flames were close to those of oxy-methane counterflow flames, which was consistent with the conclusion in Ref. [14].

Kim et al. [16] also investigated the effects of chemical and radiation on oxy-methane counterflow flame extinction by numerical simulation. They found that the effect of radiative heat loss on the critical dilutent mole factor for flame extinction was not significant, which agreed with the statement in Ref. [15]. In particular, they studied the flame extinction limit under a high strain rate. Through a scaling analysis, they proposed a correlation to quantify the relationship between the extinction stretch rate, the oxidizer Lewis number, and a specific chemical kinetic term. They suggested that the intermediate radical, OH, was a good indicator for flames, not only in oxy-fuel combustion but also in air-firing scenes.

The extinction characteristics of oxy-methane counterflow flames under elevated pressures were reported by Ref. [17]. Their study was carried out with the aid of a lab-scale experimental apparatus. Numerical simulation was also conducted to check and explain their experimental data. The authors observed that the conventional C-shape extinction curves caused by the upstream conduction loss at a lower stretch rate could also be observed under low pressures (0.1–0.2 MPa for their research schematic configuration) in oxy-methane counterflow combustion. However, under high pressures (0.5–0.7 MPa), the extinction characteristics of oxy-methane counterflow flames were different from their air-firing counterpart, and their extinction curves illustrated a monotonously broadening trend. The authors believed the difference could be attributed to radiation reabsorption, which would play a significant role at high-pressure conditions.

Hu et al. [18] investigated the flammability limits of oxy-methane flames experimentally, as well as numerically. Their experimental measurements were based on a cylinder reactor. Their results showed that CO_2 had a dramatic decrease effect on the upper flammability limits but had only a slight influence on the lower flammability limits.

The flammability limit of oxy-methane jet flames was studied in Ref. [19]. Their experimental research was carried out with the aid of a nonpremixed oxy-methane jet in a lab-scale furnace. For a jet flame, the flammability can be identified by the flame stabilization map, which consists of three parts: the attached flame region, the lifted flame region, and the blow-off region. The authors drew the flame stabilization map according to their experimental data. They observed that the addition of CO_2 had a strong effect on the flammability of oxy-methane jet flames, especially when the concentration of CO_2 in the reactants was less than 20%.

The influences of the addition of steam on the flammability limits of oxy-methane flames were reported by Ref. [20]. According to their numerical simulation of oxy-methane counterflow combustion, the flammability limits broadened with the addition of steam because the molecular specific heat capacity of steam is lower than that of CO_2.

It is well known that the flame instability is determined by three mechanisms: hydrodynamic instability, diffusive-thermal instability, and buoyancy-driven instability [21]. To date only a few publications are available about the effects of the first two mechanisms on oxy-methane flames.

Ditaranto and Hals [22] investigated experimentally the oxy-methane flame instability in a sudden expansion combustor system. The flame instability was attributed to the hydrodynamic instability. The authors observed that for the oxy-methane flames, a higher oxygen concentration was required to avoid the instability. Furthermore, they claimed that there were several different instability modes for their research prototype, depending on the ratio of flame speed to inlet flow velocity.

The instability of a cellular oxy-methane flame was investigated in Ref. [23]. It was observed that the hydrodynamic instability was responsible for the onset of flame instability, while the diffusive-thermal instability affected the flame instability at large standoff distances. In the lean oxy-methane flames, the diffusive-thermal effect tended to enhance the flame instability, whereas in the stoichiometry and rich oxy-fuel flames, it suppressed the flame instability.

Xie et al. [10] investigated experimentally the flame instability of stretched spherical oxy-methane flame. They observed that the addition of CO_2 suppressed the hydrodynamic instability, while enhancing the diffusive-thermal instability. The overall effect was that the addition of CO_2 could suppress the flame instability.

2.2.4 Flame Radiation in O_2/CO_2 Atmosphere

Flame radiation characteristics are also critically important for the design of a furnace, besides combustion featuress chamber. As previously mentioned, most fundamental parameters of combustion, such as the burning rate and maximum flame temperature, depend closely on flame radiation characteristics. Tan et al. [24] measured the temperature profiles in a pilot-scale furnace, with different dilution gases. Their experimental results indicated that the temperature distributions were quite different between oxy-fuel combustion and air-firing.

Xie et al. [10] compared the methane flame radiative heat flux in different dilution atmosphere. They found in general that the flame radiative heat flux of oxy-methane flames was much greater than its air-firing counterpart. Ditaranto and Oppelt [25] investigated the radiative heat flux characteristics of oxy-methane flames. Their experimental observations underlined that the radiative heat flux was closely dependent on the oxygen concentration and inlet Reynolds number.

Because measuring the flame radiative heat flux in large-scale furnaces is very challenging, numerical simulation is indispensable for providing the related necessary information. Since the radiative characteristics of CO_2, a kind of triatomic gas molecule, are quite different from those of N_2, the popularly used gas radiation models for air-firing should be carefully checked and revised. Consequently, the efforts on gas radiation models in O_2/CO_2 atmosphere simulation have become a "hotspot" in the oxy-fuel combustion community. Rajhi et al. [26] made a comprehensive comparison of the commonly used gas radiation models. Their research covered the simple gray gas model, the exponential wide band model, the Leckner and the Perry models, and the weighted-sum-of-gray-gases (WSGG) model. Through the comparison between numerical predictions and experimental measurements of a full-scale furnace fueled by methane diluted by CO_2, they concluded that the Leckner model and the Perry model overpredicted the temperature compared to the exponential wide band model. The WSGG model could predict accurate results compared to the benchmark model; however, its parameters should be revised according to different ratios of H_2O and CO_2 in the furnace. One of the limitations of the WSGG and Perry models is that they were not applicable for pressures other than one atmosphere. The exponential wide band model and the Leckner model both were valid for use for various pressures and ratios of H_2O and CO_2. Now a consensus has been reached that the WSGG model is a good option for oxy-fuel combustion simulation because of the balance between numerical accuracy and computational cost. Abdul-Sater and Krishnamoorthy [27] conducted an assessment of gray and nongray formulations of a WSGG model for oxy-methane combustion simulation. They focused on the effects of the thermal boundary conditions, soot, and turbulence radiation interactions. They found that the nongray model's prediction provided a lower radiant fraction than the gray model calculations. Recently, Guo et al. proposed a full spectrum *k*-distribution-based

WSGG model for oxy-fuel combustion simulation [28]. In their strategy, which was different from previous efforts where new parameters for modified WSGG models came from data fitting, the modified parameters were obtained by a full spectrum k-distribution model. The latest efforts on modeling oxy-methane flame radiation was summarized by Abdul-Sater et al. [29].

2.3 EFFECTS OF CO_2 ON REACTION KINETICS

The purpose for revealing the effects of CO_2 on the reaction kinetics of oxy-methane combustion are threefold: (1) to construct reduced reaction mechanisms, as currently it is still impossible to adopt the detailed reaction mechanism of methane to full-scale furnace simulation; (2) to understand pollution-generating mechanisms in O_2/CO_2 atmosphere; and (3) to explain the chemical effects of CO_2 on oxy-fuel combustion characteristics, such as burning rate and ignition delay. Because we previously discussed the third topic, this section focuses on the latest progress of the first two.

Andersen et al. [30] proposed a global reaction mechanism for oxy-methane combustion simulation. They compared the hydrocarbon oxidation mechanisms by Westbrook and Dryer (WD) and by Jones and Lindstedt (JL). Andersen et al. found that the CO concentration could not be predicted accurately via these mechanisms, after which they proposed an improved version to solve the problem.

With the aid of the software CHEMKIN, the present authors compared the influences of various dilution atmospheres on the oxidized pathways of methane. The research prototype was a counterflow burner, and the strain rate was fixed at $100 s^{-1}$ [31]. As shown in Fig. 2.3A, for air-firing scenes, the major reaction pathway consisted of two C1-branch paths and one C2-branch path: (1) $CH_3 \rightarrow CH_2OH \rightarrow CH_2O \rightarrow HCO \rightarrow CO \rightarrow CO_2$, (2) $CH_3 \rightarrow CH_2(s) \rightarrow CH_2 \rightarrow CH \rightarrow CH_2O \rightarrow HCO \rightarrow CO \rightarrow CO_2$, and (3) $CH_3 \rightarrow C2H_6 \rightarrow C_2H_5 \rightarrow C_2H_4 \rightarrow C_2H_3 \rightarrow C_2H_2 \rightarrow HCCO \rightarrow CO \rightarrow CO_2$. C1-branch paths represent the low temperature oxidization processes, while the C2-branch path indicates the high-temperature oxidation processes. The ratio between C1- and C2-branch paths was 1.23, which implied that in air-firing conditions, the C1-branch path was the dominant one. For OCDR cases, illustrated in Fig. 2.3B, the ratio between C1- and C2-branch paths increased to 1.88 as the contributions of $CH_3 \rightarrow CH_2OH, CH_3 \rightarrow CH_2O, CH_3 \rightarrow CH_2(s)$ increased. The C1-branch path played a more important role. Another difference, compared with its AC counterpart, was that in the OCDR condition, the contribution of $CH_2(s) \rightarrow CH_2OH$ increased from 15% to 61%, which indicated that the chain from CH_4 to CO was shortened. In the OCWR condition, shown in Fig. 2.3C, the ratio between C1- and C2-branch paths was the highest. The role of the intermediate radical H was suppressed, while that of OH was enhanced. Table 2.2 shows the detailed information.

Based on the preceding analyses, Wang et al. [31] and Chen et al. [2] proposed a global reaction mechanism for oxy-methane combustion simulation, respectively. Their results showed that their reduced reaction mechanism could reproduce sufficiently accurate numerical predictions in many scenarios.

Most of the available research on the effect of CO_2 addition on the generation of pollution focused on NOx, as air leakage is inevitable for any realistic combustion system. Park et al. [32] numerically investigated NO emission behavior in oxy-methane counterflow combustion. They claimed that the Fenimore mechanism mainly contributed to NO destruction and that the dominant source of NO formation in oxy-fuel combustion was

$$H + NO + M \leftrightarrow HNO + M \tag{2.8}$$

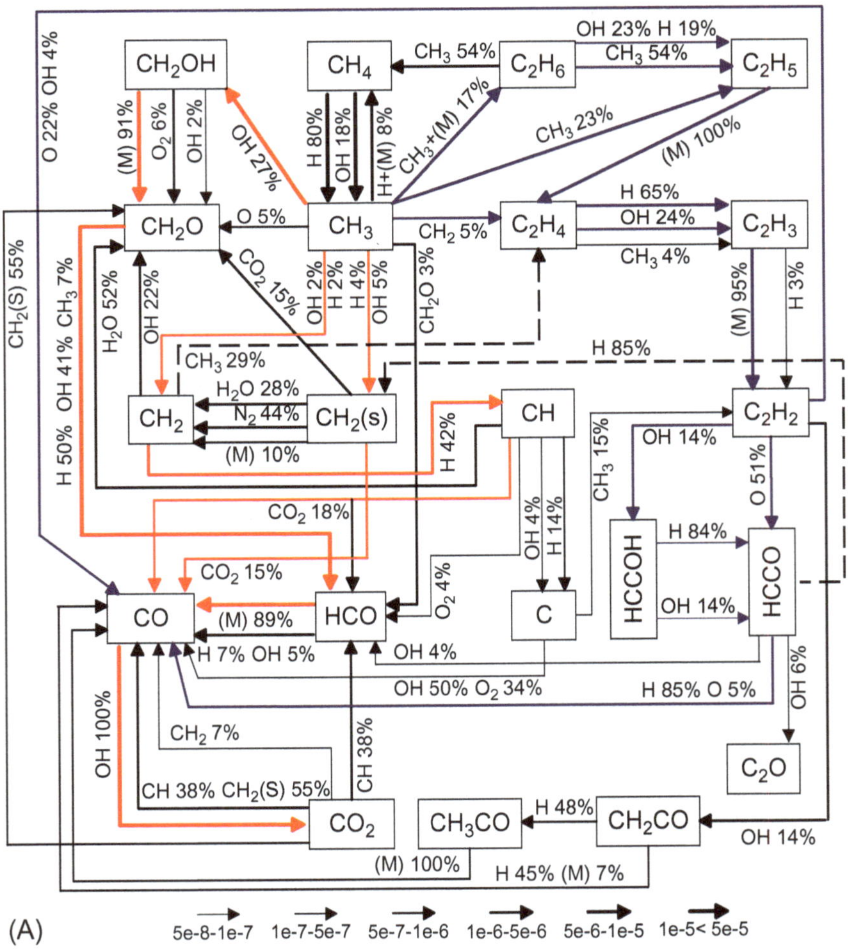

FIG. 2.3 Oxidized pathways of methane in different dilution atmosphere: (A) AC, (B) OCDR, and (C) OCWR. *(Reprinted with permission from Wang L, Liu Z, Chen S, Zheng C, Li J. Physical and chemical effects of CO_2 and H_2O additives on counterflow diffusion flame burning methane. Energy Fuels 2013;27:7602–11. Copyright (2013) American Chemical Society.)*

Mendiara and Glarborg [33] experimentally studied the NO reduction mechanism in O2/CO2 atmosphere. They observed that in stoichiometric and fuel-lean conditions, the NO reduction in oxy-fuel combustion was higher than its air-firing counterpart. The deduction was attributed to the suppression of the O/H radical pool in O_2/CO_2 atmosphere. The authors noted that the recirculation of flue gas had a significant influence on NO formation and destruction. The effect of elevated pressures on NO generation in oxy-methane combustion was reported by Seepana and Jayanti [11]. Their numerical results showed that NO concentration increased with a high pressure. The increase was attributed to the fact that the flame temperature increased with pressure, which enhanced the thermal and prompt route of NO formation.

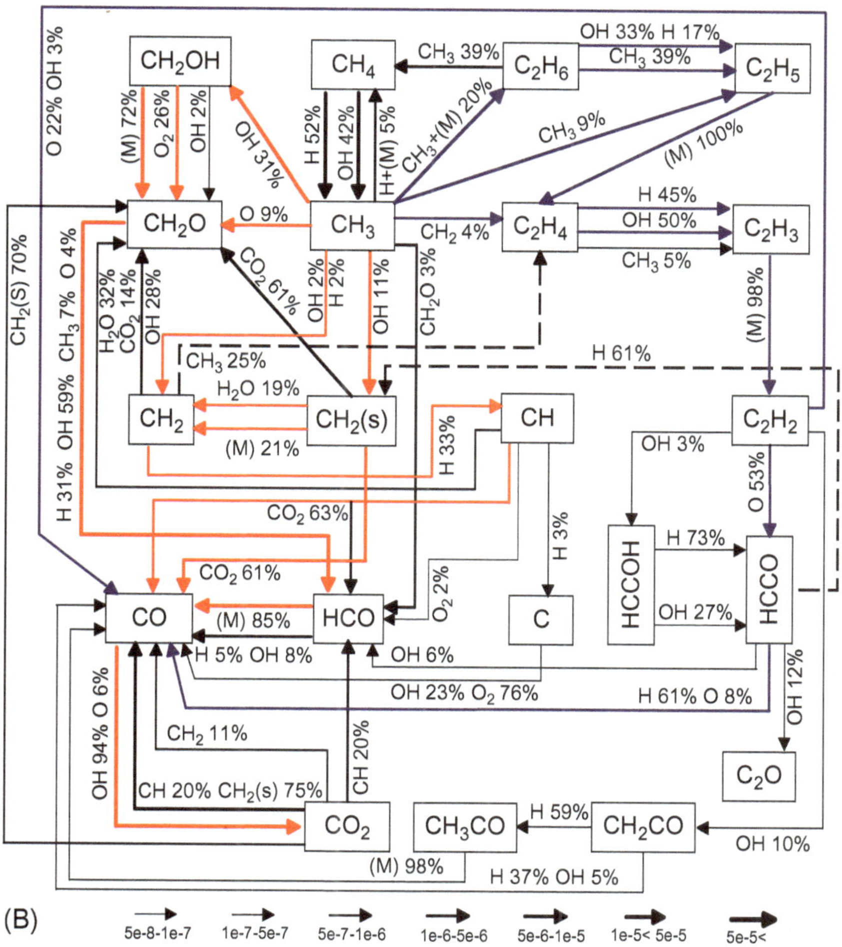

FIG. 2.3, CONT'D

(Continued)

In year 2015, a study on SOx formation in oxy-methane combustion was published [34]. The aim of Ref. [34] was to explore the possibility of utilizing "sour" natural gas resources, which consist of highly concentrated H_2S. Their numerical predictions indicated that the addition of CO_2 increased SOx generation. But the authors emphasized that further experimental data are needed to validate their predictions.

2.4 FACTORS AFFECTING OXY-FUEL COMBUSTION BEHAVIOR

Realistic oxy-fuel combustion systems, their performance and behavior may be closely dependent on two factors, namely composition of reactants and hydrodynamics of combustible mixture.

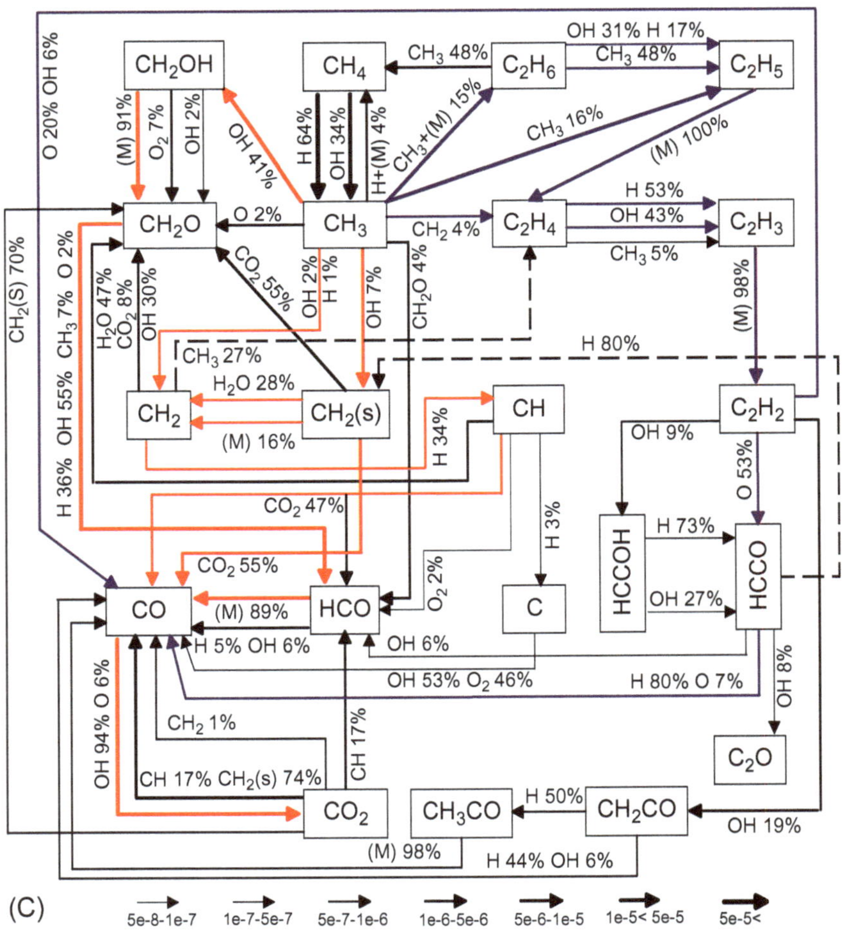

FIG. 2.3, CONT'D

TABLE 2.2 Contribution Weight of Intermediate Radical Species in Different Dilution Atmosphere

	Radical Species	AC	OCDR	OCWR
$CH_4 \rightarrow CH_3$	H	80%	52%	64%
	OH	18%	42%	34%
$CH_3 \rightarrow CH_2OH$	OH	27%	31%	41%
$CH_3 \rightarrow CH_2O$	O	5%	9%	2%
$CH_3 \rightarrow CH_2(s)$	H	4%	–	–
	OH	5%	11%	7%
$CH_2OH \rightarrow CH_2O$	M	91%	72%	91%
	O_2	6%	26%	7%
$CH_2O \rightarrow HCO$	H	50%	31%	36%
	OH	41%	59%	55%
	O	–	7%	7%
	CH3	7%	4%	2%
$HCO \rightarrow CO$	M	89%	85%	89%
	H	7%	5%	5%
	OH	5%	8%	6%
$CH_2(s) \rightarrow CO$	CO_2	15%	61%	55%
$CH \rightarrow CO$	CO_2	18%	63%	47%
$CH_3 \rightarrow C_2H_6$	CH_3+M	17%	20%	15%
$CH_3 \rightarrow C_2H_5$	CH_3	23%	9%	16%
$CH_3 \rightarrow C_2H_4$	CH_2	5%	4%	4%
$HCCO \rightarrow CO$	H	85%	61%	80%
$C_2H_6 \rightarrow C_2H_5$	OH	23%	33%	31%
	H	19%	17%	17%
	CH_3	54%	39%	48%
$C_2H_4 \rightarrow C_2H_3$	OH	24%	45%	53%
	H	65%	50%	43%
	CH_3	4%	5%	4%
$CO \rightarrow CO_2$	OH	100%	94%	94%
	O	–	6%	6%

For oxy-methane combustion, the composition of reactants mainly depends on the injection rate of oxygen and the recirculation rate/mode of flue gases. Many fundamental combustion characteristics, such as pollution generation and flame radiation, are influenced significantly by the reactants composition. These issues were previously covered, so they are not repeated here.

The key to organizing good hydrodynamics is in the design of the burner and furnace chamber. Unfortunately, to the knowledge of this books authors, the open studies on this topic are very sparse, partially because of commercial competition. For oxy-methane combustion, Tan et al. [24] investigated experimentally the influence of a swirl burner on flame behavior in a pilot-scale furnace "CANMET." The observed NO generation was sensitive to the swirl number. Hwang and Gore studied the effects of the inlet Reynolds number on oxy-methane flames [35]. Through the change of the inlet Reynolds number and of the oxidant inject configuration, four different types of flames were formed with the burner that they used. The radiation intensities of various flames were distinctly different. With the aid of numerical simulation, Habib et al. [36] investigated a typical natural gas-fired package boiler operated in O_2/CO_2 atmosphere. Their numerical results indicated that the mass and heat transfer characteristics in oxy-fuel combustion were completely different from its air-firing counterpart. Unfortunately, guidelines on how to organize better hydrodynamics through burner and chamber optimization are still absent.

2.5 SUMMARY

In this chapter, using methane as an example, some fundamentals of oxy-fuel combustion were briefly discussed. Because the characteristics of oxy-fuel combustion are definitely different from its air-firing counterpart, to make this new clean combustion strategy strongly competitive with other clean technologies, combustion experts and practitioners need to dig deeper into their understanding of it. Although the results of numerous efforts related to oxy-fuel combustion have been published, our knowledge of many important aspects of oxy-fuel combustion is still deficient, such as burner and furnace chamber design. In addition, more experimental data are needed in order to check and improve certain theoretical achievements. The deployment of oxy-fuel combustion technology will be boosted only when its potential advantages are completely harnessed. That goal can be achieved only by thoroughly understanding the fundamentals of oxy-fuel combustion technology.

REFERENCES

[1] Wang L, Liu Z, Chen S, Zheng C, Li J. Physical and chemical effects of CO_2 and H_2O additives on counterflow diffusion flame burning methane. Energy Fuel 2013;27:7602–11.

[2] Chen S, Zheng C. Counterflow diffusion flame of hydrogen-enriched biogas under MILD oxy-fuel condition. Int J Hydrog Energy 2011;36:15403–13.

[3] Mendiara T, Glarborg P. Ammonia chemistry in oxy-fuel combustion of methane. Combust Flame 2009;156:1937–49.

[4] Zhu DL, Egolfopoulos FN, Law CK. Experimental and numerical determination of laminar flame speeds of methane/(Ar, N_2, CO_2)-air mixtures as function of stoichiometry, pressure, and flame temperature. Symp Combust 1989;22:1537–45.

[5] Liu F, Guo H, Smallwood GJ. The chemical effect of CO_2 replacement of N_2 in air on the burning velocity of CH_4 and H_2 premixed flames. Combust Flame 2003;133:495–7.

[6] Heil P, Toporov D, Förster M, Kneer R. Experimental investigation on the effect of O_2 and CO_2 on burning rates during oxyfuel combustion of methane. Proc Combust Inst 2011;33:3407–13.

[7] Oh J, Noh D. Laminar burning velocity of oxy-methane flames in atmospheric condition. Energy 2012;45:669–75.

[8] Mazas A, Lacoste D, Schuller T. Experimental and numerical investigation on the laminar ame speed of CH_4/O_2 mixtures diluted with CO_2 and H_2O, In Proceedings of ASME turbo expo 2010: power for land, sea and air, United Kingdom. 2010, pp. GT2010-22512.

[9] Mazas A, Fiorina B, Lacoste DA, Schuller T. Effects of water vapor addition on the laminar burning velocity of oxygen-enriched methane flames. Combust Flame 2011;158:2428–40.

[10] Xie Y, Wang J, Zhang M, Gong J, Jin W, Huang Z. Experimental and numerical study on laminar flame characteristics of methane oxy-fuel mixtures highly diluted with CO_2. Energy Fuel 2013;27:6231–7.

[11] Seepana S, Jayanti S. Flame structure and NO generation in oxy-fuel combustion at high pressures. Energy Convers Manag 2009;50:1116–23.

[12] Koroglu B, Pryor OM, Lopez J, Nash L, Vasu SS. Shock tube ignition delay times and methane time-histories measurements during excess CO_2 diluted oxy-methanec ombustion. Combust Flame 2016;164:152–63.

[13] Sabia P, Lavadera ML, Giudicianni P, Sorrentino G, Ragucci R, de Joannon M. CO_2 and H_2O effect on propane auto-ignition delay times under mild combustion operative conditions. Combust Flame 2015;162:533–43.

[14] Seepana S, Jayanti S. Flame structure investigations of oxy-fuel combustion. Fuel 2012;93: 52–8.

[15] Li X, Jia L, Onishi T, Grajetzki P, Nakamura H, Tezuka T, et al. Study on stretch extinction limits of CH_4/CO_2 versus high temperature O_2/CO_2 counterflow non-premixed flames. Combust Flame 2014;161:1526–36.

[16] Kim TH, Park JW, Park HY, Park J, Park JH, Lim IG. Chemical and radiation effects on flame extinction and NOx formation in oxy-methane combustion diluted with CO_2. Fuel 2016;177:235–43.

[17] Maruta K, Abe K, Hasegawa S, Maruyama S, Sato J. Extinction characteristics of CH_4/CO_2 versus O_2/CO_2 counterflow non-premixed flames at elevated pressures up to 0.7 MPa. Proc Combust Inst 2007;31:1223–30.

[18] Hu X, Yu Q, Sun N, Qin Q. Experimental study of flammability limits of oxy-methane mixture and calculation based on thermal theory. Int J Hydrog Energy 2014;39:9527–33.

[19] Oh J, Noh D. The effect of CO_2 addition on the flame behavior of a non-premixed oxy-methane jet in a lab-scale furnace. Fuel 2014;117:79–86.

[20] Liu Y, Chen S, Yang B, Liu K, Zheng C. First and second thermodynamic-law comparison of biogas MILD oxy-fuel combustion moderated by CO_2 or H_2O. Energy Convers Manag 2015;106:625–34.

[21] Law CK. Combustion physics. Cambridge: Cambridge University Press; 2006.

[22] Ditaranto M, Hals J. Combustion instabilities in sudden expansion oxy-fuel flames. Combust Flame 2006;146:493–512.

[23] Yu JF, Yu R, Fan XQ, Christensen M, Konnov AA, Bai XS. Onset of cellular flame instability in adiabatic $CH_4/O_2/CO_2$ and CH_4/air laminar premixed flames stabilized on a flat-flame burner. Combust Flame 2013;160:1276–86.

[24] Tan Y, Douglas MA, Thambimuthu KV. CO_2 capture using oxygen enhanced combustion strategies for natural gas power plants. Fuel 2002;81:1007–16.

[25] Ditaranto M, Oppelt T. Radiative heat flux characteristics of methane flames in oxy-fuel atmospheres. Exp Thermal Fluid Sci 2011;35:1343–50.

[26] Rajhi MA, Ben-Mansour R, Habib MA, Nemitallah MA, Andersson K. Evaluation of gas radiation models in CFD modeling of oxy-combustion. Energy Convers Manag 2014;81:83–97.

[27] Abdul-Sater H, Krishnamoorthy G. An assessment of radiation modeling strategies in simulations of laminar to transitional, oxy-methane, diffusion flames. Appl Therm Eng 2013;61:507–18.

[28] Guo J, Li X, Huang X, Liu Z, Zheng C. A full spectrum *k*-distribution based weighted-sum-of-gray-gases model for oxy-fuel combustion. Int J Heat Mass Transf 2015;90:218–26.

[29] Abdul-Sater H, Krishnamoorthy G, Ditaranto M. Predicting radiative heat transfer in oxy-methane flame simulations: an examination of its sensitivities to chemistry and radiative property models. J Comb 2015;2015:1–20.

[30] Andersen J, Rasmussen CL, Giselsson T, Glarborg P. Global combustion mechanisms for use in CFD modeling under oxy-fuel conditions. Energy Fuel 2009;23:1379–89.

[31] Wang L, Liu Z, Chen S, Zheng C. Comparison of different global combustion mechanisms under hot and diluted oxidation conditions. Combust Sci Technol 2012;184:259–76.

[32] Park J, Park JS, Kim HP, Kim JS, Kim SC, Choi JG, et al. NO emission behavior in oxy-fuel combustion recirculated with carbon dioxide. Energy Fuel 2007;21:121–9.

[33] Mendiara T, Glarborg P. Reburn chemistry in oxy-fuel combustion of methane. Energy Fuel 2009;23:3565–72.

[34] Bongartz D, Ghoniem AF. Chemical kinetics mechanism for oxy-fuel combustion of mixtures of hydrogen sulfide and methane. Combust Flame 2015;162:544–53.

[35] Hwang S, Gore J. Characteristics of combustion and radiation heat transfer of an oxygen-enhanced fame burner. J Power Energy 2002;216:379–86.

[36] Habib M, Ben-Mansour R, Badr HM, Ahmed SF, Ghoniem AF. Computational fluid dynamic simulation of oxyfuel combustion in gas-fired water tube boilers. Comput Fluids 2012;56:152–65.

Chapter 3

Coal Ignition in Oxy-fuel Combustion Environment

Chun Zou, Lei Cai
Huazhong University of Science and Technology, Wuhan, China

3.1 INTRODUCTION

Coal ignition is closely associated with coal devolatilization, the homogeneous combustion of volatiles, and the heterogeneous combustion of char and coal. Coal can be homogeneously, heterogeneously, and hetero-homogeneously ignited based on the interaction of these processes. Homogeneous ignition is generally thought of as a two-stage process. The primary step consists of the initial ignition of volatiles accumulated around the coal particle. Subsequently, the homogeneous combustion of volatiles ignites the char particle. Heterogeneous ignition can occur via the direct attack of oxygen on the coal or char particle [1]. In hetero-homogeneous ignition, char and volatiles ignite simultaneously. Although the definition of ignition differs according to the experimental facilities and conditions adopted in the studies [2–4], the particle diameter, heating rate, oxygen mole fraction, coal type, and ambient gas temperature are widely accepted as having significant effects on the ignition mechanism [1,5–8]. Homogeneous ignition usually occurs when large particles are heated slowly and/or in low oxygen fractions, while heterogeneous ignition mostly takes place when small particles are heated quickly or the oxygen mole fraction is high. Heterogeneous ignition is more likely to occur in low-rank coals at high ambient temperatures [1].

Many researchers have examined the ignition of coal in oxy-fuel combustion. Khatami et al. [9] investigated the ignition mechanism of single particle coal and char particles in O_2/N_2 and O_2/CO_2 environments with oxygen mole fractions in the 20%–100% range. They claimed that heterogeneous ignition is likely to occur in the O_2/CO_2 environments, and homogeneous ignition in the O_2/N_2 environments. In contrast to air combustion, ignition is commonly thought to be delayed in oxy-fuel combustion at an O_2 concentration of 21% because of the higher thermal capacity (ρc_p) of CO_2. Using US eastern bituminous coal and an entrained flow reactor, Molina and Shaddix [10] found that high concentrations

Oxy-fuel Combustion. http://dx.doi.org/10.1016/B978-0-12-812145-0.00003-7

of CO_2 had no evident effect on the time of volatile combustion from ignition to burnout under a homogeneous gas-phase ignition regime. Further studies by Shaddix et al. [11] indicated that the presence of CO_2 does not affect the volatile fuel consumption rate, but slightly impacts the volatile cloud combustion rate because of a slightly lower diffusivity of volatiles in CO_2 than in N_2. Liu et al. [12] studied the ignition delay of a pulverized coal stream using identical experimental facilities and found that the ignition time strongly depends on the furnace gas temperature and coal particle size. Zhang et al. [13] imaged the ignition of a low-volatile bituminous coal particle in two O_2/CO_2 mixtures. They observed the homogeneous ignition sequence using a high-speed camera and claimed that the volatiles that accumulated on the char surface mainly converted to CO and that coal ignition in ~30% O_2/~70% CO_2 essentially matched that in air. Jovanovic et al. [14] experimentally and numerically studied the ignition of coal in drop tube furnace (DTF). The ignition distances predicted by their modified sequential coal model agreed with the experimental results. Their later studies [15] indicated that the FG model can successfully predict the ignition distance.

The aforementioned works mainly qualitatively examined the ignition phenomena in O_2/N_2 and O_2/CO_2 environments. Few studies have attempted to quantify the interaction between devolatilization, the combustion of volatiles, and coal or char combustion in the ignition process, which are significant for understanding the ignition mechanism of coal and the effects of these processes on the ignition mechanisms in O_2/CO_2 environments.

Because the physical and chemical properties of CO_2 are very different from those of N_2, the characteristics and emissions of oxy-fuel combustion are significantly different from air combustion [9–11,16–18]. In the studies of many researchers [9–11], they found that the ignition delay occurred in the O_2/CO_2 environment compared to O_2/N_2 environment with an identical O_2 mole fraction. They claimed that the higher specific heat of CO_2 led to the ignition delay. CO_2 can take part in some combustion reactions. Hirotatsu et al. [6] found that a high concentration of CO_2 greatly improved the reaction rate of the reaction $CO_2+H \Leftrightarrow CO+OH$. Song et al. [7] compared the temperature profiles of CH_4 combustion in air and in O_2/CO_2 environments using numerical simulation in a counterflow flame, and their results indicated that the chemical effects of CO_2 on the combustion temperatures were significant and even comparable to the physical effects. Consequently, the chemical properties of CO_2 can impact the ignition process of coal in O_2/CO_2 environments. However, few studies have been conducted on the chemical effects of CO_2 on the coal ignition in O_2/CO_2 environments.

At state-of-the-art oxy-fuel combustion, the flue recycle system is necessary in order to moderate the high temperature produced by the combustion of fuel in pure O_2. According to the different locations of recycled flue gas, two modes are proposed: dry flue gas recycle and wet flue gas recycle. In practical applications of oxy-fuel combustion, at least some wet flue gas recycle will be utilized [19]. Moreover, the wet flue gas recycle approach is simpler and lowers the potential for air ingress because the flue gas is not needed to pass through a condenser or wet scrubbers. The wet flue gas recycle approach offers slightly higher efficiency than the dry flue gas recycle approach [19,20]. Some studies have reported experimental results of coal in a wet flue gas recycle system [21,22]. However, the effect of different concentrations of steam on the ignition of coal particles in $O_2/CO_2/H_2O$ environments has not been studied in detail. In other words, the physical and chemical effects of steam on the ignition are not clear.

In this chapter, the homogeneous and heterogeneous processes in O_2/CO_2 environments are studied in detail by numerical simulation. The interactions among the rate of devolatilization, volatiles combustion, and char combustion are revealed. The effect of CO_2 and H_2O on the ignition of coal is particularly analyzed in this chapter.

3.2 NUMERICAL SIMULATION OF THE IGNITION PROCESS OF PULVERIZED COAL IN O_2/CO_2 ENVIRONMENTS

In this section, an ignition mechanism model is proposed to state the ignition process of coal in O_2/CO_2 environments. Jovanovic's ignition experiments were adopted to verify the simulated results [14,15].

3.2.1 Modeling Approach

ANSYS Fluent version 12 was adopted to simulate the ignition process of coal. The coal combustion process was modeled in the Eulerian-Lagrangian frame [23]. The standard $\kappa-\varepsilon$ double equations model was used to model the dynamics of the flow [14,15]. The random walk model was used to simulate the coal particle velocity and trajectories [14,24]. The discrete ordinate (DO) radiation model was used to simulate the radiation heat transfer during pulverized coal combustion or oxy-coal combustion [24,25]. The gas mixture absorption coefficient was calculated by using the weighted-sum-of-gray-gases model (WSGGM).

In this work, a single rate kinetic model was adopted [15,24,25], and the devolatilization rate was expressed by Eq. (3.1),

$$\frac{-1}{W}\frac{dW}{dt} = \mathfrak{R}_{\text{dvol}} = K = A\exp\left(\frac{-E_{\text{a}}}{R_{\text{gas}}T_{\text{p}}}\right) \tag{3.1}$$

where K is the devolatilization rate constant, A is the preexponential factor, E_a is the activation energy, and W is the mass of volatiles. The kinetic parameters (A, E_a) are shown in Table 3.1 [14].

The kinetic/diffusion model was used for devolatilized char combustion and nondevolatilized coal combustion. The char and coal combustion rates were represented by Eq. (3.2),

$$\mathfrak{R}_{\text{het}} = -A_{\text{p}}\frac{\rho R_{\text{gas}}T_{\text{ox}}f_{\text{ox}}}{M_{\text{ox}}}\frac{D_0 \cdot R}{D_0 + R} \tag{3.2}$$

TABLE 3.1 Combustion Parameters Inputs for Model

Parameters	
Devolatilization kinetic parameters	
Preexponential factor	1.34E4 (s^{-1})
Activation energy	2.41E07 (J/kmol)
Char combustion kinetic parameters	
Preexponential factor	4.50E−05 ($kg/m^2/s/P$)
Activation energy	6.90E07 (J/kmol)
Coal combustion kinetic parameters	
Preexponential factor	1.60E−04 ($kg/m^2/s/P$)
Activation energy	1.49E08 (J/kmol)
Volatile combustion	
Preexponential factor	3.88E08 ($kg/m^2/s/Pa$)
Activation energy	1.59E08 (J/kmol)

where $\mathfrak{R}_{het}$ represents the rate of nondevolatilization coal combustion ($\mathfrak{R}_{het,coal}$) or the rate of devolatilization char combustion ($\mathfrak{R}_{het,char}$). R is the chemical rate and is expressed by Eq. (3.3):

$$R = A_{het} \exp\left(-\frac{E_{het}}{R_{gas}Tp}\right) \tag{3.3}$$

D_0 is the bulk molecular diffusion rate in the expression

$$D_0 = C_1\left\{\left[\left(T_P + T_\infty\right)/2\right]^{0.75} / d_P\right\} \tag{3.4}$$

C_1 is the mass diffusion coefficient, defined as follows:

$$C_1(T,p) = \frac{ShD(T_0)M_{ox}}{T_0^{1.75}} \tag{3.5}$$

The kinetic parameters used for devolatilized char and nondevolatilized coal combustion are different and are shown in Table 3.1 [14].

In this section, the char-CO_2 and char-H_2O gasification reactions were not considered in the simulation [20,26].

In this section, for homogeneous combustion, that is, volatiles combustion, a one-step global reaction with a single hypothetic component of volatile matter, $C_xH_yO_z$, was used: [14,27]

$$C_xH_yO_z + \alpha O_2 = \beta CO_2 + \gamma H_2O \tag{3.6}$$

Here the coefficients x, y, and z are obtained based on the proximate and ultimate analyses of coal, as shown in Table 3.2. The rate of the volatile combustion reaction is expressed as

$$\mathfrak{R}_{\mathrm{hom,vol}} = M_{\mathrm{vol}} A_{\mathrm{hom,vol}} \exp\left(\frac{-E_{\mathrm{hom,vol}}}{R_{\mathrm{gas}} T_{\mathrm{gas}}}\right) C_{\mathrm{vol}}^{a} C_{O_2}^{b} V_{\mathrm{cell}} \tag{3.7}$$

where a and b are rate exponents.

The turbulence-chemistry interaction was modeled by the finite-rate/eddy-dissipation combustion model [28].

The inflection point of particle temperature versus distances was deemed as the ignition point in the simulation [14].

TABLE 3.2 Characteristics of Russian High-Volatile Bituminous Coal

Coal Type	High-Bituminous
Proximate analysis (%)	
Moisture	3.1
Volatile matter	30
Ash	14.9
Fixed carbon	52
Ultimate analysis (%)	
C	66.04
H	4.18
N	1.99
S	2.06
O	7.07

Three ignition mechanisms were identified based on the conditions, as follows [14,29]:

1. For a homogeneous ignition mechanism, the rate of devolatilization must be greater than that of nondevolatilization coal combustion ($\mathfrak{R}_{devol} > \mathfrak{R}_{het,coal}$), and the combustion rate of volatiles must be greater than that of devolatilization char ($\mathfrak{R}_{hom,vol} > \mathfrak{R}_{het,char}$).
2. For a heterogeneous ignition mechanism, the rate of devolatilization must be greater than that of nondevolatilization coal combustion ($\mathfrak{R}_{devol} > \mathfrak{R}_{het,coal}$), and the combustion rate of devolatilization char must be greater than that of volatiles ($\mathfrak{R}_{het,char} > \mathfrak{R}_{hom,vol}$).
3. For a heterogeneous-homogeneous combustion mechanism, the rate of nondevolatilization coal combustion must be greater than that of devolatilization ($\mathfrak{R}_{het,coal} > R_{devol}$).

The ignition mechanisms were determined at the position of particle ignition based on three conditions. The three conditions were implemented in the FLUENT code by a user-defined function (UDF). If the ignition of coal is controlled by homogeneous mechanisms ($\mathfrak{R}_{devol} > \mathfrak{R}_{het,coal}$ and $\mathfrak{R}_{hom,vol} > \mathfrak{R}_{het,char}$), devolatilization and volatiles combustion will occur, while devolatilization char combustion will be stopped. If the ignition mechanism is heterogeneous ($\mathfrak{R}_{devol} > \mathfrak{R}_{het,coal}$ and $\mathfrak{R}_{het,char} > \mathfrak{R}_{hom,vol}$), devolatilization and devolatilization char combustion will occur, while volatiles combustion will be stopped. If the ignition mechanism is heterogeneous-homogeneous ($\mathfrak{R}_{het,coal} > \mathfrak{R}_{devol}$), devolatilization, volatiles combustion, and devolatilization char combustion will occur [14].

3.2.2 Validation of the Model

The prediction results of the ignition distance were compared with the results of the experiment. Because the submodel of the devolatilization of volatiles significantly affects the ignition of coal [15,30], the results of the present model were also compared with the results of the CFD model with the default devolatilization submodels in FLUENT, CPD, and FG [14]. Fig. 3.1 shows that the present CFD model successfully predicted the ignition distance during air and oxy-fuel combustion, and the results of the present CFD model were comparable with those of the CFD model with the CPD and FG submodel. Therefore the present CFD model could be used to investigate the ignition mechanism of coal in oxy-fuel combustion.

3.2.3 Ignition Mechanisms

To evaluate the ignition mechanisms, the rates of devolatilization, volatiles combustion, char combustion, and nondevolatilization coal combustion at the ignition point were calculated. The results for the environments of 21%

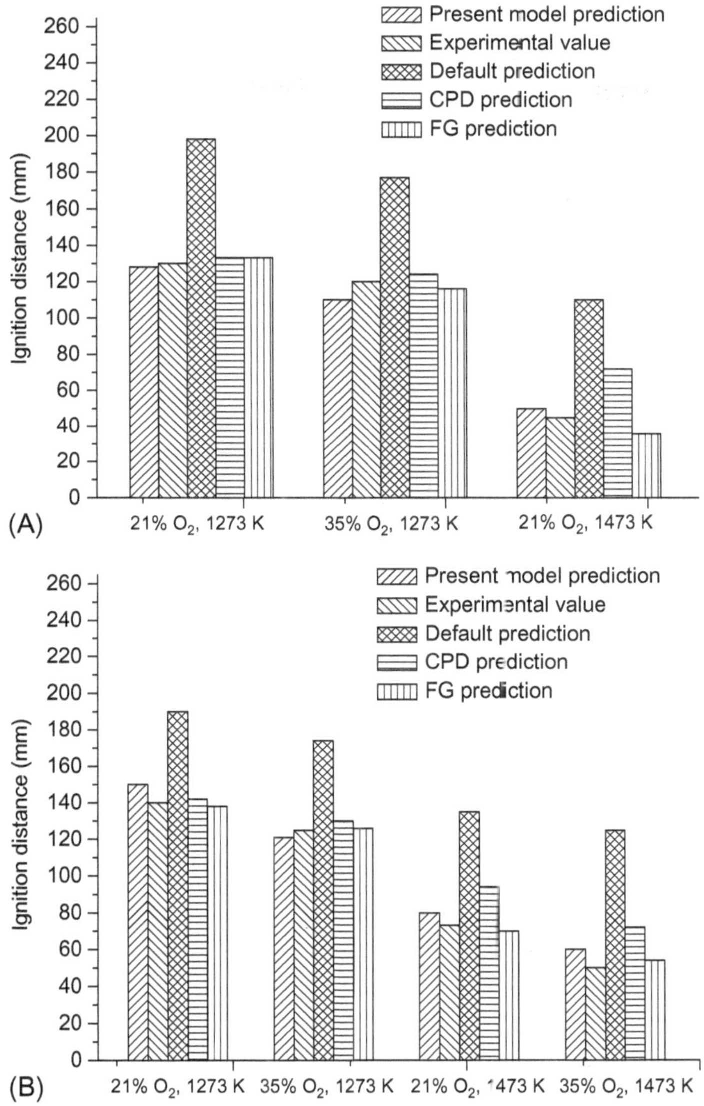

FIG. 3.1 Comparison of ignition distance of present CFD model, experiment, and CFD model with default devolatilization model in FLUENT, CPD, and FG submodel. (A) O_2/N_2 mixtures and (B) O_2/CO_2 mixtures.

O_2/79% CO_2 and 35% O_2/65% CO_2 are shown in Fig. 3.2. For the environment of 21% O_2/79% CO_2, the rates of devolatilization and volatiles combustion were greater than those of coal combustion and char combustion, respectively, for secondary air temperatures between 1073 and 1473 K. According to the criteria of the ignition mechanism, coal particles ignite homogeneously. Shaddix [11] and Zhang [13] observed homogeneous ignition at similar

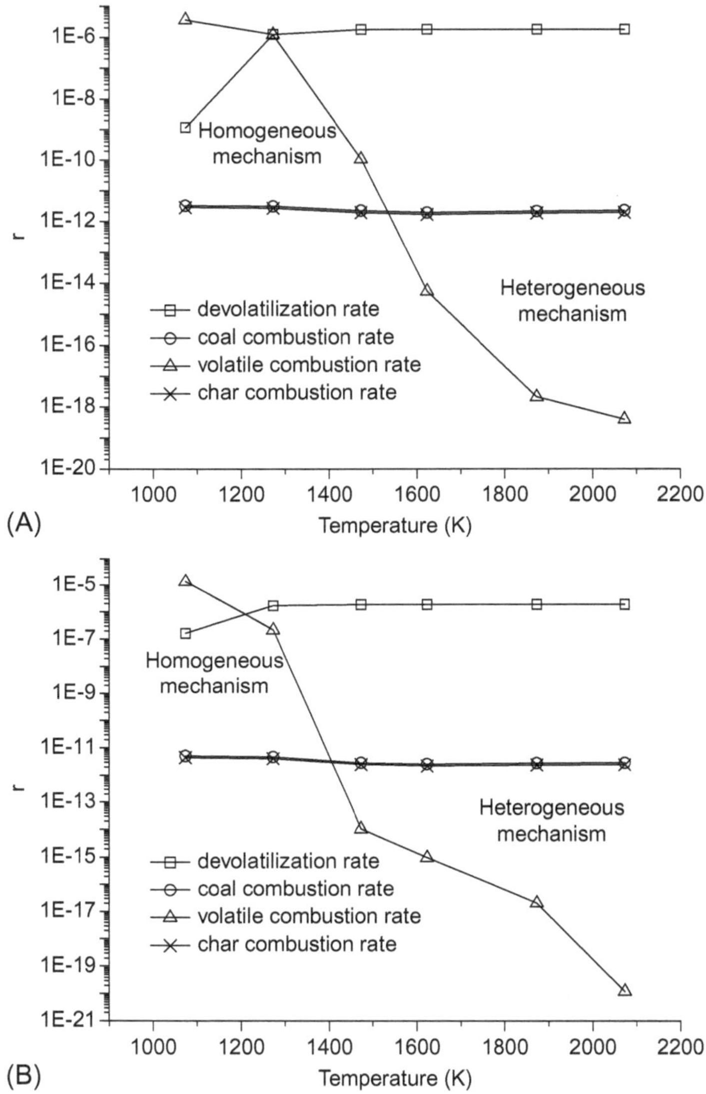

FIG. 3.2 Rates of devolatilization, volatiles, and char and coal combustion at the ignition point for different secondary oxidant temperatures. (A) 21% O_2/79% CO_2 and (B) 35% O_2/65% CO_2.

conditions. When the secondary air temperature exceeded 1473 K, the rate of char combustion was higher than that of volatiles combustion, which indicated that coal ignition occurred heterogeneously based on the criteria of the ignition mechanism. Khatami also reported that ignition of coal was heterogeneous at this furnace temperature range in O_2/CO_2 environments [9]. As the secondary temperature increased, the reduction of the mole fraction of volatiles, caused by the reduction of ignition delay time, decreased the volatiles combustion rate, whereas the char combustion rate increased. When the secondary temperature exceeded a critical temperature, the char combustion rate was higher than the volatiles combustion rate. In other words, char ignited before volatiles ignited, which implied heterogeneous ignition. Fig. 3.2B shows that homogeneous ignition occurred at secondary air temperatures of 1073 and 1273 K for an environment of 35% O_2/65% CO_2, while heterogeneous ignition occurred at temperatures greater than 1273 K according to the criteria of the ignition mechanism.

A comparison of Fig. 3.2A and B shows that when the mole fraction of O_2 in CO_2 increased from 21% to 35%, (1) the rates of char and coal combustion at the ignition point increased by one order of magnitude, which favored the onset of heterogeneous ignition; and (2) the rate of volatiles combustion at the ignition point decreased by more than one order of magnitude. Therefore the ignition mechanism of the coal particle changed from homogeneous ignition to heterogeneous ignition at the identical secondary air temperature of 1473 K when the O_2 mole fraction increased from 0.21 to 0.35 in CO_2.

In light of the preceding analysis, the ignition mechanism changed because of (1) the significant reduction of the rate of devolatilization and (2) the increase of the char combustion rate.

3.2.4 Homogeneous Ignition Process

The homogeneous ignition process was examined in detail at a secondary air temperature of 1273 K and an environment of 35% O_2/65% CO_2. For this case, the ignition delay time was 32 ms, and the ignition distance was 130 mm down from the exit of coal particles. Fig. 3.3A reveals that the rate of devolatilization changed with time. Based on Fig. 3.3A, the process of devolatilization began at 30 ms, 2 ms after which homogeneous ignition occurred. During the volatiles combustion process, the rate of devolatilization drastically reduced because of the reduction of the coal particle mass. The devolatilization process ended at 49 ms. Fig. 3.3B shows that the volatiles combustion rate varied versus time in this case. Fig. 3.3B also shows that the volatiles combustion rate did not obviously vary until 45 ms. At this time, the mole fraction of volatiles around the coal particle was maximized, as shown in Fig. 3.3C. The high mole fraction of volatiles accelerated the homogeneous combustion rate to its maximum at 58 ms. The volatiles combustion process terminated at 63 ms because the volatiles were entirely consumed. The duration of the volatiles combustion process was 31 ms.

Fig. 3.3C shows that the mole fraction of volatiles around the coal particles changed over time. The volatiles around the coal particle stemmed from devolatilization and survived without consumption via volatiles combustion. Fig. 3.3C shows that the curve was symmetrical and Arrhenius-like, which indicates that the devolatilization and volatiles combustion processes were not disturbed by the char combustion process. This phenomenon arose because of the high rate

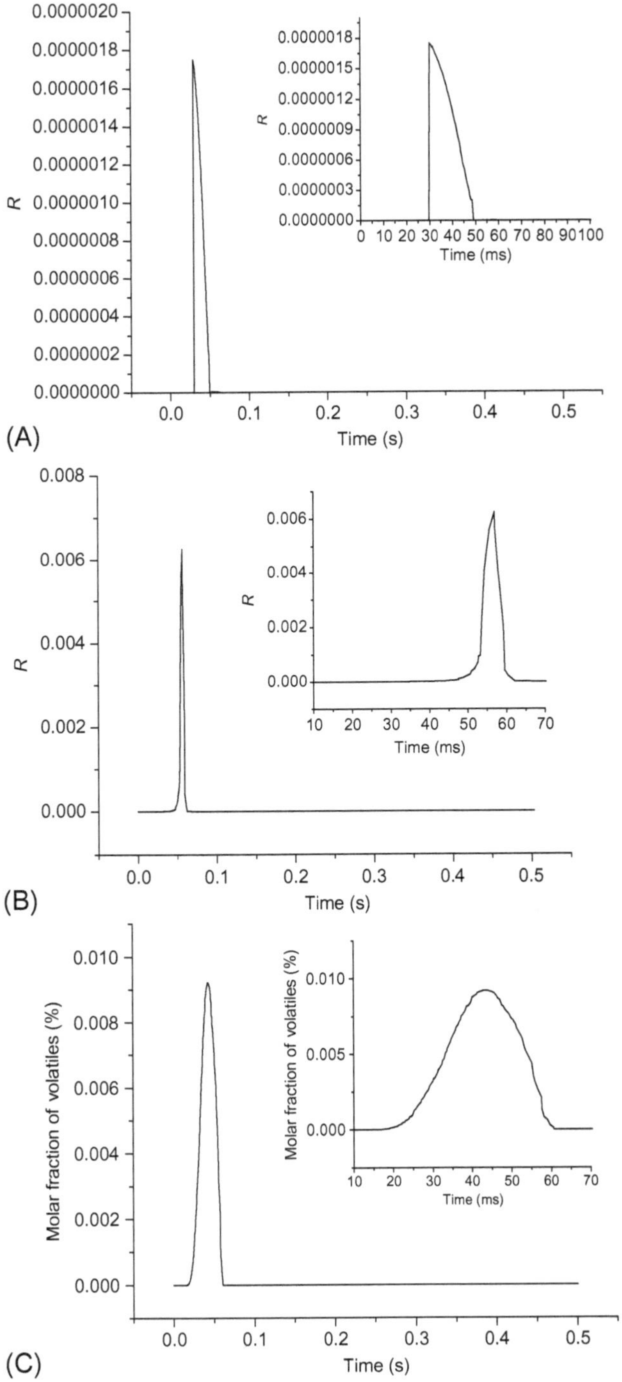

FIG. 3.3 (A) Rate of devolatilization, (B) rate of volatiles combustion, (C) mole fraction of volatiles around the surface of the char particle,

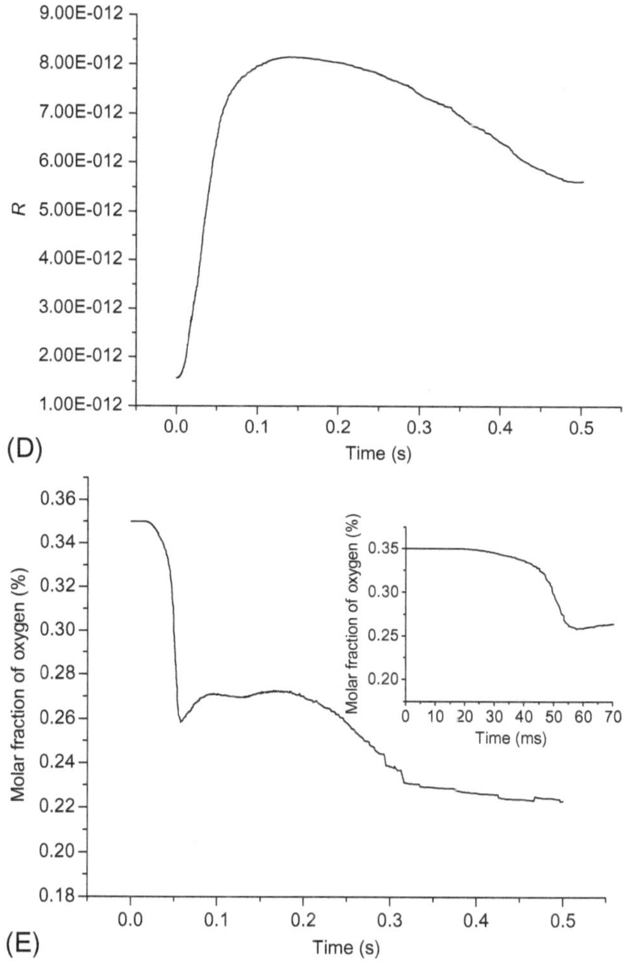

FIG. 3.3, CONT'D (D) rate of char combustion, and (E) mole fraction of oxygen around the surface of char particle, and versus time during the homogeneous ignition process.

of volatiles combustion, which is five orders of magnitude greater than that of char combustion at the time of the char ignition.

During the simulation of the homogeneous ignition process, the ignition of char was determined based on the condition that the rate of char combustion change over time was 10% of its maximum. According to this condition, char ignited at 45 ms, at which time, the mole fraction of volatiles around the particle was maximized, and the rate of volatiles combustion markedly increased. The heat released from volatiles combustion was sufficient to induce char ignition. Fig. 3.3D indicates that the rate of char combustion steadily increased from 45 ms to 63 ms. In other words, char combustion was not affected by the devolatilization or volatiles combustion during the homogeneous ignition process of coal.

Fig. 3.3E shows the variance of the mole fraction of O_2 around the surface of coal particle versus time. After homogeneous ignition, the O_2 mole fraction steadily decreased until the char ignited at 45 ms, after which the O_2 mole fraction drastically decreased. The O_2 mole fraction was minimized when volatiles combustion ended and then rapidly increased to reach a second maximum. This phenomenon arose because volatiles combustion overlapped char combustion between 45 and 63 ms, which increased the consumption of O_2. Subsequently, the O_2 mole fraction increased because of the end of volatiles combustion. The amount of rebound from the minimum to the second maximum was approximately 0.07, which also demonstrated that the amount of O_2 consumed by the char during the homogeneous ignition process was small.

3.2.5 Heterogeneous Ignition Process

Heterogeneous ignition was examined in an environment of 35% O_2/65% CO_2 and at a secondary air temperature of 1473 K. According to Section 3.2, the coal particle ignited at 15 ms and 60 mm down from the exit of primary air at this condition. After ignition, the rate of char combustion steadily increased until it reached 43 ms (Fig. 3.4A) and was subsequently minimized at 45 ms. This phenomenon was associated with the combustion of volatiles that subsequently occurred. During the simulation of heterogeneous ignition, the ignition of volatiles was determined based on the fact that the rate of volatiles combustion at the ignition point was 10% of its maximum. According to this condition, the ignition of volatiles occurred at 43 ms, as shown in Fig. 3.4B, which was responsible for the reduction of the rate of char combustion at the time. Fig. 3.4B indicates that the rate of volatiles combustion was maximized at 45 ms, which corresponded to the minimum char combustion rate. These findings show that the combustion of volatiles strongly affected the combustion of chars. During the heterogeneous ignition process, a significant amount of oxygen was consumed by volatiles because of their high combustion rate. As a result, the rate of char combustion decreased during the volatiles combustion process. Volatiles combustion ended at 50 ms after 7 ms, which was considerably shorter than the homogeneous ignition process.

The devolatilization process is discussed based on Fig. 3.4C. Volatiles were released slightly at the time of heterogeneous ignition. This finding demonstrated that coal ignites heterogeneously. Devolatilization started at 26 ms and ended at 47 ms, just 2 ms before the end of volatiles combustion. The devolatilization process lasted 21 ms and was longer than that for homogeneous ignition.

When volatiles ignited, the mole fraction of volatiles was approximately 0.01, which was near its maximum and greater than during homogeneous ignition (approximately 0.006) (Fig. 3.4D). This difference may be attributed to the limited amount of heat released from char combustion, which resulted in a higher mole fraction of volatiles required for ignition.

Fig. 3.4E shows that the mole fraction of oxygen was minimized when volatiles combustion ended, which was similar to homogeneous ignition. The rebound amount of the oxygen mole fraction was much greater than that in homogeneous ignition. This difference implied that a certain amount of the char had been consumed when the volatiles combustion finished in the heterogeneous ignition process. This process is crucial to the formation of c-NOx in coal combustion.

The interaction between the volatiles combustion and char combustion was analyzed for the heterogeneous ignition process from Fig. 3.4A–D. Because the rate of volatiles combustion was much higher than that of char combustion, volatiles combustion depressed the char combustion because of the competition for O_2 around the coal particle. Therefore the curve of the rate of char combustion

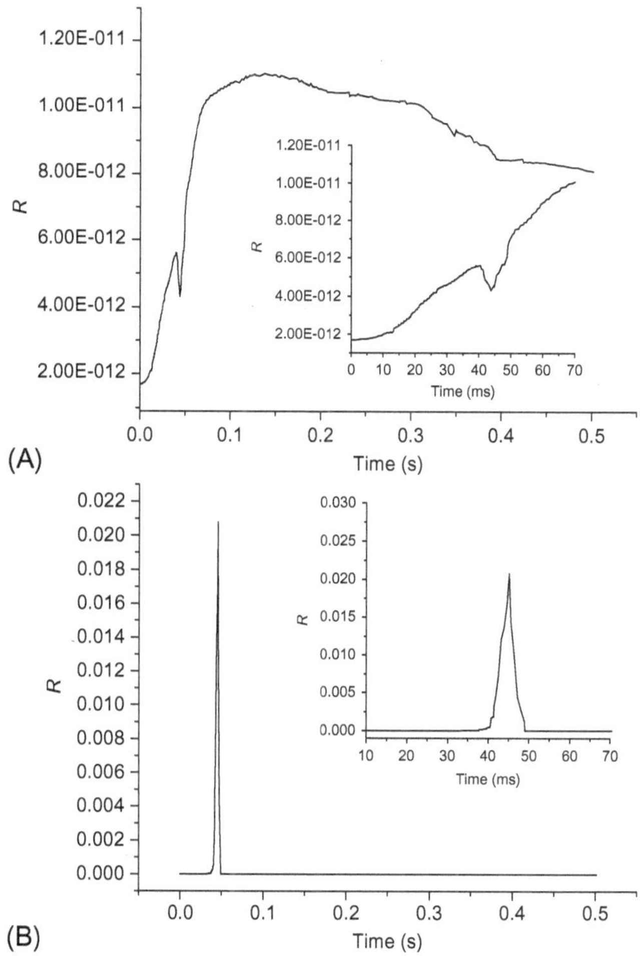

FIG. 3.4 (A) Rate of char combustion, (B) rate of volatiles combustion,

(Continued)

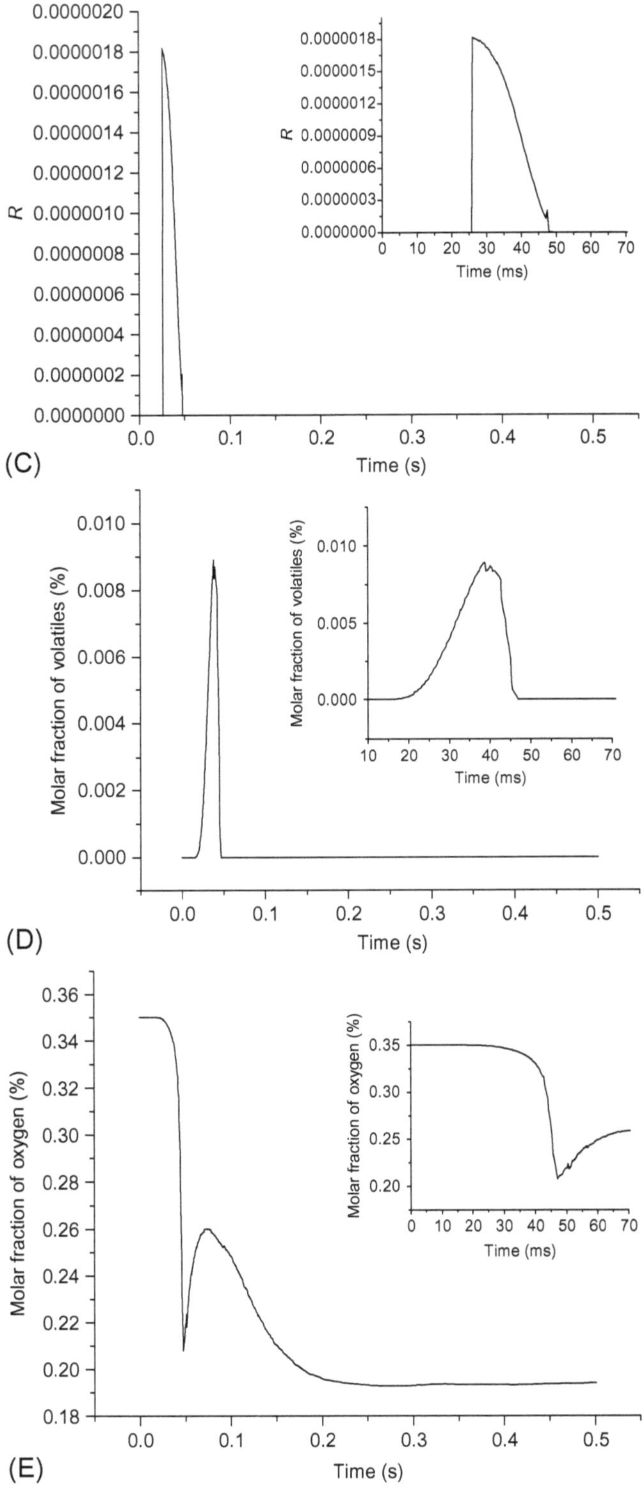

FIG. 3.4, CONT'D (C) rate of devolatilization, (D) mole fraction of volatiles around the surface of the char particle, and (E) mole fraction of oxygen around the surface of char particle, and versus time during the heterogeneous ignition process.

showed a v-notch because of the volatiles combustion process, as shown in Fig. 3.4A. After the volatiles ignited, the rate of char combustion decreased to its minimum as the rate of volatiles combustion increased to its maximum; it then increased with the reduction of the volatiles combustion rate. Conversely, the rate of volatiles combustion increased slowly after volatiles ignition because of the competition for O_2 by char combustion. When the mole fraction of volatiles passed through its maximum, the rate of volatiles combustion drastically decreased because of the sharp reduction of the mole fraction of volatiles. As a result, the curve of the rate of volatiles combustion was skewed. Many researchers have observed that char combustion is extinguished for the heterogeneous ignition process. This phenomenon did not occur in the simulation. However, it is likely to occur when the mole fraction of O_2 around the coal particle is very low, which is caused by the combustion of great amounts of volatiles, which cannot sustain char combustion.

3.3 EFFECTS OF CO_2 ON THE IGNITION OF PULVERIZED COAL

In this section, numerical simulation was used to analyze the effects of the physical and chemical properties of CO_2 on the homogeneous ignition and heterogeneous ignition processes of coal in O_2/CO_2 environments based on Jovanovic's experiments. The artificial materials *X*, *Y*, and *Z*, whose physical properties can be set according to needs and chemical properties are inert, were used to isolate the physical properties, such as the volume specific heat, mass diffusivity, and radiation. GRI 3.0 was adopted to analyze the chemical effects of CO_2 on the homogeneous ignition of coal, and the CO_2 gasification was considered. The other submodels were shown in Section 3.2.

3.3.1 Validation of the Model

The comparison of the results of the present model with those of experiments, as shown in Fig. 3.5, showed that the present CFD model could accurately predict the position of coal ignition points during both O_2/N_2 and oxy-fuel conditions. Thus the present CFD model could be used to investigate the coal ignition mechanism in O_2/CO_2 combustion.

The ignition mechanism was shown in Section 3.2, and the cases of 21% O_2/79% CO_2 and 35% O_2/65% CO_2 in 1473 K were selected to state the effect of CO_2 on the ignition of coal in homogeneous and heterogeneous ignition mechanism.

3.3.2 The Chemical Effect of CO_2 on the Homogeneous Ignition of Pulverized Coal at O_2/CO_2 Environments

The physical properties and chemical properties of CO_2 on the ignition of coal were separated during simulation. An artificial material *X*, which had the same thermochemical, transport, and radiation properties as CO_2 and was chemically

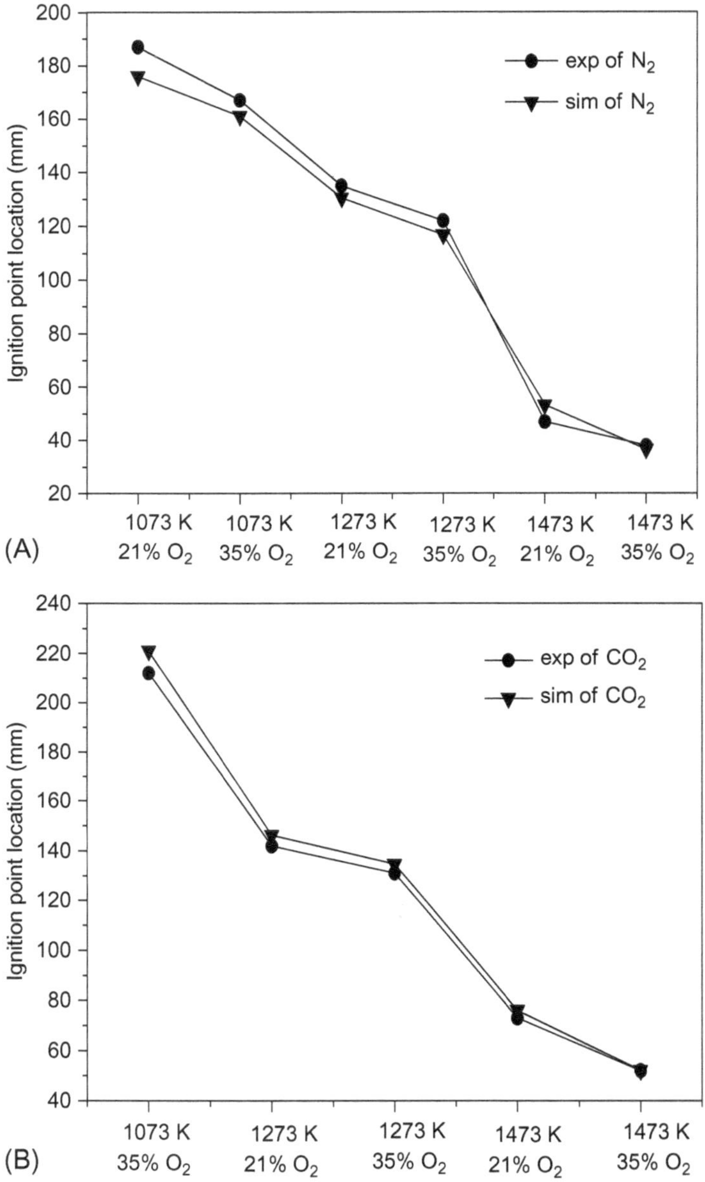

FIG. 3.5 Comparison of ignition distance of present model and experiment. (A) O_2/N_2 mixtures and (B) O_2/CO_2 mixtures.

inert, was introduced in the present study, as shown in Table 3.3. Consequently, the differences of the position of ignition between the *X* case and the O_2/CO_2 case showed the effects of the chemical properties of CO_2 on the ignition.

Fig. 3.6 shows the ignition distances of coal in CO_2 in 21% O_2/79% CO_2 and 1473 K, in which pulverized coal ignited homogeneously, and corresponding *X*

TABLE 3.3 Physical Properties of Different Materials at 1473 K, 0.1 MPa

Gas	N_2	CO_2	*X*	*Y*	*Z*	*W*
C_p (J/mol K)	27.38	53.24	53.24	27.38	53.24	53.24
Absorption coff.(1/m)	0	0.43	0.43	0.43	0.43	0
Mass diff. (m^2/s)	4.37e−4	2.97e−4	2.97e−4	2.97e−4	4.37e−4	2.97e−4

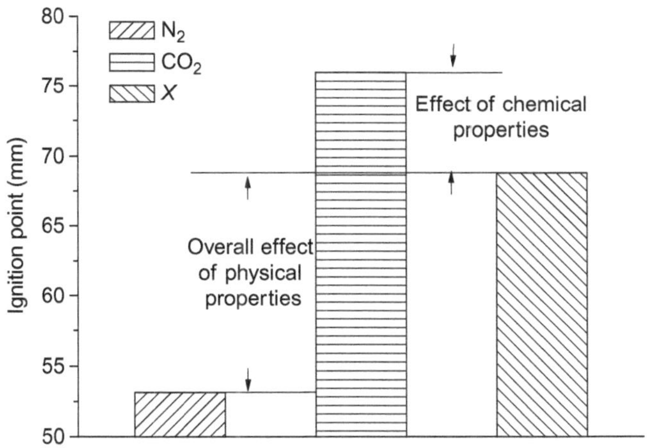

FIG. 3.6 Comparison of ignition point position of O_2/N_2, O_2/CO_2 and O_2/X environments on 1473 K, 21% O_2 condition.

cases. The interval between CO_2 and *X* case with identical O_2 concentration represented the effect of global chemical properties of CO_2 on the ignition of coal. Fig. 3.6 shows that ignition point positions in the CO_2 environments were much higher than those in the *X* environments, implying that the chemical properties of CO_2 also resulted in the delay of ignition of coal in CO_2 environments compared with in N_2 environments. Moreover, the effects of the chemical properties of CO_2 were comparable to those of the physical properties.

Methane was generally used as a surrogate for the volatile of coal in many studies. Thus the ignition time delays of methane in the 21% O_2/79% N_2, 21% O_2/79% *X*, and 21% O_2/79% CO_2 environments were calculated with the perfect stirred reactor (PSR) by CHEMKIN. Fig. 3.7 shows that ignition delay time in the CO_2 environment was significantly longer compared to the *X* and N_2 cases. Moreover, the heat release rate difference (HRRD) of each elementary reaction between the *X* case and the corresponding CO_2 case was introduced to study the chemical effects of CO_2 on the ignition time and was defined as $\triangle H_i = H_{i;X} - H_{i;CO2}$, where $H_{i;X}$ and $H_{i;CO2}$ represent the heat release rate of elementary reaction *i* in O_2/X and O_2/CO_2 environments, respectively. The top ten $\triangle H_i$ are

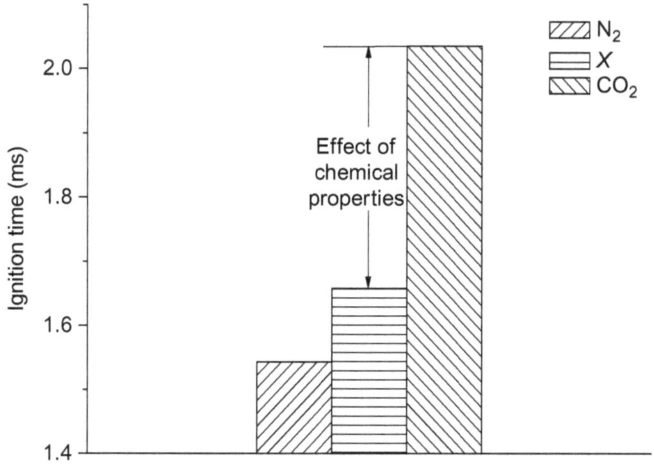

FIG. 3.7 Comparison of ignition time of O_2/N_2, O_2/X and O_2/CO_2 environments on 1473 K, 21% O_2 condition.

TABLE 3.4 Elementary Reactions of Temperature Sensitivity Coefficient

Reaction Number	Elementary Reactions	$\triangle H_i = H_{i;X} - H_{i;CO2}$
R99	$OH + CO \Leftrightarrow H + CO_2$	1.129E+10
R10	$O + CH_3 \Leftrightarrow H + CH_2O$	9.898E+09
R38	$H + O_2 \Leftrightarrow O + OH$	−6.782E+09
R284	$O + CH_3 \Leftrightarrow H + H_2 + CO$	6.534E+09
R153	$CH_2(S) + CO_2 \Leftrightarrow CO + CH_2O$	−6.352E+09
R52	$H + CH_3(+M) \Leftrightarrow CH_4(+M)$	2.323E+09
R58	$H + CH_2O \Leftrightarrow HCO + H_2$	2.001E+09
R168	$HCO + O_2 \Leftrightarrow HO_2 + CO$	1.779E+09
R84	$OH + H_2 \Leftrightarrow H + H_2O$	1.004E+09
R101	$OH + CH_2O \Leftrightarrow HCO + H_2O$	−3.833E+08

listed in Table 3.4, and corresponding elementary reactions are shown in Fig. 3.8. Table 3.4 shows that $\triangle H_{R99}$ ($CO_2 + H \Leftrightarrow OH + CO$) are the most among these ten reactions because R99 is an exothermic reaction in the *X* case, while it is an endothermic reaction in the CO_2 case because of the high concentration of CO_2, as shown in Fig. 3.8. Consequently, the high concentration of CO_2 resulted in an increase in the amount of heat absorption in the chemistry reaction system,

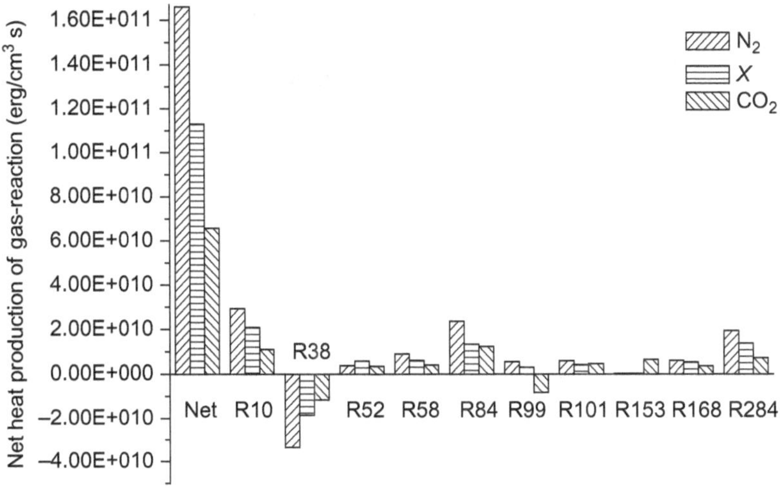

FIG. 3.8 The net heat production of gas-reaction at O_2/N_2, O_2/X and O_2/CO_2 environments at ignition time.

leading to the ignition delay in the O_2/CO_2 environment compared to the O_2/N_2 environment.

To study the effects of the specific heat, thermal radiation, and mass diffusion coefficient of CO_2 on the ignition, three artificial materials that were chemically inert were introduced. The artificial material *Y* had the specific heat of N_2 and the thermal radiation and mass diffusion coefficient of CO_2. The artificial material *Z* had the mass diffusion coefficient of N_2 and the specific heat and thermal radiation of CO_2. The artificial material *W* had the thermal radiation of N_2 and the mass diffusion coefficient of CO_2, as shown in Table 3.3. Consequently, the differences between the ignition points of *X* case and *Y* case showed the effect of the specific heat of CO_2 on the ignition, the differences between *X* case and *Z* case showed the effect of the mass diffusion coefficient of CO_2, and the differences between *X* case and *W* case showed the effect of the thermal radiation of CO_2.

The ignition distances in the CO_2, *X*, *Y*, *Z*, and *W* cases are shown in Fig. 3.9. The higher specific heat of CO_2 resulted in a longer ignition delay on the homogeneous ignition, which is regarded as the reason for the ignition delay in O_2/CO_2 environments by many researchers. The higher thermal radiation of CO_2 was also beneficial to the ignition delay because of more heat loss by thermal radiation. Moreover, as shown in Fig. 3.9, the lower mass diffusion coefficient caused the further ignition delay in O_2/CO_2 environments. Fig. 3.9 also shows that the effects of higher thermal radiation and lower mass diffusion coefficient on the ignition delay were much smaller than that of higher specific heat in the homogeneous ignition of coal. Consequently, the reasons for the homogeneous ignition delay of coal in the O_2/CO_2 environment compared to the corresponding O_2/N_2 environment were the higher specific heat and chemical properties of CO_2.

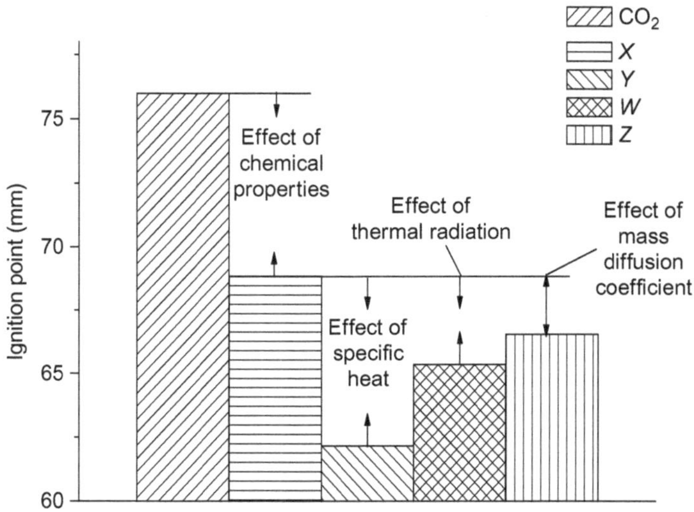

FIG. 3.9 Ignition distances in O_2/CO_2, O_2/X, O_2/Y, O_2/W, and O_2/Z environments on 1473 K, 21% O_2 condition.

3.3.3 The Effect of CO_2 on the Heterogeneous Ignition of Pulverized Coal at O_2/CO_2 Environments

According to the previous study, the case of the temperature of 1473 K in a 35% O_2/65% CO_2 environment was chosen to discuss the effects of CO_2 on heterogeneous ignition. Fig. 3.10 shows that the gap between CO_2 and X cases was very slight, implying that the effects of the chemical properties of CO_2 on

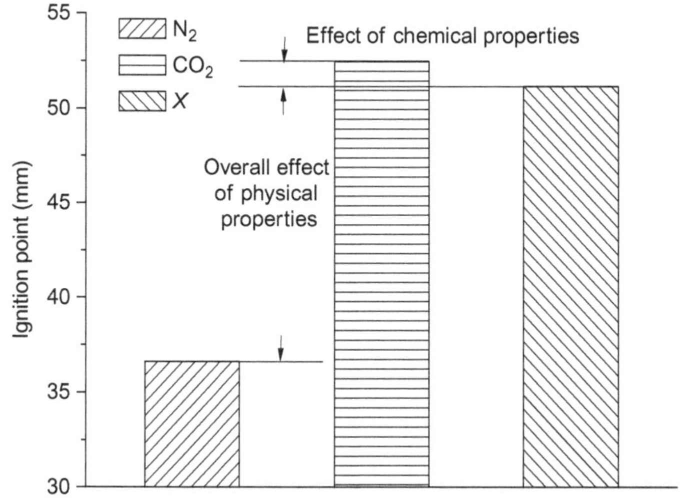

FIG. 3.10 Ignition distances in O_2/N_2, O_2/CO_2, and O_2/X environments with 35% O_2 in 1473 K.

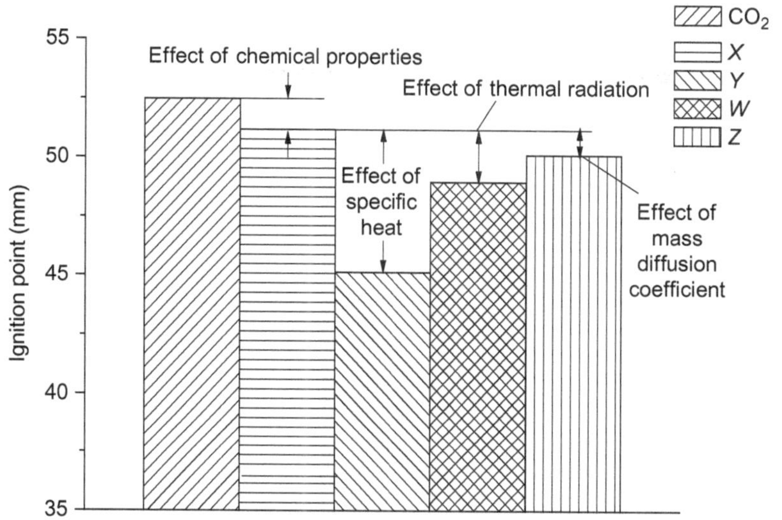

FIG. 3.11 Ignition distance of O_2/CO_2, O_2/X, O_2/Y, O_2/W, and O_2/Z environments on 1473 K 35% O_2 condition.

ignition delay were far weaker than those of the physical properties of CO_2 in the heterogeneous ignition.

Fig. 3.11 shows that the higher specific heat of CO_2 obviously influenced ignition delay on the heterogeneous ignition. It also shows that the higher thermal radiation of CO_2 and the lower mass diffusion coefficient led to significant ignition delay because the coal particle lost more heat due to radiation and because of the difficulty of O_2 diffusing to the coal particle surface, respectively. Fig. 3.11 also shows that the extent of the effects of the specific heat of CO_2 on the heterogeneous ignition delay of coal was comparable with the sum of the effects of the thermal radiation and mass diffusion coefficient of CO_2. In other words, the effects of the thermal radiation and mass diffusion coefficient of CO_2 were the other two physical reasons for the heterogeneous ignition delay of coal in O_2/CO_2 environments. Consequently, the reasons for the ignition delay in the heterogeneous ignition of pulverized coal in O_2/CO_2 environments compared with O_2/N_2 environments were mainly the higher specific heat, higher thermal radiation, and lower mass diffusion coefficient of CO_2 compared to N_2.

3.4 EFFECTS OF H_2O ON THE IGNITION OF PULVERIZED COAL

In this section, the ignition of bituminous coal particles was studied by experimental and numerical methods under oxy-fuel conditions with the addition of different concentrations of H_2O. The experiments were conducted in a drop-tube furnace utilizing a high-speed camera, and the numerical simulations were performed by the previously proposed model to conduct a detailed investigation of the effect of H_2O on the ignition of coal particles in O_2/CO_2 mixtures.

3.4.1 Experimental

A Chinese bituminous coal was employed in this study. The coal was air-dried, ground, and sieved to a size cut of 97–105 μm. The proximate and ultimate analysis of the coal sample is shown in Table 3.5.

The experiments were carried out in an electrically heated DTF at a constant wall temperature of 1373 K, as shown in Fig. 3.12. The furnace was a transparent quartz tube 1.0 m in length, which consisted of two layers

TABLE 3.5 Properties of the Coal Sample

Proximate analysis (%w)	Moisture 6.40	Volatile matter 36.95	Ash 2.90	Fixed carbon 53.75
Ultimate analysis (daf, %w)	Carbon 72.82	Hydrogen 4.88	Nitrogen 0.93	Oxygen + sulfur 21.37

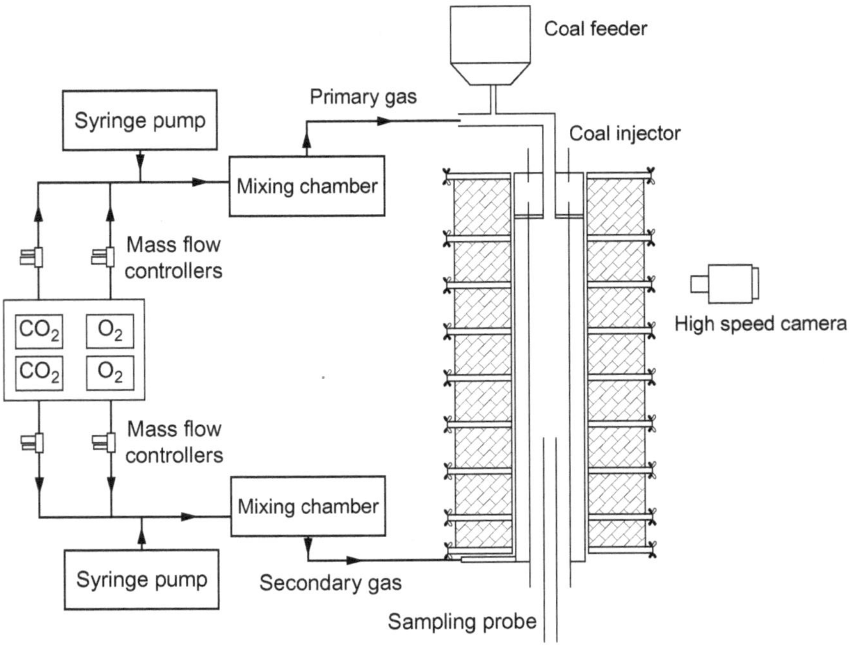

FIG. 3.12 Schematic structure of the DTF.

with inner diameters of 50 and 80 mm, respectively. The furnace was fitted for this work in this manner: Viewports with a diameter of 20 mm were arranged along the furnace for nonintrusive measurement at eight levels, as illustrated in Fig. 3.12. Pulverized coal particles were introduced from the scraper feeder and entrained by a primary gas flow of 1.0 L/min (STP) to the inner chamber through the injector with a diameter of 10 mm. The secondary gas that was used for combustion was carried from the bottom of the reactor into the annulus at a rate of 9.0 L/min (STP). The temperature of the secondary gas attained the wall temperature at the top of the reactor, passed through a quartz flow straightener, and then mixed with the coal and primary gas. The flow rates of CO_2 and O_2 were controlled by mass flow controllers. The steam was generated by the mixing chamber, in which distilled water was heated up to 473 K by an electric resistance wire to ensure complete vaporization. Liquid water was controlled by a high-precision syringe pump. Stainless steel tubing connected the mixing chamber to the reactor, which was wrapped with heating tape and insulated to maintain a constant temperature of 423 K.

Nonintrusive observation of the ignition characteristics was achieved through the viewports that were arranged along the furnace with a diameter of 20 mm. A FASTCAM SA4 high-speed camera was used in all experiments at a frame speed of 1000 fps. The focus of the camera was set to view the central line of the furnace, where the particles ideally appeared. In each case, the exposure time was set to approximately 1000 μs, and the recording was made at least three times. Several coal injectors with different lengths were used in the experiments. An appropriate injector was used in a specific environment to ensure that the ignition images occurred in the field of the viewports. Thus the ignition characteristics could be recorded relatively more accurately and continuously.

The furnace was heated up to 1373 K at 15°C min^{-1} for approximately 1.5 h, and then a constant temperature was maintained. In the ignition tests, the O_2/CO_2 environments were applied with three different oxygen mole fractions (21%, 30%, and 40%). To evaluate the effect of steam on the ignition characteristics in O_2/CO_2 conditions, the mole fractions of 5%, 10%, 20%, and 30% of steam were tested as a substitution of CO_2 in each O_2/CO_2 environment.

3.4.2 Experimental Results

The ignition distance was defined as the distance between the ignition point and the feeding point that could be detected for a certain environment [9,14]. For a given environment, approximately 100 ignition samples of single coal particles were selected to estimate the average ignition distance.

The experimental results of ignition distance are shown in Fig. 3.13. Fig. 3.13 shows that the effects of steam on the ignition of coal in oxy-steam

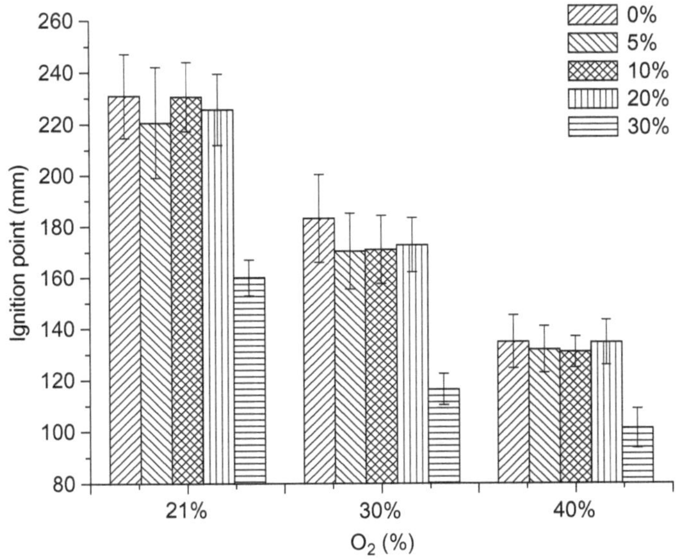

FIG. 3.13 Experimental ignition distances in O_2/CO_2 environments with and without the addition of H_2O.

combustion changed with the mole fractions of O_2 and steam in the environment. At the mole fractions of O_2 of 21% and 30%, all the distances of ignition in the steam cases were shorter than those in the case without steam with an identical mole fraction of O_2, implying that the addition of steam was conducive to the ignition of coal, resulting in ignition occurring sooner. Moreover, in the range of the mole fraction of steam from 5% to 20%, the distances of the ignition of coal had slight differences between each other for an identical mole fraction of O_2. At the mole fraction of steam of 30%, however, the distances of ignition dropped dramatically compared to the others, indicating that the high concentration of steam strongly impacted the ignition. When the mole fraction of O_2 increased to 40%, the distances of the ignition in the range of the mole fraction from 5% to 20% were very close to the case without steam, showing that the effects of steam were not obvious on the ignition of coal in these environments. However, the distance of the ignition at the mole fraction of steam of 30% was evidently shorter than the other cases with an identical mole fraction of O_2. Moreover, the extent of drop was significantly lower than the corresponding cases of the mole fraction of O_2 of 21% and 30%. These behaviors showed that the effects of steam are significantly weakened with increasing O_2 concentration. This novel phenomenon may be accounted for by the different physical and chemical properties between steam and CO_2, such as the specific heat, thermal radiation, diffusivity, and reactivity. Therefore the role of steam in the ignition processes of coal particles in $O_2/CO_2/H_2O$ environments should be comprehensively analyzed through the numerical method.

3.4.3 Modeling

The detailed submodels were described in Section 3.2. The steam gasification reaction was taken into account in the simulations to assess the effect of steam on the ignition behaviors in the oxy-fuel environments [20].

3.4.4 Validation of the Model

Fig. 3.14 depicts the comparisons between the experimental and numerical ignition distances in O_2/CO_2 environments with and without the addition of steam. The relative errors between the numerical and experimental results were less than 5%. Consequently, the CFD model could predict well the ignition process of coal in O_2/CO_2 environments with or without the addition of steam. Thus the model was used to analyze in detail the effect of steam in O_2/CO_2 environments on the ignition of coal particles. The physicochemical properties of steam are significantly different from those of CO_2, thus the physical and chemical effect of steam on the ignition of coal will be depicted in the following sections.

3.4.5 Effect of Steam Physical Properties

To evaluate the effect of the physical properties of steam on the ignition of coal in $O_2/CO_2/H_2O$ environments, the chemical properties of steam were isolated in the simulations. An artificial material *X* was introduced to substitute steam in the simulations. The physical properties of *X* were set as the same as those of

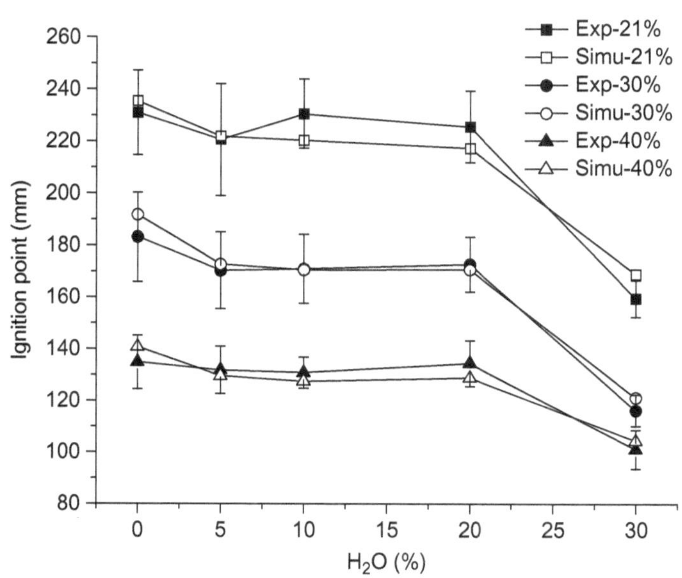

FIG. 3.14 Experimental and simulation results of ignition distance with and without the addition of H_2O.

steam, but *X* had the same chemical properties as CO_2. The higher thermal radiation, lower specific heat, and higher diffusivity of steam, compared to CO_2, had different effects on the ignition of coal during the combustion process. Therefore three artificial materials *Y*, *Z*, and *W* were displayed therein. The thermal radiation and chemical properties of *Y* were set with the same values of CO_2, and the other physical properties of *Y* were kept at the same values of steam. The specific heat and chemical properties of *Z* were set with the same values of CO_2, and the other physical properties of *Z* were kept at the same values of steam. The diffusivity and chemical properties of *W* were set with the same values of CO_2, and the other physical properties of *W* were kept at the same values of steam.

Fig. 3.15 shows the results of the ignition distance reflecting the effects of the physical properties of steam on the ignition of coal in $O_2/CO_2/H_2O$ environments. The gap between the cases of O_2/CO_2 environments and $O_2/CO_2/X$ environments represented the global effect of the physical properties of steam on the ignition of coal. It is obvious that the physical properties of steam led to the ignition delay shown in Fig. 3.15. The gap between $O_2/CO_2/X$ and $O_2/CO_2/Y$ represented the effect of the thermal radiation of steam on the ignition compared to CO_2. The ignition distance in the cases of $O_2/CO_2/Y$ was shorter than that in the case of $O_2/CO_2/X$ at an identical steam concentration, which implied that higher thermal radiation of steam compared to CO_2 resulted in ignition delay. This phenomenon also meant that a higher thermal radiation of steam would lead to ignition delay because steam would absorb more heat than

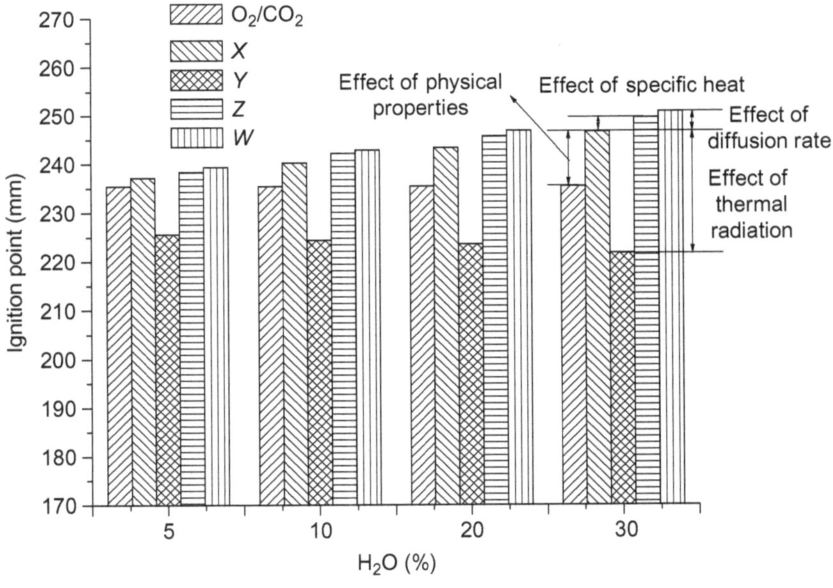

FIG. 3.15 Prediction results of ignition distance considering the physical properties of steam with oxygen mole fraction of 21%.

CO_2. The gaps between $O_2/CO_2/X$ and $O_2/CO_2/Z$ and $O_2/CO_2/W$ represented the effect of specific heat and diffusivity of steam on the ignition compared to CO_2, respectively. The values of $O_2/CO_2/Z$ and $O_2/CO_2/W$ were larger than that of $O_2/CO_2/X$ at an identical steam concentration, which demonstrated that lower specific heat and higher diffusivity of steam compared to CO_2 resulted in the ignition occurring sooner. Consequently, the physical property of steam leading to the ignition delay was its thermal radiation.

As seen in Fig. 3.15, the gap between the case of $O_2/CO_2/X$ and $O_2/CO_2/Y$ was larger than the gap between the case of $O_2/CO_2/X$ and $O_2/CO_2/Z$ and the gap between the case of $O_2/CO_2/X$ and $O_2/CO_2/W$ at an identical steam concentration, implying that the pulverized coal ignition was affected more by the thermal radiation of steam than by the specific heat and diffusivity of steam. Moreover, the gap between the case of $O_2/CO_2/X$ and $O_2/CO_2/Z$ was slightly smaller than that between the cases of $O_2/CO_2/X$ and $O_2/CO_2/W$, showing that the effect of the diffusivity of steam on the pulverized coal was comparable to that of the specific heat of steam. Finally, the gaps among different physical properties increased with increasing concentrations of steam. These phenomena show that the effect of the physical properties of steam on the ignition of coal increased with increasing steam concentrations.

3.4.6 Effect of Steam Chemical Properties

The effects of steam on coal ignition were positive, as shown in Fig. 3.13, and the physical properties of steam had a negative effect on the ignition of coal in $O_2/CO_2/H_2O$ environments; consequently, the chemical properties of steam had a positive effect on the ignition of coal. Two factors of the chemical effect of steam on the ignition of coal were considered in this study. One was the effect of the steam shift reaction, and the other was the effect of the steam gasification reaction. To assess the detailed effect of the two reactions on the ignition of coal in $O_2/CO_2/H_2O$ environments, two artificial materials U and V were introduced into the simulations to replace steam. In the cases of $O_2/CO_2/U$, the steam gasification reaction was cancelled in the simulations, and in the cases of $O_2/CO_2/V$, the steam shift reaction was cancelled in the simulations.

Fig. 3.16 shows the comparisons among the predicted results of ignition distance considering the effect of chemical properties of steam on the ignition of coal in $O_2/CO_2/H_2O$ environments with an oxygen mole fraction of 21%. The predicted results of $O_2/CO_2/U$ and $O_2/CO_2/V$ represented the effect of the steam gasification reaction and the steam shift reaction, respectively. The results indicated that the effect of the two reactions resulted in the advanced ignition of coal in $O_2/CO_2/H_2O$ environments. The gap between the results of $O_2/CO_2/H_2O$ and $O_2/CO_2/U$ was smaller than that between $O_2/CO_2/H_2O$ and $O_2/CO_2/V$ for comparable steam concentrations, as demonstrated in Fig. 3.16. This phenomenon revealed that the effect of the steam shift reaction resulted in the advanced ignition of coal particles more significantly than that of the steam gasification reaction.

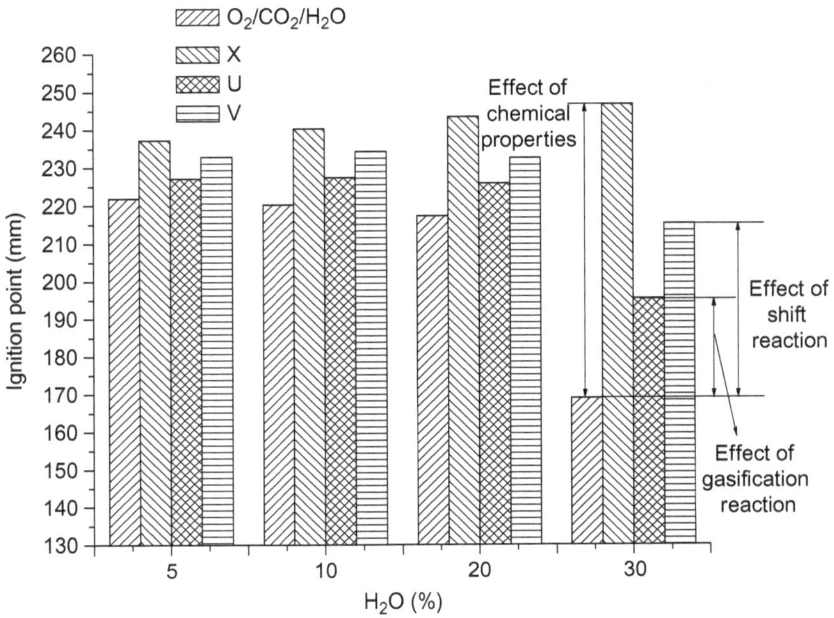

FIG. 3.16 Prediction results of ignition distance considering the chemical properties of steam with oxygen mole fraction of 21%.

The numerical simulations of the ignition mechanism in 30% O_2 are shown in Fig. 3.17 (the ignition mechanism of other O_2 mole fractions were also conducted, and the results followed a similar trend so are not shown here). In the case of 5%, 10%, and 20% steam, the rate of devolatilization was higher than that of coal combustion, and the rate of volatiles combustion was higher than that of char combustion. According to the criteria of the ignition mechanism mentioned in Section 3.2, the ignition mechanism was homogeneous. In the case of 30% steam, the rate of devolatilization was higher than that of coal combustion, and the rate of volatiles combustion was lower than that of char combustion, so the ignition mechanism was heterogeneous. Obviously, the different mole fractions of steam resulted in the discrepancy of the ignition mechanism. The coal ignited homogeneously when a low mole fraction of steam (5%, 10%, and 20%) was added to the O_2/CO_2 environment. Thus the ignition distances barely changed at low mole fractions of steam. However, when steam was increased to 30%, the coal ignited heterogeneously, and the ignition distances decreased significantly.

3.5 SUMMARY

1. For homogeneous ignition in O_2/CO_2 environments, the char generally ignites at the maximum mole fraction of volatiles near the surface of the coal

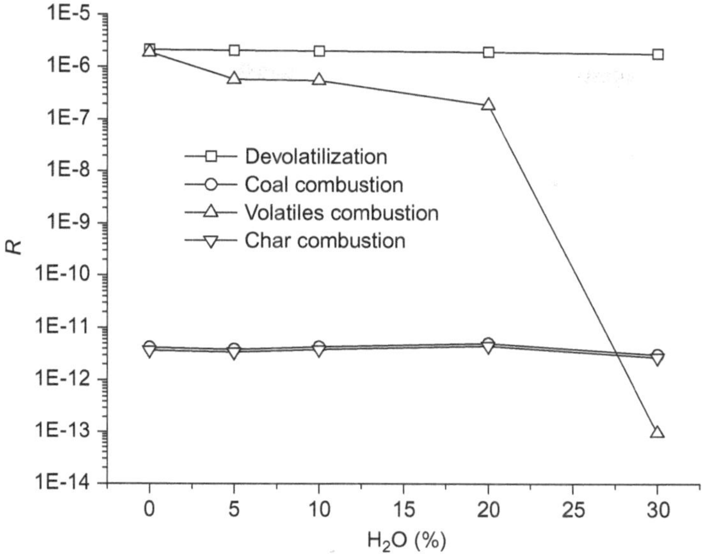

FIG. 3.17 Rates of devolatilization, volatiles, and char and coal combustion at the ignition point location with and without the addition of H_2O.

particle. The char combustion process does not disturb the devolatilization and volatiles combustion processes. For heterogeneous ignition in O_2/CO_2 environments, volatiles generally ignite at the maximum mole fraction of volatiles near the surface of the coal particle. The interaction between the combustion of volatiles and char combustion is significant. The curve of the rate of char combustion exhibits a v-notch because of the volatiles combustion process, and the curves of the rate of volatiles combustion and mole fraction of volatiles are skewed because of the char combustion process.

2. In the homogeneous ignition mechanism of coal in the O_2/CO_2 environment, the reasons for the ignition delay compared to the corresponding O_2/N_2 environment were the higher specific heat and the chemical properties of CO_2. The chemical properties of CO_2 resulted in the ignition delay during the homogeneous ignition of pulverized coal in O_2/CO_2 environments. The high concentration of CO_2 caused an increase in the amount of heat absorption in the chemistry reaction system, especially R99 ($CO_2 + H \Leftrightarrow OH + CO$), which was an exothermic reaction in the *X* case and an endothermic reaction in the CO_2 case. The higher specific heat, higher thermal radiation, and lower mass diffusion coefficient of CO_2 played important roles in the ignition delay in the heterogeneous ignition of pulverized coal in O_2/CO_2 environments, as compared to O_2/N_2 environments.
3. The overall effects of the physical properties of steam resulted in the ignition delay of coal in O_2/CO_2 environments with the addition of steam. Compared to CO_2, the lower specific heat and higher diffusivity of steam resulted in

the advanced ignition of coal, and the higher radiation thermal of steam resulted in the ignition delay. The chemical properties of steam resulted in the advanced ignition of coal. Compared to the steam gasification reaction, the steam shift reaction was more favorable to the ignition of coal.

REFERENCES

[1] Essenhigh RH, Misra MK, Shaw DW. Ignition of coal particles: a review. Combust Flame 1989;77:3–30.

[2] Zhang D-k, Wall TF. Ignition of coal particles: the influence of experimental technique. Fuel 1994;73(7):1114–9.

[3] Wall T, Gupta R, Gururajan V, Zhang D-k. The ignition of coal particles. Fuel 1991;70(9):1011–6.

[4] Wall T, Gururajan V. Combustion kinetics and the heterogeneous ignition of pulverized coal. Combust Flame 1986;66(2):151–7.

[5] Howard J, Essenhigh R. The mechanism of ignition of pulverized coal. Combust Flame 1965;9(3):337–9.

[6] Howard J, Essenhigh R. Mechanism of solid-partical combustion with simultaneous gas-phase volatiles combustion. Proc Combust Inst 1967;11:399–408.

[7] Annamalai K, Durbetaki P. A theory on transition of ignition phase of coal particles. Combust Flame 1977;29:193–208.

[8] Du X, Annamalai K. The transient ignition of isolated coal particle. Combust Flame 1994;97(3):339–54.

[9] Khatami R, Stivers C, Levendis YA. Ignition characteristics of single coal particles from three different ranks in O_2/N_2 and O_2/CO_2 atmospheres. Combust Flame 2012;159:3554–68.

[10] Molina A, Shaddix CR. Ignition and devolatilization of pulverized bituminous coal particles during oxygen/carbon dioxide coal combustion. Proc Combust Inst 2007;31(2):1905–12.

[11] Shaddix CR, Molina A. Particle imaging of ignition and devolatilization of pulverized coal during oxy-fuel combustion. Proc Combust Inst 2009;32(2):2091–8.

[12] Liu Y, Geier M, Molina A, Shaddix CR. Pulverized coal stream ignition delay under conventional and oxy-fuel combustion conditions. Int J Greenhouse Gas Control 2011;5:S36–46.

[13] Zhang L, Binner E, Qiao Y, Li C-Z. High-speed camera observation of coal combustion in air and O_2/CO_2 mixtures and measurement of burning coal particle velocity. Energy Fuel 2009;24(1):29–37.

[14] Jovanovic R, Milewska A, Swiatkowski B, Goanta A, Spliethoff H. Numerical investigation of influence of homogeneous/heterogeneous ignition/combustion mechanisms on ignition point position during pulverized coal combustion in oxygen enriched and recycled flue gases atmosphere. Int J Heat Mass Transf 2011;54(4):921–31.

[15] Jovanovic R, Milewska A, Swiatkowski B, Goanta A, Spliethoff H. Sensitivity analysis of different devolatilisation models on predicting ignition point position during pulverized coal combustion in O_2/N_2 and O_2/CO_2 atmospheres. Fuel 2012;101:23–37.

[16] Huang X, Jiang X, Han X, Wang H. Combustion characteristics of fine-and micro-pulverized coal in the mixture of O_2/CO_2. Energy Fuel 2008;22(6):3756–62.

[17] Bejarano PA, Levendis YA. Single-coal-particle combustion in O_2/N_2 and O_2/CO_2 environments. Combust Flame 2008;153:270–87.

[18] Gil M, Riaza J, Álvarez L, Pevida C, Pis J, Rubiera F. Oxy-fuel combustion kinetics and morphology of coal chars obtained in N_2 and CO_2 atmospheres in an entrained flow reactor. Appl Energy 2012;91:67–74.

[19] Tan Y. Oxy-fuel combustion for power generation and carbon dioxide (CO_2) capture. UK: Woodhead Publishing Ltd; 2011. p. 35–53.
[20] Hecht ES, Shaddix CR, Geier M, Molina A, Haynes BS. Effect of CO_2 and steam gasification reactions on the oxy-combustion of pulverized coal char. Combust Flame 2012;159:3437–47.
[21] Riaza J, Álvarez L, Gil M, Pevida C, Pis J, Rubiera F. Effect of oxy-fuel combustion with steam addition on coal ignition and burnout in an entrained flow reactor. Energy 2011;36:5314–9.
[22] Marek E, Świątkowski B. Experimental studies of single particle combustion in air and different oxy-fuel atmospheres. Appl Therm Eng 2014;66:35–42.
[23] FLUENT6.3. User's guide. United States: Fluent Inc.; 2006.
[24] Álvarez L, Gharebaghi M, Pourkashanian M, Williams A, Riaza J, Pevida C, et al. CFD modelling of oxy-coal combustion in an entrained flow reactor. Fuel Process Technol 2011;92(8):1489–97.
[25] Álvarez L, Gharebaghi M, Jones J, Pourkashanian M, Williams A, Riaza J, et al. CFD modeling of oxy-coal combustion: prediction of burnout, volatile and NO precursors release. Appl Energy 2013;104:653–65.
[26] Jones J, Pourkashanian M, Waldron D, Williams A. Prediction of NOx and unburned carbon in ash in highly staged pulverised coal furnace using overfire air. J Energy Inst 2010;83(3):144–50.
[27] Hashimoto N, Kurose R, Hwang S-M, Tsuji H, Shirai H. A numerical simulation of pulverized coal combustion employing a tabulated-devolatilization-process model (TDP model). Combust Flame 2012;159(1):353–66.
[28] Shi S-P, Zitney S, Shahnam M, Syamlal M, Rogers W. Modelling coal gasification with CFD and discrete phase method. J Energy Inst 2006;79(4):217–21.
[29] Ponzio A, Senthoorselvan S, Yang W, Blasiak W, Eriksson O. Ignition of single coal particles in high-temperature oxidizers with various oxygen concentrations. Fuel 2008;87(6):974–87.
[30] Khare S, Wall T, Farida A, Liu Y, Moghtaderi B, Gupta R. Factors influencing the ignition of flames from air-fired swirl pf burners retrofitted to oxy-fuel. Fuel 2008;87(7):1042–9.

Chapter 4

Pulverized Coal Combustion Characteristics in Oxy-fuel Atmospheres

Liqi Zhang*, Baojun Yi†
**Huazhong University of Science and Technology, Wuhan, China*
†Huazhong Agricultural University, Wuhan, China

4.1 INTRODUCTION

Oxy-fuel combustion is considered to be the most promising CO_2 capture technology [1,2]. Changing from N_2 to CO_2 can alter combustion characteristics, flame propagation, and boiler heat transfer. Among these factors, the most significant changes are those related to combustion characteristics [3–9], which are important in the design of boilers. Therefore discovering the influence of characteristic temperatures is critical.

4.2 THE RELATIONSHIP BETWEEN PULVERIZED COAL PROPERTIES AND CHARACTERISTIC TEMPERATURES

To investigate the relationship between pulverized coal properties and characteristic temperatures, 28 kinds of Chinese coals were chosen for this research. All of the coals were investigated in three O_2 atmospheres in order to determine the relationship between the combustion characteristics and the coal properties under various O_2 concentrations and to help achieve maximum economic benefits for the kinds of coal that favour the presence of O_2.

4.2.1 Samples and Experimental Process

Twenty-eight kinds of Chinese coals were selected for the study of the combustion characteristics of pulverized coal under oxy-fuel conditions. Twenty-two of the coals were Xinzhuang (XZ), Liuzhi (LZ), Tiangou (TG), Xiyangcun (XYC), Shanxi (SX1), Shangping (SP), Sichuan (SC), Ma'anshan (MAS), Xiwujiang (XWJ), Sijiaotian (SJT), Shanxi (SX2), Liupanshui (LPS), Datong (DT), Wudong (WD), Dahebian (DHB), Lingnancang (LNC), Shenhua (SH), Cuijiazhai (CJZ), Tianchi (TC), Shenfu (SF), Xiaolongtan (XLT), and Dananhu (DNH). The origins of the remaining six coals were unknown, as shown in Table 4.1.

Oxy-fuel Combustion. http://dx.doi.org/10.1016/B978-0-12-812145-0.00004-9

TABLE 4.1 Proximate Analysis, Ultimate Analysis, and LHV of Pulverized Coal

		Proximate Analysis (wt%.ad)				Ultimate Analysis (wt%.daf)					
No	Sample	*M*	*VM*	*A*	*FC*	*C*	*H*	O^a	*N*	*S*	LHV(MJ/kg)
1	XZ	1.12	11.16	29.67	58.05	89.57	3.12	3.30	0.98	3.03	22.98
2	LZ	1.07	12.08	35.04	51.81	81.64	4.96	5.43	1.33	6.65	19.79
3	TG[b]	0.76	13.24	22.32	63.68	63.60	3.62	24.34	0.91	7.53	26.00
4	XYC	1.14	14.61	21.98	62.27	90.07	4.06	3.58	1.61	0.68	26.89
5	SX1	0.78	15.42	20.85	62.95	89.71	3.89	4.39	1.44	0.56	27.55
6	SP	0.99	15.58	20.72	62.71	90.34	3.69	4.15	1.34	0.47	26.81
7	SC	0.96	16.16	25.16	57.72	84.33	4.74	4.03	1.13	5.77	25.11
8	MAS	1.12	17.25	16.23	65.40	89.63	4.65	2.48	2.17	1.06	28.93
9	XWJ	1.17	17.86	52.27	28.70	78.81	4.06	10.90	1.68	4.56	13.24
10	SJT	0.88	20.36	27.92	50.84	76.94	4.51	10.53	1.27	6.75	22.24
11	SX2	0.89	23.24	24.07	51.8	86.15	5.06	6.41	1.54	0.84	26.47
12	–	1.06	26.03	25.63	47.28	84.32	4.94	5.56	1.49	3.70	25.19
13	LPS	1.29	26.31	26.97	45.43	80.92	4.73	8.97	1.44	3.95	23.67
14	DT	1.52	28.59	27.02	42.87	77.17	4.81	14.41	1.31	2.31	21.72
15	–	1.09	29.55	14.64	54.72	83.45	4.61	8.76	2.14	1.04	27.01

16	WD	2.11	30.95	14.46	52.48	80.34	5.04	12.27	1.15	1.20	27.80
17	–	1.21	31.14	28.55	39.10	88.86	4.37	2.90	1.35	2.52	21.03
18	–	1.60	31.26	15.04	52.10	81.31	4.56	11.00	2.13	1.00	26.67
19	–	1.56	31.94	14.29	52.21	81.42	4.67	10.85	2.07	0.99	26.73
21	DHB	1.14	31.96	28.38	38.52	81.38	5.18	8.67	1.52	3.24	23.28
21	LNC	2.70	32.26	25.43	39.61	79.79	6.08	11.38	1.68	1.07	22.05
22	SH	2.05	32.46	17.14	48.35	76.37	4.65	17.14	1.20	0.64	25.60
23	CJZ	2.65	35.24	17.26	44.85	71.17	3.91	23.46	1.09	0.36	22.19
24	TC	3.11	36.53	6.16	54.20	81.49	4.81	11.84	0.89	0.97	26.62
25	–	2.81	38.12	4.65	54.42	76.77	4.44	17.27	1.00	0.52	27.59
26	SF	2.34	42.28	5.46	49.92	78.69	4.15	15.12	1.29	0.75	27.69
27	XLT	2.58	45.24	25.84	26.34	66.11	5.15	25.95	1.88	0.91	18.26
28	DNH	2.85	46.28	22.94	27.93	61.71	4.22	32.77	1.23	0.07	14.95

[a]*By difference.*
[b]*Tiangou coal.*

The ignition stage, the stage of maximum reaction rate, and the burnout stage were considered during pulverized coal combustion. Combustion experiments were conducted by NETZSCH STA 409C thermogravimetric analysis. A 10 ± 0.2 mg sample was placed in an alumina crucible. The heating program ranged from room temperature to 900°C at 20°C/min. The furnace atmosphere was 21% O_2/79% CO_2, 30% O_2/70% CO_2, and 100% O_2. The gas flow rate was 100 mL/min.

4.2.2 Combustion Curves of Different Coals in O_2/CO_2 Atmosphere

The DTG curves of pulverized coals under three O_2 concentrations are shown in the Fig. 4.1. Coal increased in accordance with the volatile matter content. As the content of the volatile matter increased, step by step, the DTG remained close to the low temperature zone. Compared to 21% O_2 concentration, the characteristic temperature decreased in a 30% O_2 concentration. When the O_2 concentration increased to 100%, all combustion curves moved to the low temperature zone, and the characteristic temperature decreased greatly. From the two curves of 21% O_2 concentration to 30% O_2 concentration (as shown in Fig. 4.1), it was determined that several coals had weight changes that were significantly different from the adjacent volatile matter content of coal. This state meant that volatile matter had the most effect on pulverized coal combustion, although other factors affect the combustion characteristics.

4.2.3 Effect of Proximate Analysis, Ultimate Analysis, and Mineral Elements on Characteristic Temperature

Fig. 4.2 depicts the effect of characteristic temperature varied by proximate analysis. The points in the figure show that the H_2O content of different coals concentrated in the vicinity of 1% and 3% but that it concentrated less at 2%, which indicated that various coals had different H_2O absorption capacities. The volatile matter distributed mainly in the vicinity of 15% and 30%, the fixed carbon had a uniform distribution between 40% and 65%, and the content of ash distributed mainly in the vicinity of 15%–30%. On the one hand, these findings are related to the general characteristics of coal; on the other hand, they may partially reflect the characteristics of China's thermal coal. The characteristic temperature decreased as the moisture and volatile matter content increased, whereas the opposite trend was observed with fixed carbon and ash. Moisture and volatiles had a high reactivity at a low temperature, and their densities were relatively low. Generally speaking, their positions were more prone to forming holes, which increased the entire combustion reaction. In addition, the volatile matter combustion released a great deal of heat. Coal with a high fixed carbon content had a relatively large density that was equivalent to denser carbon. The reactivity was lower. A high ash content during the combustion reaction,

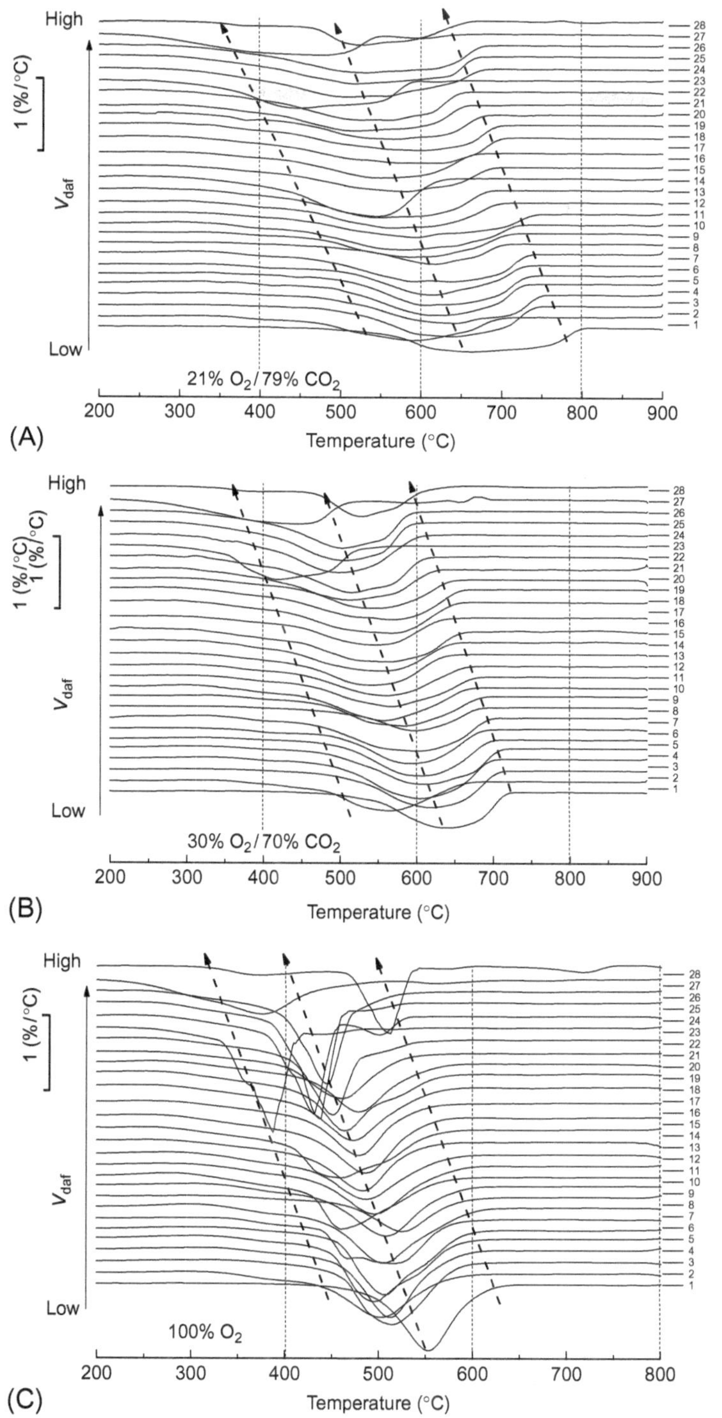

FIG. 4.1 The DTG curve of 28 kinds of pulverized coal under different O_2 concentrations.

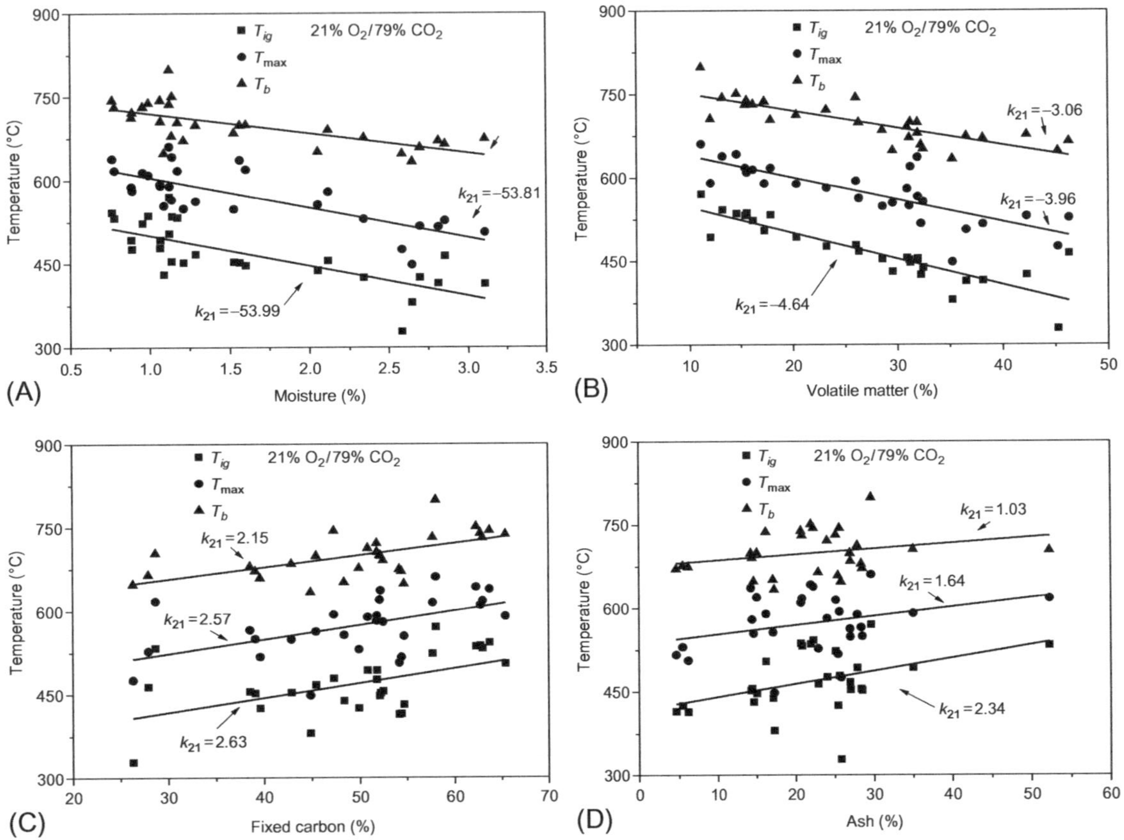

FIG. 4.2 Effect of proximate analysis on the characteristic temperature. (Solid line is the fitted line of the filled symbols; dashed line is the fitted line of the open symbols.)

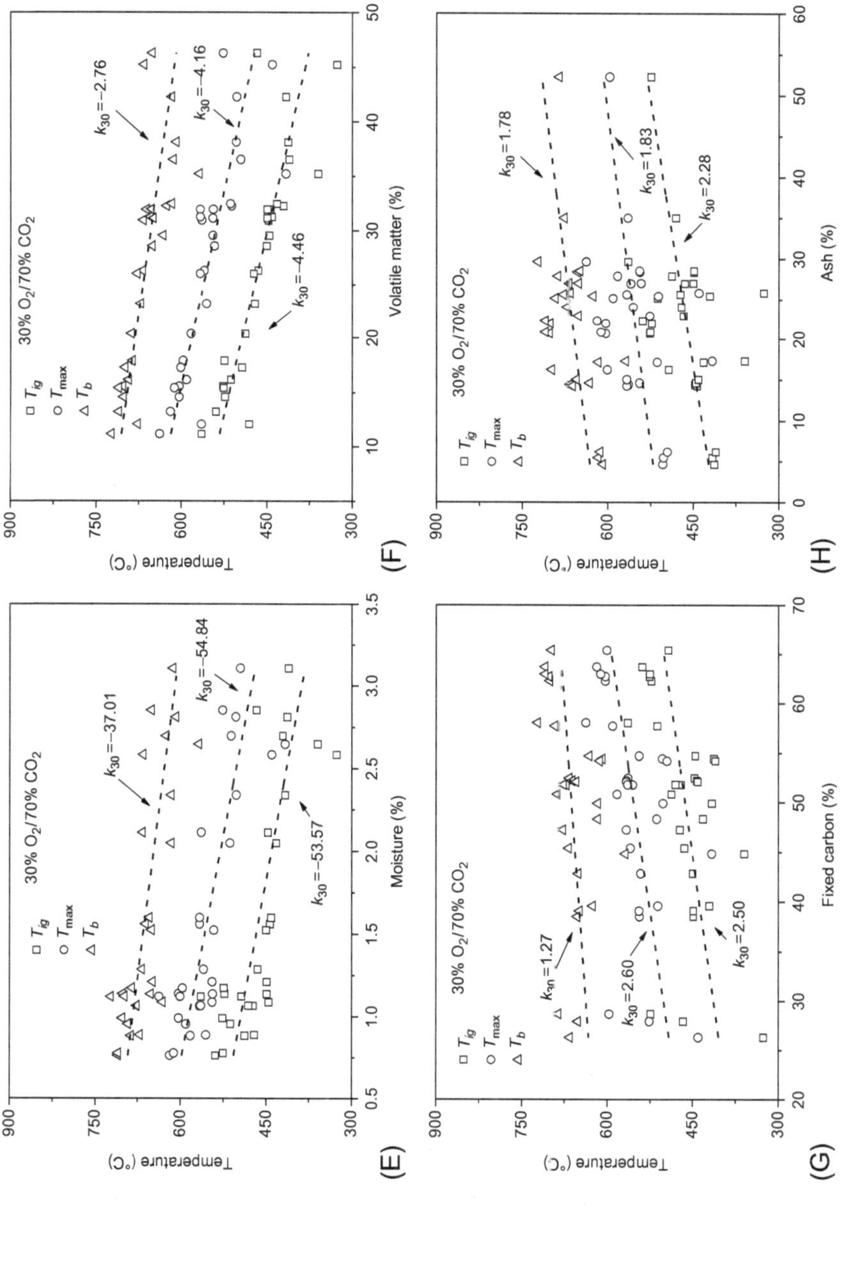
(E)
30% O2/70% CO2
Tig
Tmax
Tb
k30=−37.01
k30=−54.84
k30=−53.57
Temperature (°C)
Moisture (%)
(F)
30% O2/70% CO2
k30=−2.76
k30=−4.16
k30=−4.46
Volatile matter (%)
(G)
30% O2/70% CO2
k30=1.27
k30=2.60
k30=2.50
Fixed carbon (%)
(H)
30% O2/70% CO2
k30=1.78
k30=1.83
k30=2.28
Ash (%)

FIG. 4.2, CONT'D

especially for high-speed reactions, will prevent C–O contact, thus hindering the whole reaction process. Compared to a 21% O_2 concentration, on average, T_{ig}, T_{max}, and T_b with 30% O_2 decreased by 6°C, 20°C, and 37°C. As the reaction progressed, the effect of O_2 concentration gradually increased, indicating relatively scarce O_2 at the burnout stage.

The kinds of coal which are easily affected by O_2 concentration were determined by comparing the k values under two O_2 concentrations, see reference [10]. According to the slope, O_2 strongly affected the combustion characteristic of coals with high concentrations of moisture, fixed carbon, and low volatile matter.

This phenomenon occurred because the moisture in pulverized coal evaporated and was released during combustion, which consumed energy. The high moisture content consumed more energy, which was more significant with the high O_2 concentration. In addition, the volatile matter contained energy, and its reaction rate was much faster than for solid char. So the increased combustion reaction was more obvious. High-grade coals usually have a high fixed-carbon content. The values of the characteristic temperature in the study were high. The high fixed carbon content also required a long burnout time because the outer layer of ash produced by the reaction hindered the internal reactions of unburned C [11]. In the highly concentrated O_2 condition, the O_2 easily penetrated the interior of the char. Therefore increasing the O_2 concentration had a greater effect on the burnout stage.

4.2.4 Conclusions

From the combustion characteristic of various pulverized coals in an oxy-fuel condition, it can be concluded that the characteristic temperature decreases as the content of moisture and volatile matter increases and that the characteristic temperature increases as the content of fixed carbon and ash increases. As the O_2 concentration increases, the overall trend does not change, but the characteristic temperature decreases. Coals with high concentrations of moisture, fixed carbon, and low volatile matter combustion are strongly affected by O_2.

4.3 EFFECT OF H_2O ON THE COMBUSTION CHARACTERISTICS OF PULVERIZED COAL IN AN OXY-FUEL ATMOSPHERE

Many studies focus on the influence of the replacement of N_2 with CO_2. The steam content in an oxy-fuel combustion atmosphere is significantly higher than that in conventional air combustion, as reported in the literature [12–14]. Such a high concentration of H_2O inevitably leads to greater changes in pulverized coal combustion behavior. H_2O in the atmosphere has a strong impact on mass transfer [15]. Nearly 5% of carbon consumption under dry recycling and 9% under wet recycling can be attributed to H_2O gasification reactions [16]. Adding 5% to approximately 20% of H_2O to a O_2/CO_2 atmosphere can increase the ignition temperature and decrease the carbon burnout rate [17]. All of the studies showed that H_2O strongly affected the combustion of pulverized coal under oxy-fuel conditions. However, these studies did not provide details on the impact, and there is a lack of research on the effect of high H_2O concentrations. This section investigates in detail the effect of H_2O on the combustion characteristics of pulverized coal in oxy-fuel atmospheres.

4.3.1 Coal Sample Preparation and Experimental Process

Three differently ranked Chinese coals were chosen from those shown in Table 4.1: Xiyangcun (XYC) anthracite coal, Lingnancang (LNC) bituminous coal, and Xiaolongtan (XLT) brown coal. Coal samples were crushed and sieved from 45 μm to approximately 75 μm.

The reactivity experiments were conducted in a NETZSCH STA449F3 water vapor furnace. 10 ± 0.2 mg of the coal sample were placed in an alumina crucible. The temperature program involved heating from room temperature to 150°C at 10°C/min, insulating for 5 min, and then heating to 900°C at 20°C/min. The furnace atmosphere was O_2/CO_2 mixed with 180°C steam. The total gas flow rate was 100 mL/min.

To evaluate the combustion performance of pulverized coal, comprehensive combustibility index S was defined as [18,19]

$$S = \frac{(dw/dt)_{\max} \times (dw/dt)_{av}}{T_{ig}^2 \times T_b} \tag{4.1}$$

where $(dw/dt)_{\max}$ is the maximum combustion rate and $(dw/dt)_{av}$ is the average weight loss rate. Generally, a higher S value indicates a higher reactivity of the pulverized coal combustion process.

4.3.2 Effect of H_2O on the Combustion Behaviors of Coal

The characteristic temperatures and combustion characteristic indexes under 21% O_2/79% CO_2 and 21% O_2/49% CO_2/30% H_2O atmospheres are plotted in Fig. 4.3. The atmosphere containing H_2O led to an increase in the T_{ig} and the

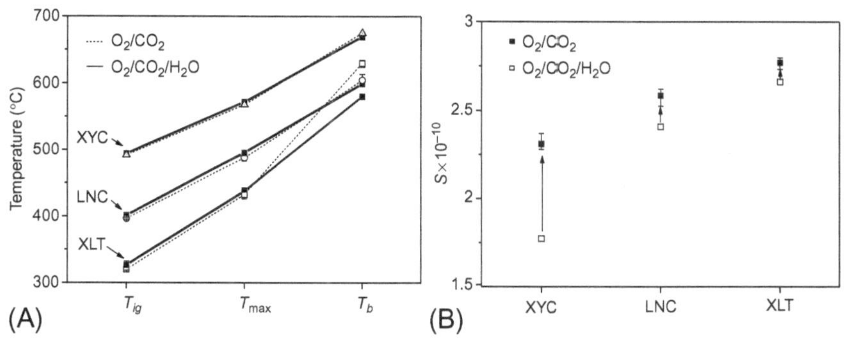

FIG. 4.3 Effect of H_2O on the combustion behaviors of different types of coal: (A) characteristic temperature and (B) S.

T_{max} but a decrease in the T_b. Generally, with a higher specific heat of CO_2 than that of N_2, the T_{ig} of CO_2 atmosphere is higher than the N_2 atmosphere [20], and the specific heat of H_2O is higher than that of CO_2. So the T_{ig} in the H_2O atmosphere should be delayed similarly. The thermal diffusivity of H_2O was greater than that of CO_2, and the heat released from combustion was taken away more easily. So the combustion was hindered. The higher emissivity of H_2O had a similar result. As the reaction progressed, the content of carbon was gradually reduced, and the effect of the combustion reaction was gradually enhanced. The diffusion rate of O_2 in H_2O atmosphere was also higher than in CO_2 atmosphere. So the partial pressure of O_2 on the char surface was higher in the atmosphere containing H_2O than in an atmosphere without H_2O with the same O_2 concentration. Finally, the H_2O-containing atmosphere was conducive to the combustion reactivity of pulverized coal. The comprehensive combustibility index S of coal was evidently higher in an $O_2/CO_2/H_2O$ atmosphere than in an O_2/CO_2 atmosphere, as depicted in Fig. 4.3B. Compared to XYC anthracite coal and LNC bituminous coal, XLT, a higher-volatility brown coal, showed little change between the atmospheres without H_2O and the atmospheres containing H_2O.

The effect of 30% H_2O on the combustion at the same O_2 concentration has been studied. To convenient the result of H_2Os effect on coal combustion in more oxy-fuel atmosphere conditions, $O_2/CO_2/H_2O$ atmospheres with 30% H_2O under various O_2 concentrations were studied. The result is plotted in Fig. 4.4. The characteristic temperature decreased as the O_2 concentration increased. There was a difference between the atmosphere containing H_2O and the atmosphere without H_2O. In the O_2/CO_2 atmosphere, the characteristic temperature was decreased linearly from an atmosphere of 21%–50% O_2 concentration. However, in the $O_2/CO_2/H_2O$ atmosphere, the characteristic temperature decreased from 21% to 30%, and the rate of decline slowed down above a 30% O_2 concentration. It can be concluded that the impact of steam was even more significant in the low O_2 concentration. S had a similar trend.

Based on the previous analysis, 30% H_2O in an $O_2/CO_2/H_2O$ atmosphere benefited coal combustion in all O_2 concentrations. However, the effect of other H_2O

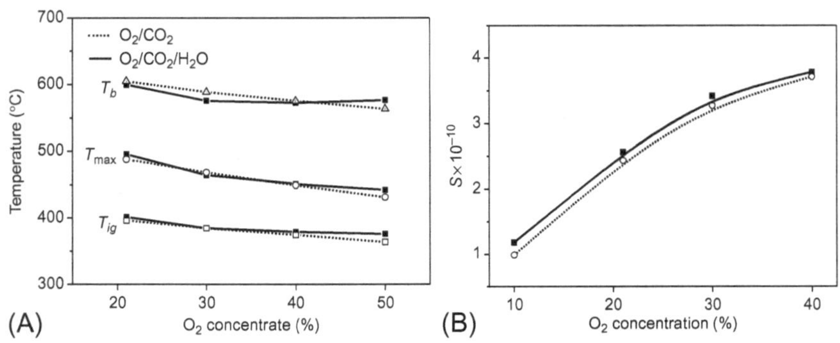

FIG. 4.4 Effect of 30% H_2O concentration on coal combustion behavior under different O_2 concentrations: (A) characteristic temperature and (B) S.

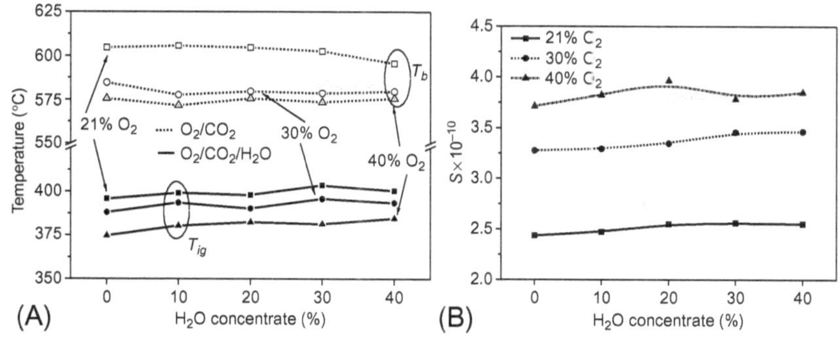

FIG. 4.5 Effect of H_2O concentration on coal combustion behavior under different O_2 concentrations: (A) T_{ig} and (B) S.

concentrations in different O_2 and CO_2 atmospheres is still unclear. The impact of H_2O concentrations on coal combustion under various O_2 atmospheres has been analyzed, and the results are shown in Fig. 4.5. When switching combustion atmospheres from those without H_2O to those with H_2O in all the O_2 concentrations, the T_{ig} increased. Under an $O_2/CO_2/H_2O$ atmosphere, the change of trend for T_{ig} was not obvious for the change of H_2O concentration. At the burnout stage, the T_b decreased obviously as the H_2O concentration increased from 0% to 40% at a 21% O_2 condition. As the O_2 concentration increased to 30%, the decreasing tendency of the T_b gradually slowed, and it was essentially constant at a 40% O_2 condition. At low O_2 concentration, the reaction rates is low, at which condition internal diffusion is the crucial factor affecting reaction rate. For a high diffusion rate of O_2 in H_2O atmosphere, H_2O have a significant impact on T_b at low O_2 concentration. In a high O_2 concentration, at which condition internal diffusion is not crucial factor affecting reaction rate, so H_2O have no impact on T_b. Fig. 4.5B shows the changes of S with a H_2O concentration. The combustion reactivity increased with an increase of the H_2O concentration under a 21% O_2 concentration and a 30% O_2 concentration, but the combustion reactivity increased first and then decreased for a 40% O_2 concentration.

In previous studies, the impact of H_2O concentration on combustion under O_2/CO_2 atmosphere was unclear [17,21]. The problem was that increasing the H_2O concentration resulted in the CO_2 concentration decreasing, and which is more important is difficult to estimate. Atmospheres containing constant concentrations of O_2 and H_2O as well as N_2 instead of CO_2 were used to study how CO_2 affects the coal combustion under $O_2/CO_2/H_2O$ atmosphere. As can be seen in Fig. 4.6, the replacement of CO_2 in $O_2/CO_2/H_2O$ by N_2 decreased both T_{ig} and T_b. CO_2 hindered the coal combustion at both the ignition and the burnout stages. The T_{ig} greatly affected the low H_2O concentration, and the T_b greatly influenced the high H_2O concentration. S is shown in Fig. 4.6B, and the trend confirmed that H_2O replacement of CO_2 was not conducive to combustion and was more pronounced in a high O_2, low H_2O atmosphere.

The effect of CO_2 in an $O_2/CO_2/H_2O$ atmosphere can be obtained. Replacing CO_2 with H_2O reduces the CO_2 ratio, which can improve the ignition.

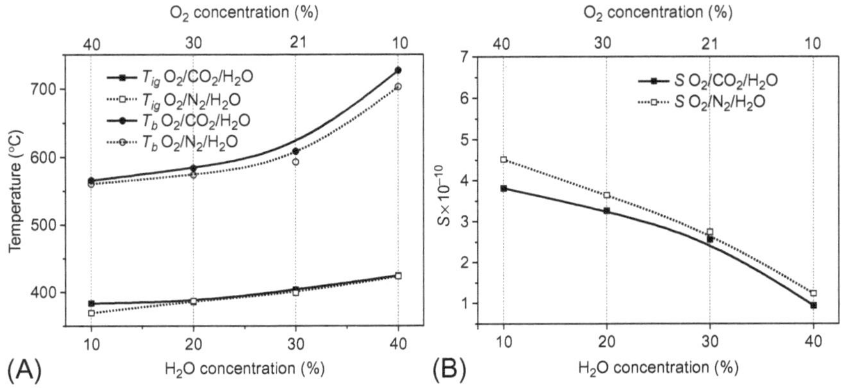

FIG. 4.6 Effect of CO_2 on coal combustion behavior under different O_2 concentrations with H_2O: (A) T_{ig} and T_b and (B) S.

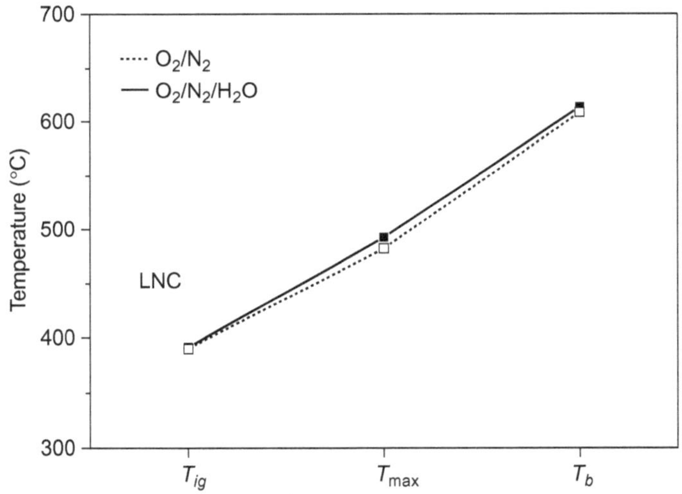

FIG. 4.7 Effect of H_2O on characteristic temperature of coal at O_2/N_2 atmosphere.

Theoretically, the T_{ig} should decrease, but the actual result is the opposite. The reason may be that the effect of H_2O increases on T_{ig} to a great extent. As for T_b, partially replacing CO_2 with H_2O can reduce the T_b, and the decreasing T_b may be affected by the decreasing CO_2 concentration. Furthermore, replacing H_2O with CO_2 can improve the combustion because of the decreasing impact of CO_2. Concerning S, the replacement can improve both values, which is consistent with the role of CO_2 in coal combustion. Taken together, it is clear that inclusion of H_2O in O_2/CO_2 can markedly postpone the ignition, but how it affects the burnout stage is still unclear. To further confirm the role of H_2O, the combustion characteristics of LNC coal under O_2/N_2 and $O_2/N_2/H_2O$ atmospheres were studied. As shown in Fig. 4.7, H_2O did not help coal ignition but was capable of promoting the combustion reaction rate after ignition.

4.3.3 Conclusions

In $O_2/CO_2/H_2O$ combustion atmospheres, CO_2 hinders the coal combustion under both ignition and the burnout stage. The presence of H_2O leads to a longer ignition delay, faster burnout, and a greater comprehensive reactivity in various O_2 concentrations.

4.4 INFLUENCE OF EVOLUTION OF COAL CHAR STRUCTURE ON CHARACTERISTIC TEMPERATURE

Compared to conventional pulverized coal combustion, gasification influences in oxy-fuel combustion are more complicated, especially in a high H_2O concentration under $O_2/CO_2/H_2O$ conditions [22]. The influence of CO_2 and H_2O on characteristic temperature was introduced in Section 4.3. Also, a number of studies showed that partial replacement of CO_2 by H_2O in coal combustion resulted in raising the ignition temperature, decreasing the burnout temperature, and comprehensively improving reactivity [17,21,23,24]. Char structure caused by N_2 to CO_2 in the atmosphere has also attracted researchers attention. O_2/CO_2 char has a thick surface and compact pores, and the consumption rate of organic functional groups is slower or delayed [25]. CO_2 char has a higher reactivity than N_2 char in both air and oxy-fuel combustion atmospheres, and the combustion atmosphere exerts less influence on the reactivity of CO_2 char compared to N_2 char [26]. Generally, the structural characteristics of pyrolysis char affect combustion characteristics, and coal with a higher surface area and better dispersion of ash phase leads to higher combustion reactivity [27]. In conclusion, CO_2 has a significant impact on the pulverized coal char structure during the combustion process, so will the influence of highly reactive H_2O be greater?

Because of a lack of research on char gasification reactions in oxy-fuel atmospheres containing a high H_2O concentration, this section covers the effect of the char structure after a gasification reaction on the combustion characteristic in an atmosphere with a high H_2O concentration.

4.4.1 Sample Preparation and Experimental Methods

Six typical Chinese coals were selected from those in Table 4.1: Xiyangcun (XYC) lean coal, Xiwujiang (XWJ) anthracite, Datong (DT) bituminous coal, Lingnancang (LNC) bituminous coal, Xiaolongtan (XLT) lignite, and Dananhu (DNH) lignite. Pyrolysis and gasification experiments were carried out in a drop-tube furnace, which was introduced in the previous [28]. The furnace was heated to the experimental temperature of 1000°C. Four kinds of atmosphere were designed: N_2, 50% N_2/50% H_2O, CO_2, and 50% CO_2/50% H_2O. The total gas flow rate was 5 L/min, the primary airflow rate was 1 L/min, and the secondary airflow rate was 4 L/min. The amount of the sample was set at 0.1 g/min. Proximate analysis was used to calculate carbon content, carbon burnout rate, and char reactivity.

4.4.2 Analysis Instruments and Carbon Burnout Rates of Char

A Sirion 200 field emission scanning electron microscopy was used to observe the surface topography of the sample. A LabRAM HR800 laser confocal Raman spectroscopy was adopted to analyze the carbon structure. An Accelerated Surface Area and Porosimetry System (ASAP 2020) was used for analyzing the specific surface area and pore structure. Combustion reactivity was tested on a STA449F3 steam thermal analyzer produced by NETZSCH (Germany)—the procedure was described in Section 4.3.1.

In order to study the process of coal combustion, a calculation method of carbon burnout rate was proposed based on the ash tracer method, [29]

$$B = \left(1 - \frac{A_0}{1 - A_0} \times \frac{1 - A_i}{A_i}\right) \tag{4.2}$$

where 0 is raw coal, *i* is *i* position or *i* moment of char, *A* is ash content of coal, and char is proximate analysis.

The carbon burnout rate was calculated by the ash tracer method, as shown in Table 4.2. The carbon burnout rate increased as the volatile content increased, and this phenomenon was shown in all of the atmospheres. The carbon content of XYC was only 8% in N_2 atmosphere and only 24% to approximately 28% in CO_2 and H_2O atmospheres, which indicated a slower pyrolysis and gasification reaction rate. The reaction rate did not change greatly for XWJ, DT, and LNC between pyrolysis and gasification. For lignite XLT and DNH, the carbon burnout rate under N_2 pyrolysis atmosphere was 50%–70%. But the carbon burnout rate was 90% or more in CO_2 and H_2O atmospheres, and even close to 100% in some atmospheres, which was a higher pyrolysis and gasification reactivity. In addition, the carbon burnout rates of the three pulverized coals (XYZ, XWJ, and DT) in CO_2/H_2O atmospheres were higher than those in N_2/H_2O atmospheres, but LNC, XLT, and DNH were the opposite. In general, gasification reactions containing CO_2 had a higher reactivity than the no-gasification reaction of N_2, but the reactivity of pulverized coal in an atmosphere mix of CO_2 and H_2O

TABLE 4.2 Carbon Burnout Rate of Pulverized Coal Under Various Atmospheres

Atmosphere	XYC	XWJ	DT	LNC	XLT	DNH
N_2	7.93	36.58	42.75	45.61	50.58	68.80
N_2/H_2O	27.15	47.05	58.61	55.94	92.99	97.04
CO_2	24.30	33.89	44.86	43.99	98.54	99.26
CO_2/H_2O	24.78	41.43	54.43	64.13	95.80	97.99

was not necessarily the result of a superposition of an individual CO_2 or H_2O atmosphere. Depending on the characteristics of the carbon burnout rate, three typical coals (XYC, LNC, and XLT) were selected from six kinds of pulverized coal, which were, respectively, for the ignition stage, the rapid reaction stage, and the burnout stage.

4.4.3 SEM Analysis

Fig. 4.8 shows a single particle image enlarged to 1200 or 1300 in order to highlight the surface structure. As shown by the distribution of raw coal and char, the surface of the XYC particle was flat, while the surfaces of LNC and XLT were three-dimensional. The particle shape of the XYC char is similar to that of coal. Compared to raw coal, the edge of the LNC char sample was relatively smooth. The bubble-like structures in the surface may have been a result of the volatile release and the great porosity. For XLT, more pores of char than raw coal were observed only in an N_2 atmosphere. In other cases, the most obvious characteristic of XLT char is the small particle size, which in this case resulted mainly from the low carbon content and high ash content. The structure of the raw coal was dense, while the pore structure of the char was obviously richer than raw coal, as reflected in the bubble-like structure and blasting pores on the surface. Char particles from an H_2O-containing atmosphere were surrounded by fewer fine particles than the char particles in the atmosphere without H_2O, because fine coal particles in the H_2O-containing atmosphere were easily reacted and taken away from the airflow.

4.4.4 Pore Structure Analysis

The SEM image showed the topography of the particle surface. The gas adsorption method was used to analyze the pore structure. Fig. 4.9A shows the surface area of three coals and their corresponding char samples. The specific surface area of coal was small. Except for XLT, there was no significant difference in the specific surface area between N_2 char and raw coal. The sharp increase of the specific surface area for XLT N_2 char was related to the dramatic change of the carbon conversion rate, indicating that the impact of the specific surface area was limited during the pulverized coal pyrolysis process. For XYC and LNC, the specific surface area of N_2/H_2O char was larger than that of N_2 char. The same happened to the specific surface area of CO_2/H_2O char when compared with CO_2 char. It showed the H_2O-enhanced gasification reaction of pulverized coal in both pyrolysis and gasification processes. Specific surface areas of two pulverized coals in the N_2/H_2O atmosphere were smaller than those in the CO_2/H_2O atmosphere, indicating that gasification reactions in coexisting CO_2 and H_2O were more intense than when CO_2 and H_2O exist separately.

Changes in the pore volume of three coals were similar to the changes in the specific surface area. The pore size trend was different among three coals. The big pore size of raw coal resulted in a small specific surface area and pore

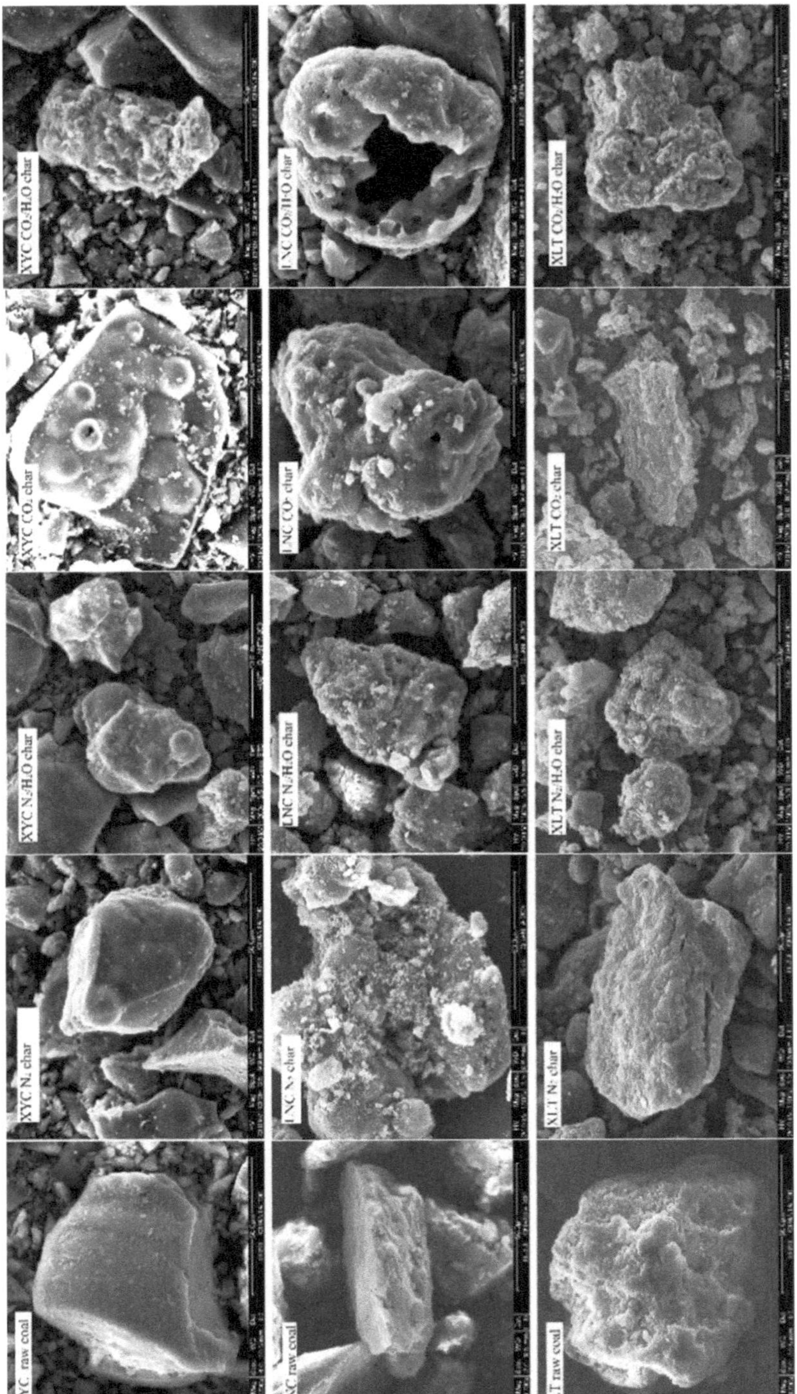

FIG. 4.8 SEM image of the surface of single coal and char particles.

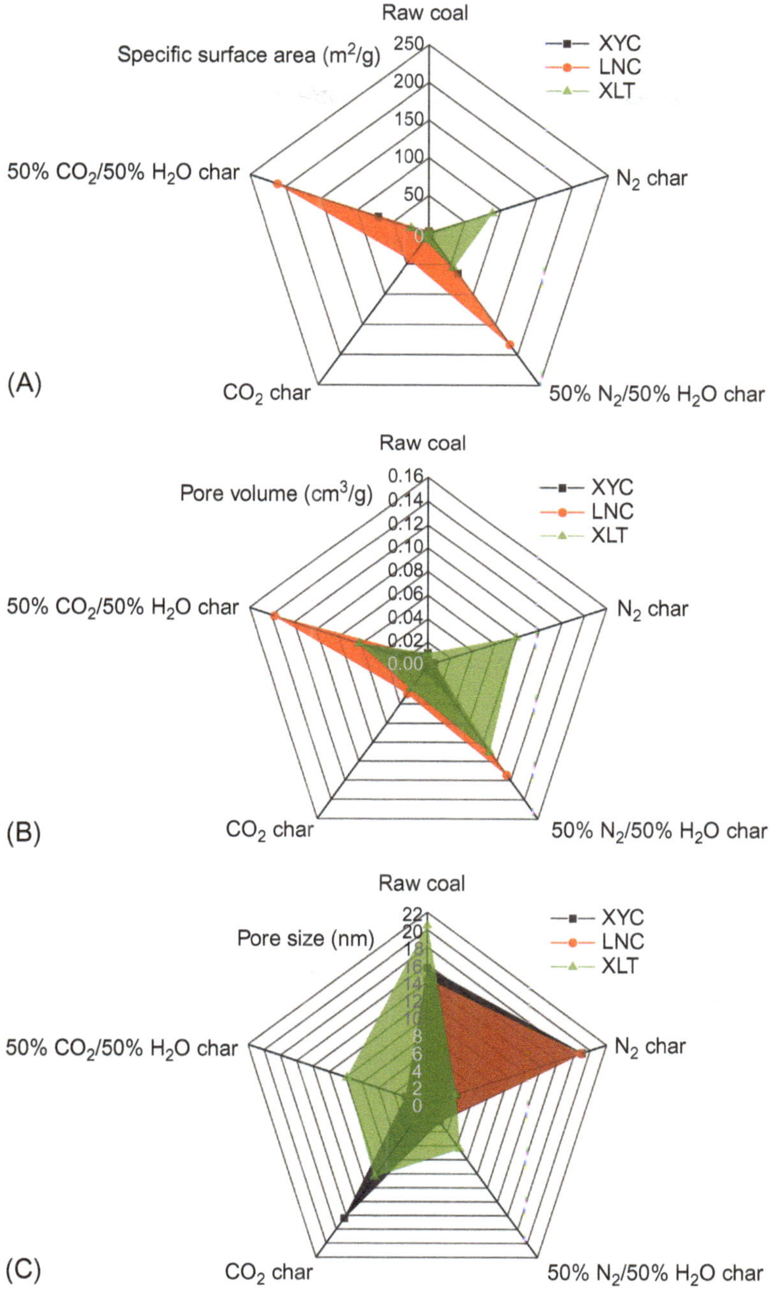

FIG. 4.9 Three kinds of coals and chars pore structure parameters.

volume. With the pyrolysis and gasification process, a large number of micropores were produced, and the char pore structure became more loosely porous, which was obvious for XYC and LNC under H_2O-containing atmospheres. The high reactivity of XLT related to the short coal-formation time and a specific coal-formation environment. The remarkable reduction of char pores under an atmosphere containing CO_2, H_2O and other gasification agents may have been caused by carbon consumption during the gasification reaction, which cannot only produce structure that is more porous but also to diminish the proportion of large pores.

4.4.5 Raman Analysis

The pore structure of the pulverized coal determined the reaction sites, while the carbon crystal structure of coal affected reactivity. Char with a high degree of graphite showed a lower reactivity. Raman analysis was conducted on the carbon crystal structure.

The char was a highly disordered carbon material. Pure graphite peaked only at 1580 cm^{-1}, referred as G-peak, in the Raman spectrum. Because of lattice defects and hetero atom vibrations, a series of additional peaks occurred at 1150, 1350, 1530, and 1620 cm^{-1} [30–35]. For comparison, two parameters of I_G/I_{all} and I_{D3}/I_{all} were chosen. The G-peak was associated with the level of carbon graphite. The ratio of the G-peak area to the total peaks area was named I_G/I_{all} and was related to the degree of graphitized carbon in the coal. The greater value implied a higher degree of orderly and lower reactivity. The D_3 peak was related to amorphous carbon. The ratio of the D_3 peak area to the total peak was named I_{D3}/I_{all}. The greater value indicated higher amorphous carbon content and higher activity.

As shown in Fig. 4.10, the variation trends of I_G/I_{all} were different for various coals. The I_G/I_{all} of XYC under the two N_2 atmospheres was greater, indicating that the graphite structure was easier to form in the N_2 atmosphere. The I_G/I_{all} was smaller in the two CO_2 atmospheres, so the graphitization degree was lower. In both LNC and XLT, the I_G/I_{all} was higher in the H_2O-containing atmosphere, indicating that adding H_2O could accelerate the orderly arrangement of the carbon structure.

The distribution of amorphous carbon under different atmospheres is shown in Fig. 4.11. Compared to N_2 atmosphere char, increases of amorphous in CO_2 atmosphere for three kinds of coal presented little difference, which illustrated a restricted impact of C-CO_2 gasification on amorphous carbon. The comparison of atmosphere with and without H_2O, either for CO_2 atmosphere or N_2 atmosphere, showed that amorphous carbon in the H_2O-containing atmosphere was reduced. It showed that amorphous carbon in coal could be quickly consumed in the H_2O-containing atmosphere. On the one hand, pulverized coal had a higher reactivity in the H_2O-containing atmosphere; on the other hand, the remaining char had a lower reactivity, which explained the phenomenon that the carbon

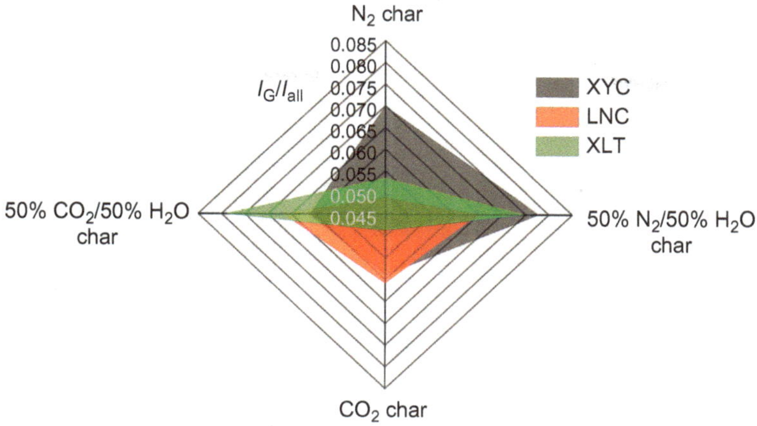

FIG. 4.10 I_G/I_{all} of three coals.

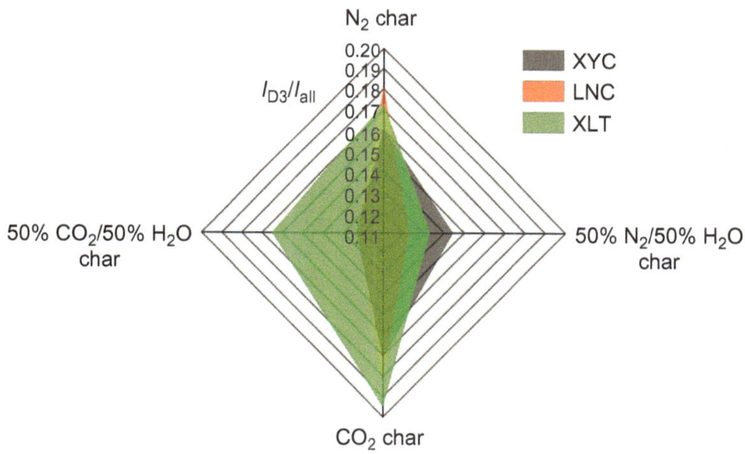

FIG. 4.11 I_{D3}/I_{all} of three coals.

burnout rate among various types of coals varied greatly for coal samples, but varied little for char samples in a H_2O-containing atmosphere.

4.4.6 Combustion Reactivity Analysis

The structural characteristic of char under various atmospheres was analyzed. The combustion characteristics of pulverized coal were tested under a 21% O_2/79% CO_2 atmosphere and a 21% O_2/49% CO_2/30% H_2O atmosphere.

The level of influence of an atmosphere containing 30% H_2O on T_{ig} is shown in Fig. 4.12. Influence of T_{ig} of char samples under a 21% O_2/49% CO_2/30% H_2O atmosphere was less than under a 21% O_2/79% CO_2 atmosphere. This means that atmospheres containing H_2O dramatically reduced the change of

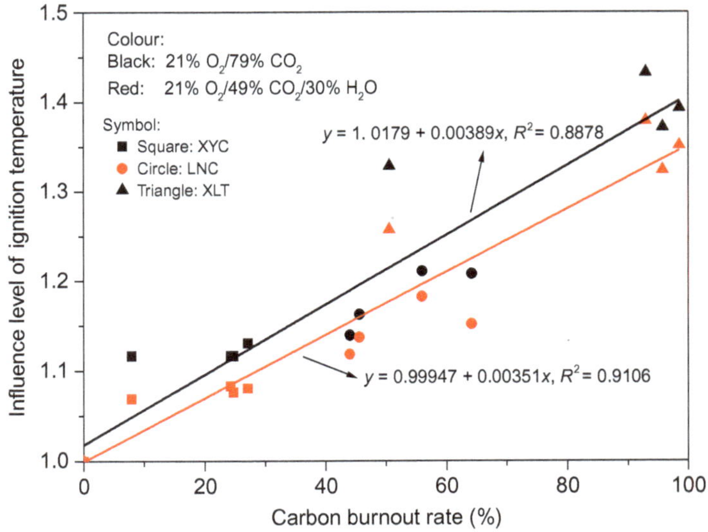

FIG. 4.12 Influence levels of ignition temperature under H_2O-containing atmosphere.

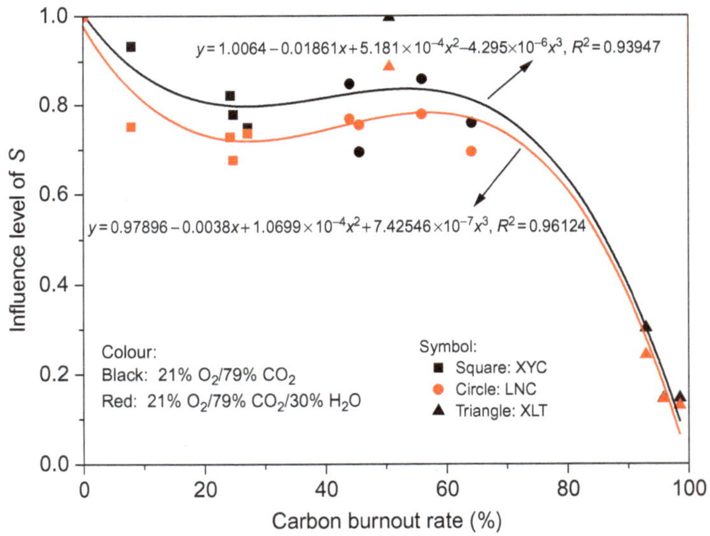

FIG. 4.13 The influence levels of comprehensive combustion characteristics index under H_2O-containing atmosphere.

char T_{ig} at 21% O_2 concentrations. From the slopes of two fitted line, the influence degree was enlarged as the carbon burnout rate increased. The method of slope calculating can be found in the literature [10]. The extent of variation was around 10%.

As demonstrated in Fig. 4.13, the influence level of *S* gradually decreased as the char burnout rate increased. In contrast to the linear change of ignition temperature, the relative value of *S* showed a cubic reduction. It can be seen that the changes were mainly in the early stage of combustion and in the burnout stage and that the changes were small between the carbon burnout rates of 20%–60%, which indicated that the char reactivity change process was in the volatile release stage and the char burnout stage. In addition, the relative value of *S* in the H_2O-containing atmosphere was lower than that in the atmosphere without H_2O, which indicated that the combustion reaction was more intense in the H_2O-containing atmosphere and that the reactivity of the char also changed significantly according to the degree of the reaction progress. In general, combustion characteristics had a significant relationship with the char structure, especially in the ignition and burnout stages.

4.4.7 Conclusion

Char properties of various atmospheres were studied in this chapter, and the impact of pyrolysis and gasification atmosphere on char formation was also investigated. Six kinds of pulverized coal were selected in the study. The results are as follows.

Char in an atmosphere containing H_2O had a higher carbon consumption rate in most cases. The gasification reaction of all pulverized coals was able to consume carbon, and H_2O-gasification reactivity was generally higher than CO_2-gasification reactivity. The gasification reaction of high-rank coal was significantly enhanced in H_2O-containing atmospheres, but for low-rank coal, H_2O atmospheres caused no significant change in gasification.

Both CO_2 chars and H_2O-containing atmosphere chars had a richer pore structure, compared, respectively, to N_2 chars and chars in atmospheres without H_2O. With the increasing content of volatile matter in coal, the porosity of char from a H_2O-containing atmosphere was significantly enhanced compared to that of an atmosphere with no H_2O. The graphitization degree increased after thermal reaction. The amorphous carbon ratio of CO_2 char was higher than that of N_2 char, but decreased in an atmosphere containing H_2O.

An atmosphere containing H_2O dramatically reduced the change of the char ignition temperature, and the relative value of *S* showed a cubic reduction. In general, the combustion characteristics had a significant relationship with the char structure, especially in the ignition and burnout stages.

REFERENCES

[1] Boot-Handford ME, Abanades JC, Anthony EJ, Blunt MJ, Brandani S, Mac Dowell N, et al. Carbon capture and storage update. Energy Environ Sci 2014;7:130–89.

[2] Chen L, Yong SZ, Ghoniem AF. Oxy-fuel combustion of pulverized coal: characterization, fundamentals, stabilization and CFD modeling. Prog Energy Combust Sci 2012;38: 156–214.

[3] Zhang L, Binner E, Qiao Y, Li C-Z. In situ diagnostics of Victorian brown coal combustion in O_2/N_2 and O_2/CO_2 mixtures in drop-tube furnace. Fuel 2010;89:2703–12.

[4] Liu H, Zailani R, Gibbs BM. Comparisons of pulverized coal combustion in air and in mixtures of O_2/CO_2. Fuel 2005;84:833–40.

[5] Khatami R, Stivers C, Joshi K, Levendis YA, Sarofim AF. Combustion behavior of single particles from three different coal ranks and from sugar cane bagasse in O_2/N_2 and O_2/CO_2 atmospheres. Combust Flame 2012;159:1253–71.

[6] Andersson K, Johnsson F. Process evaluation of an 865MW_e lignite fired O_2/CO_2 power plant. Energy Convers Manag 2006;47:3487–98.

[7] Kiga T, Takano S, Kimura N, Omata K, Okawa M, Mori T, et al. Characteristics of pulverized-coal combustion in the system of oxygen/recycled flue gas combustion. Energy Convers Manag 1997;38:S129–34.

[8] Kimura N, Omata K, Kiga T, Takano S, Shikisima S. The characteristics of pulverized coal combustion in O_2/CO_2 mixtures for CO_2 recovery. Energy Convers Manag 1995;36:805–8.

[9] S-l N, Han K-h, C-m L. Characteristic of coal combustion in oxygen/carbon dioxide atmosphere and nitric oxide release during this process. Energy Convers Manag 2011;52:532–7.

[10] Yi B, Zhang L, Huang F, Xia Z, Mao Z, Ding J, et al. Investigating the combustion characteristic temperature of 28 kinds of Chinese coal in oxy-fuel conditions. Energy Convers Manag 2015;103:439–47.

[11] Wen C. Noncatalytic heterogeneous solid-fluid reaction models. Ind Eng Chem 1968;60:34–54.

[12] Anheden M, Burchhardt U, Ecke H, Faber R, Jidinger O, Giering R, et al. Overview of operational experience and results from test activities in Vattenfall's 30 MW_{th} oxyfuel pilot plant in Schwarze Pumpe. Energy Procedia 2011;4:941–50.

[13] Kakaras E, Koumanakos A, Doukelis A, Giannakopoulos D, Vorrias I. Oxyfuel boiler design in a lignite-fired power plant. Fuel 2007;86:2144–50.

[14] Duan L, Sun H, Zhao C, Zhou W, Chen X. Coal combustion characteristics on an oxy-fuel circulating fluidized bed combustor with warm flue gas recycle. Fuel 2014;127:47–51.

[15] Yu J, Zhou K, Ou W. Mass transfer coefficients considering effects of steam in oxy-fuel combustion of coal char. Fuel 2013;111:48–56.

[16] Hecht ES, Shaddix CR, Geier M, Molina A, Haynes BS. Effect of CO_2 and steam gasification reactions on the oxy-combustion of pulverized coal char. Combust Flame 2012;159:3437–47.

[17] Riaza J, Álvarez L, Gil MV, Pevida C, Pis JJ, Rubiera F. Effect of oxy-fuel combustion with steam addition on coal ignition and burnout in an entrained flow reactor. Energy 2011;36:5314–9.

[18] Wang C, Zhang X, Liu Y, Che D. Pyrolysis and combustion characteristics of coals in oxyfuel combustion. Appl Energy 2012;97:264–73.

[19] Wang CB, Lei M, Yan WP, Wang SL, Jia LF. Combustion characteristics and ash formation of pulverized coal under pressurized oxy-fuel conditions. Energy Fuel 2011;25:4333–44.

[20] Li Q, Zhao C, Chen X, Wu W, Li Y. Comparison of pulverized coal combustion in air and in O_2/CO_2 mixtures by thermo-gravimetric analysis. J Anal Appl Pyrolysis 2009;85:521–8.

[21] Gil M, Riaza J, Álvarez L, Pevida C, Pis J, Rubiera F. A study of oxy-coal combustion with steam addition and biomass blending by thermogravimetric analysis. J Therm Anal Calorim 2012;109:1–7.

[22] Mao Z, Zhang L, Zhu X, Pan C, Yi B, Zheng C. Modeling of an oxy-coal flame under a steam-rich atmosphere. Appl Energy 2016;161:112–23.

[23] Yi B, Zhang L, Huang F, Mao Z, Zheng C. Effect of H_2O on the combustion characteristics of pulverized coal in O_2/CO_2 atmosphere. Appl Energy 2014;132:349–57.

[24] Riaza J, Álvarez L, Gil MV, Khatami R, Levendis YA, Pis JJ, et al. Ignition behavior of coal and biomass blends under oxy-firing conditions with steam additions. Greenhouse Gases Sci Technol 2013;3:397–414.

[25] Li Q, Zhao C, Chen X, Wu W, Lin B. Properties of char particles obtained under O_2/N_2 and O_2/CO_2 combustion environments. Chem Eng Process Process Intensif 2010;49:449–59.

[26] Wang B, Sun L, Su S, Xiang J, Hu S, Fei H. Char structural evolution during pyrolysis and its influence on combustion reactivity in air and oxy-fuel conditions. Energy Fuel 2012;26:1565–74.

[27] Lee D-W, Bae J-S, Park S-J, Lee Y-J, Hong J-C, Choi Y-C. The pore structure variation of coal char during pyrolysis and its relationship with char combustion reactivity. Ind Eng Chem Res 2012;51:13580–8.

[28] Yi B, Zhang L, Yuan Q, Yan S, Zheng C. The evolution of coal char structure under the oxy-fuel combustion containing high H_2O. Fuel Process Technol 2016;152:294–302.

[29] El-Samed AKA, Hampartsoumian E, Farag TM, Williams A. Variation of char reactivity during simultaneous devolatilization and combustion of coals in a drop-tube reactor. Fuel 1990;69:1029–36.

[30] Beyssac O, Goffé B, Petitet J-P, Froigneux E, Moreau M, Rouzaud J-N. On the characterization of disordered and heterogeneous carbonaceous materials by Raman spectroscopy. Spectrochim Acta A Mol Biomol Spectrosc 2003;59:2267–76.

[31] Ferrari A, Robertson J. Interpretation of Raman spectra of disordered and amorphous carbon. Phys Rev B 2000;61:14095.

[32] Sheng C. Char structure characterised by Raman spectroscopy and its correlations with combustion reactivity. Fuel 2007;86:2316–24.

[33] Shim H-S, Hurt RH. Thermal annealing of chars from diverse organic precursors under combustion-like conditions. Energy Fuel 2000;14:340–8.

[34] Sadezky A, Muckenhuber H, Grothe H, Niessner R, Pöschl U. Raman microspectroscopy of soot and related carbonaceous materials: spectral analysis and structural information. Carbon 2005;43:1731–42.

[35] Jawhari T, Roid A, Casado J. Raman spectroscopic characterization of some commercially available carbon black materials. Carbon 1995;33:1561–5.

Chapter 5

Nitrogen and Sulfur Behavior During Oxy-fuel Combustion and Its Retention

Jing Liu*, Hao Liu†, Xin Guo*, Hong Yao*, Juan Chen‡
*Huazhong University of Science and Technology, Wuhan, China
†Soochow University, Jiangsu, China
‡Shandong University, Jinan, China

5.1 INTRODUCTION

Oxy-fuel combustion has technological superiority because of its low NO_x emissions and its capability to easily remove SO_2 [1]. In oxy-fuel combustion, nitrogen is removed from the oxidant, with fuel combusting in highly concentrated oxygen rather than in a conventional air atmosphere. Also, recycled flue gas (RFG) is applied to control the flame temperature and to supplement the volume loss of N_2. The replacement of ordinary air with O_2/CO_2 leads to modified distributions of temperatures in field and product species, as well as radiation flux resulting from the property differences between CO_2 and N_2 [2]. These changes will affect the nitric and sulfur chemistries, thereby reducing the emission of NO_x by one-third and of SO_2 by 30% [3]. However, utilizing an oxygen concentration higher than 21%, compared to a conventional air atmosphere, may result in enhancing the formation of fuel NO_x and sulfur trioxide SO_3. Understanding the fate of nitric oxide and sulfur oxide during oxy-fuel combustion is critical for controlling pollution as well as for the design and operation of oxy-fuel coal-fired power plants.

5.2 NO FORMATION AND EMISSION DURING OXY-FUEL COMBUSTION

NO_x is formed through three main routes: thermal and prompt formation from N_2 and from fuel-bound nitrogen (Fuel N) [2]. Thermal NO evolves from homogeneous phase reactions between gaseous molecular N_2 and O_2, as described by the extended Zeldovich mechanism: [4,5]

$$N_2 + O \rightarrow NO + N \tag{5.1}$$

Oxy-fuel Combustion. http://dx.doi.org/10.1016/B978-0-12-812145-0.00005-0

$$N + O_2 \rightarrow NO + O \tag{5.2}$$

$$N + OH \rightarrow NO + H \tag{5.3}$$

Thermal NO is closely temperature-dependent and becomes dominant only at high temperatures (>1500°C) [2]. Prompt NO is formed through reactions between free nitrogen and hydrocarbon radicals, which occur mostly under fuel-rich conditions. Fuel N is decomposed into volatile-N and char-N during the devolatilization process [2]. The volatile-N (an intermediate gaseous compound, including HCN and NH_3) then transforms through a battery of subsequent reactions to form either NO or N_2, while char-N undergoes a set of heterogeneous reactions with the char oxidation [2].

Compared to air combustion, the amount of thermal NO_x and prompt NO in oxy-fuel combustion is decreased due to the very light concentration of N_2 in the combustor, in which N_2 is replaced by CO_2 in the RFG [6]. However, the NO emission in oxy-fuel combustion is also affected by recycled NO and high concentration CO_2 in the furnace. First, the recycled NO_x in the volatile matter release stage is reduced. Second, owing to reburning through RFG, the interactions between recycled NO_x and fuel N as well as hydrocarbons can further decrease NO_x formation [7]. The reduction level depends on the qualities of CO_2 and the recycled NO concentration, oxygen/fuel stoichiometric ratio λ, coal rank and volatility, and varied combustion conditions.

5.2.1 Influence of the Increased CO_2 Concentration

In O_2/CO_2 combustion, the CO_2 volume fraction in exhaust gases can be up to 0.95. This CO_2-rich atmosphere is likely to alter many features of the process. Examples include conversion from fuel N to NO, reduction of recycled NO in the reducing atmosphere in the combustion region of the volatile matter from coal, and interactions between recycled NO and fuel N. The NO emission in an O_2/CO_2 combustion system and contributions of various factors, including the effect of the reduction of recycled NO, the influence of the CO_2 concentration on the conversion ratio of fuel N to NO, and the interaction between recycled NO and fuel N, were investigated experimentally and analytically [8].

In order to evaluate the effect of CO_2 concentration in the recycled gas on the conversion ratio (*CR*) from fuel N to NO, experiments were conducted in an electrically heated, one-dimensional, premixed reactor under atmospheres of various CO_2 concentrations, at $T_{max} = 1450\,K$, fuel $N = 1.22\,wt\%$ (as N atom mass), and recycled NO =0 ppm [8]. When no coal was fed and only gas-phase reactions occurred, the *CR* value increased slightly with the increase of CO_2 concentration in the recycled gas. The experimental data were regressed to be a linear correlation with CO_2 concentration:

$$CR = 0.3633 + 0.0833 C_{CO_2} \quad \text{when} \quad \lambda = 0.7 \tag{5.4a}$$

$$CR = 0.563 + 0.0833 C_{CO_2} \quad \text{when} \quad \lambda = 1.0 \tag{5.4b}$$

$$CR = 0.5667 + 0.1167C_{CO_2} \quad \text{when} \quad \lambda = 1.2 \tag{5.4c}$$

However, in the presence of coal, the conversion ratio from fuel N to NO decreased with increasing CO_2 concentration:

$$CR = 0.3667 - 0.0833C_{CO_2} \quad \text{when} \quad \lambda = 0.7 \tag{5.5a}$$

$$CR = 0.5167 + 0.0833C_{CO_2} \quad \text{when} \quad \lambda = 1.0 \tag{5.5b}$$

$$CR = 0.5467 + 0.0833C_{CO_2} \quad \text{when} \quad \lambda = 1.2 \tag{5.5c}$$

The preceding difference can be explained through the fuel NO formation process, with the effect of coal burning upon the reduction of NO in the presence of CO. When only gas-phase reactions occurred, the concentration of oxygen-containing species increased with an increase in the CO_2 concentration, and consequently *CR* increased slightly. However, in the presence of coal, the lower proportion of the intermediate species containing N increased under the impact of coal plus CO. Furthermore, as the CO concentration increased with an increasing CO_2 concentration, *CR* tended to decrease. In conclusion, the conversion ratio from fuel N to NO increased with the increase of CO_2 concentration in the absence of coal, but it decreased in the presence of coal.

5.2.2 Influence of the Recycled NO

From a practical point of view, the amount of fuel N exhausted from the system as NO is important. For the whole system, the global conversion ratio from fuel N to exhausted NO is defined as:

$$CR^* = \left(\begin{matrix} N\,\text{atom number of} \\ \text{NO in exhaust gas} \end{matrix} \right) \Big/ \left(\begin{matrix} N\,\text{atom number} \\ \text{of fuel N} \end{matrix} \right) \tag{5.6}$$

Different values of the gas recirculation ratio, derived at different levels of oxygen/fuel stoichiometric ratio, were used to calculate *CR**. The different oxygen/fuel stoichiometric ratio, λ=0.7, 1.0 and 1.2, were considered. Taking λ=1.0, for example, Fig. 5.1 shows the global conversion ratio of fuel N to NO (*CR**) in various cases at an oxygen/fuel stoichiometric ratio of λ=1.0. The factors considered in the various cases are listed in Table 5.1.

The global conversion ratio of fuel N to NO (*CR**) decreased to some degree from Case B to Case C, when the replacement ratio for CH_4 by coal increased from zero to 0.2, which was due to the enhancement of NO reduction at the solid surface of the coal particle (or "heterogeneous effect" of coal). The global conversion ratio of fuel N to NO decreased remarkably from Case C to Case D when the reduction of recycled NO was included.

When the influence of CO_2 on *CR* was included, the global conversion ratio of fuel N to NO decreased from Case D to Case E. In addition, when the impact of recycled NO on *RR* was considered, the global conversion ratio of fuel N to NO decreased from Case E to Case F. The *CR** value in Case G was lower

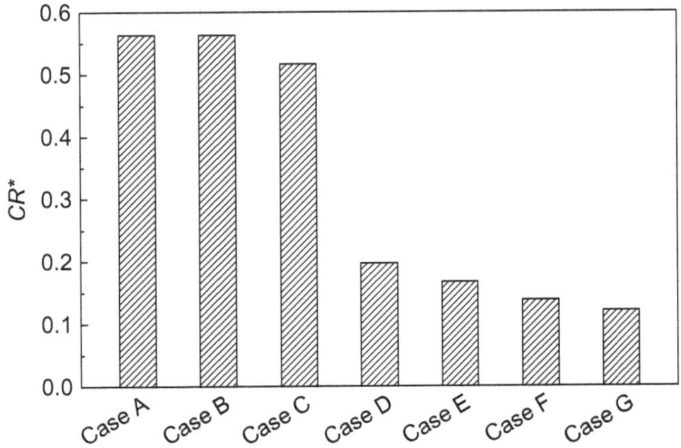

FIG. 5.1 Global conversion ratio from fuel N to exhausted NO at oxygen/fuel stoichiometric ratio $\lambda = 1.0$. *(From Liu H, Yuan Y, Yao H, Dong S, Ando T, Okazaki K. Factors affecting NO reduction during O2/CO2 combustion. Energy Fuel 2011;25(6):2487–92.)*

than that in Case F, suggesting that *CR** decreased because of the interaction between recycled NO and fuel N.

In brief, the reduction ratio of NO in recycled gas increased with the NO concentration when $\lambda = 0.7$ and 1.0, but decreased when $\lambda = 1.2$. The conversion ratio from fuel N to NO decreased with the increase of NO concentration in the recycled gas, because of the interaction between fuel N and recycled NO.

To clarify the mechanism of efficient NO reduction in O_2/CO_2 coal combustion, the contributions of various factors were quantified [8]. The contributions at oxygen/fuel stoichiometric ratio $\lambda = 1.0$ are shown in Fig. 5.2. The reduction of recycled NO was the major contributor, although not as dominant as at $\lambda = 0.7$. The contributions of various factors increased in the following order: interaction between recycled NO and fuel N, influence of recycled NO on the reduction ratio *RR*, influence of CO_2 on the *CR*, heterogeneous effect, and reduction of recycled NO.

In total, the relative importance of various factors to the low global conversion ratio from fuel N to exhausted NO in O_2/CO_2 combustion changed with oxygen/fuel stoichiometric ratio (λ). Nevertheless, under all conditions investigated, the reduction of recycled NO was the major contributor (over 70%).

5.2.3 Influence of Coal Rank

Three types of Chinese coals (lignite, bituminous, and anthracite) were used for investigating the homogeneous and heterogeneous contributions of high concentrations of CO_2 and recycled NO through a drop-tube furnace (DTF) at 1373 K [9]. The char yield and nitrogen distribution during devolatilization were studied. The yields of CO_2 chars were lower than those of N_2 chars for all three coals. The

TABLE 5.1 Factors Considered in Various Cases

	Air Combustion	Recycle of Flue Gas	Heterogeneous Effect	Reduction of Recycled NO	Influence of CO_2 on *CR*	Influence of Recycled NO on *RR*	Interaction Between Recycled NO and Fuel N
Case A	○						
Case B		○					
Case C		○	○				
Case D		○	○	○			
Case E		○	○	○	○		
Case F		○	○	○	○	○	
Case G		○	○	○	○	○	○

From Liu H, Yuan Y, Yao H, Dong S, Ando T, Okazaki K. Factors affecting NO reduction during O2/CO2 combustion. Energy Fuel 2011;25(6):2487–92.

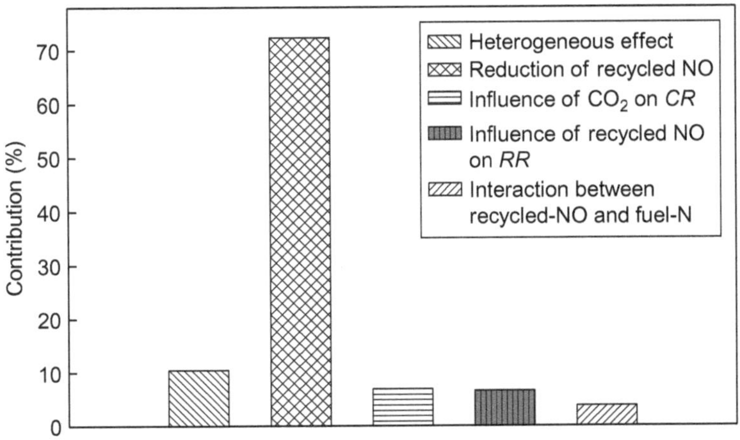

FIG. 5.2 Contributions of various factors at oxygen/fuel stoichiometric ratio $\lambda = 1.0$. *(From Liu H, Yuan Y, Yao H, Dong S, Ando T, Okazaki K. Factors affecting NO reduction during O2/CO2 combustion. Energy Fuel 2011;25(6):2487–92.)*

nitrogen content in N_2 chars was much higher than that in CO_2 chars for all three coals, which may have been the result of the weak nitrogen-containing sites in the char being attacked by the highly concentrated CO_2. Moreover, the nitrogen content in the lignite and bituminous chars was less than half of that in the raw coal, which suggested that more nitrogen was released as volatile-N during the lignite and bituminous devolatilization, while most of the coal-N remained in the anthracite chars as char-N after devolatilization. This may have been due to the different volatile contents of the three coals [9].

Fig. 5.3 shows the conversion ratios of coal-N to the exhaust NO for the lignite, bituminous, and anthracite coal combustion at different O_2 fractions of 21%, 27%, and 32% under O_2/N_2 and oxy-coal combustion conditions. The recycled ratio was set at 0.8 in O_2/RFG. Therefore the N amount in RFG was about 80%, while the N amount in outlet-NO was 20% of the total N amount of NO [9].

The conversion ratio of coal-N to NO in O_2/N_2 was highest and decreased following the $O_2/N_2 > O_2/CO_2 > O_2$/RFG order for all O_2 fractions. The conversion ratio of coal-N to NO increased with the increase of the O_2 fraction for all coals tested. The particle temperature was higher for a high O_2 fraction [9]. More NCO radicals (the important intermediate in volatile-N and char-N oxidation) was able to convert to NO at a higher temperature [9]. Reducing agents such as CH_i may have reduced NO to N_2. However, these agents can also be oxidized more quickly at a higher temperature. The increase of NO formation and decrease of NO reduction jointly resulted in a higher NO emission at a higher O_2 fraction.

Although the nitrogen content was almost the same in the lignite and bituminous coal, more NO was produced during the combustion of lignite compared to

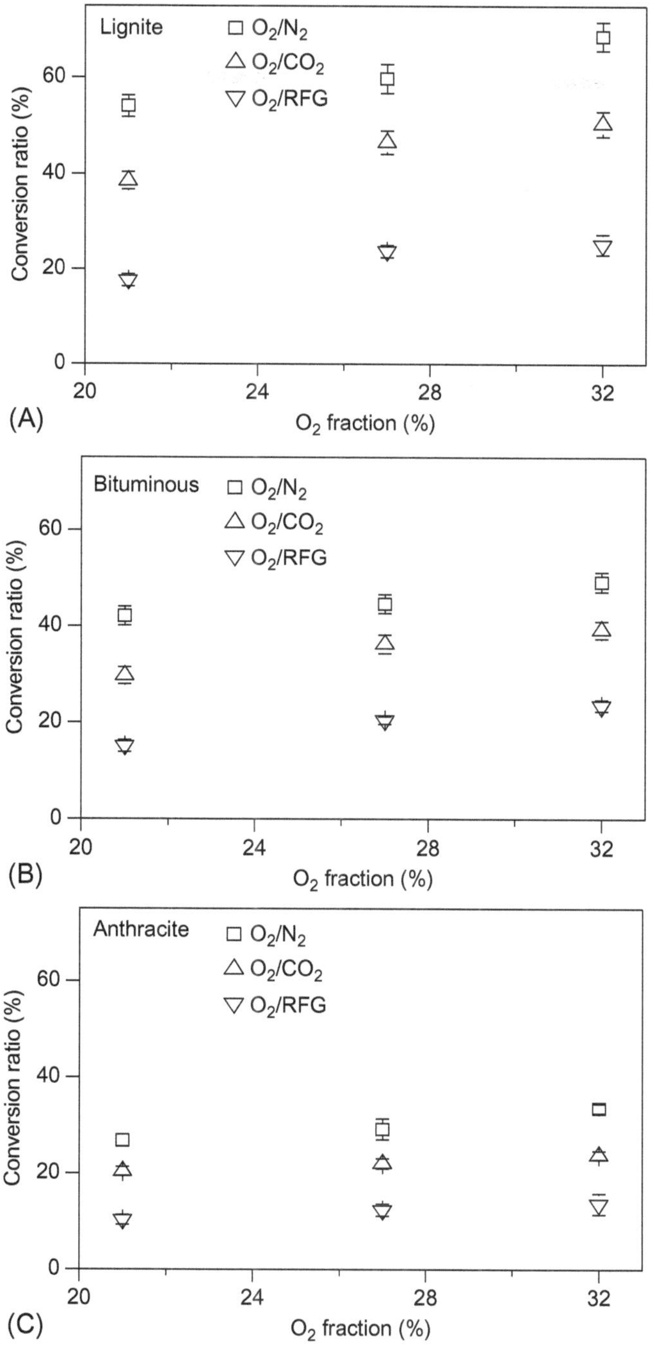

FIG. 5.3 Conversion ratios of NO from coal-N during coal combustion for the three coals. *(From Hu X, Li X, Luo G, Yao H. Homogeneous and heterogeneous contributions of CO2 and recycled NO to NO emission difference between air and oxy-coal combustion. Fuel 2016;163:1–7.)*

bituminous coal combustion. The nitrogen content was greatest in the anthracite among the three coals tested; however, the anthracite combustion produced the least NO at the same atmosphere and O_2 fraction. It can be observed that the conversion ratio of coal-N to NO increased with decreasing coal rank. Namely, the conversion ratio of coal-N to NO for the lignite was the highest, and for the anthracite, it was the lowest [9].

5.2.4 Homogeneous and Heterogeneous Contributions of CO_2 and Recycled NO

High concentrations of CO_2 and recycled NO influenced NO emissions in oxy-coal combustion through both homogeneous and heterogeneous reactions and ultimately decreased the NO release amount. The reactions between recycled NO and the reducing agents led to the difference in NO emission rates between the cases of O_2/CO_2 and O_2/RFG. For char combustion, the difference of NO conversion ratios between O_2/N_2 and O_2/CO_2 atmospheres was considered as the heterogeneous effect of CO_2 because little volatile matter was retained in the char. The contribution of heterogeneous recycled NO removal on NO emissions can also be evaluated similarly by the NO conversion ratio difference between O_2/CO_2 and O_2/RFG during char combustion [9]. Fig. 5.4 shows a flowchart estimating the effect of CO_2 and recycled NO. The detailed contributions of homogeneous and heterogeneous effects of CO_2 and recycled NO on the overall NO decrements from O_2/N_2 to O_2/RFG are shown in Fig. 5.5.

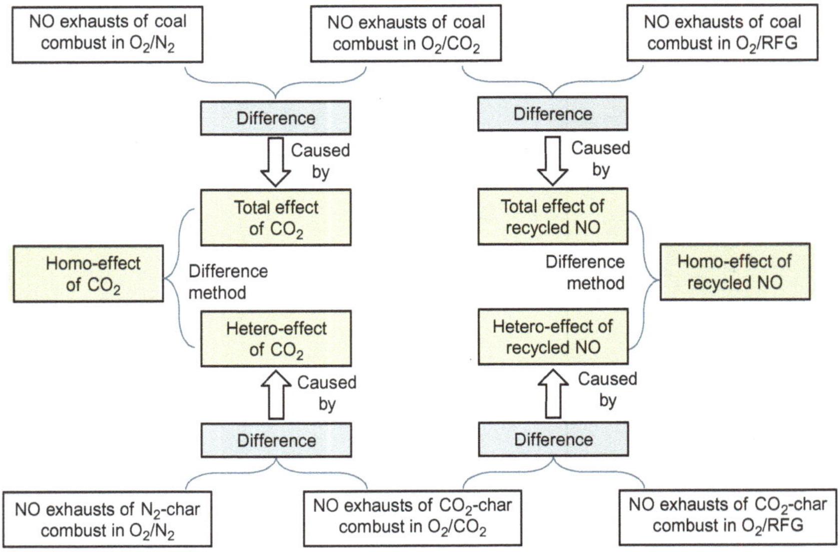

FIG. 5.4 Flowchart to estimate the detailed contributions of CO_2 and recycled NO. *(From Hu X, Li X, Luo G, Yao H. Homogeneous and heterogeneous contributions of CO2 and recycled NO to NO emission difference between air and oxy-coal combustion. Fuel 2016;163:1–7.)*

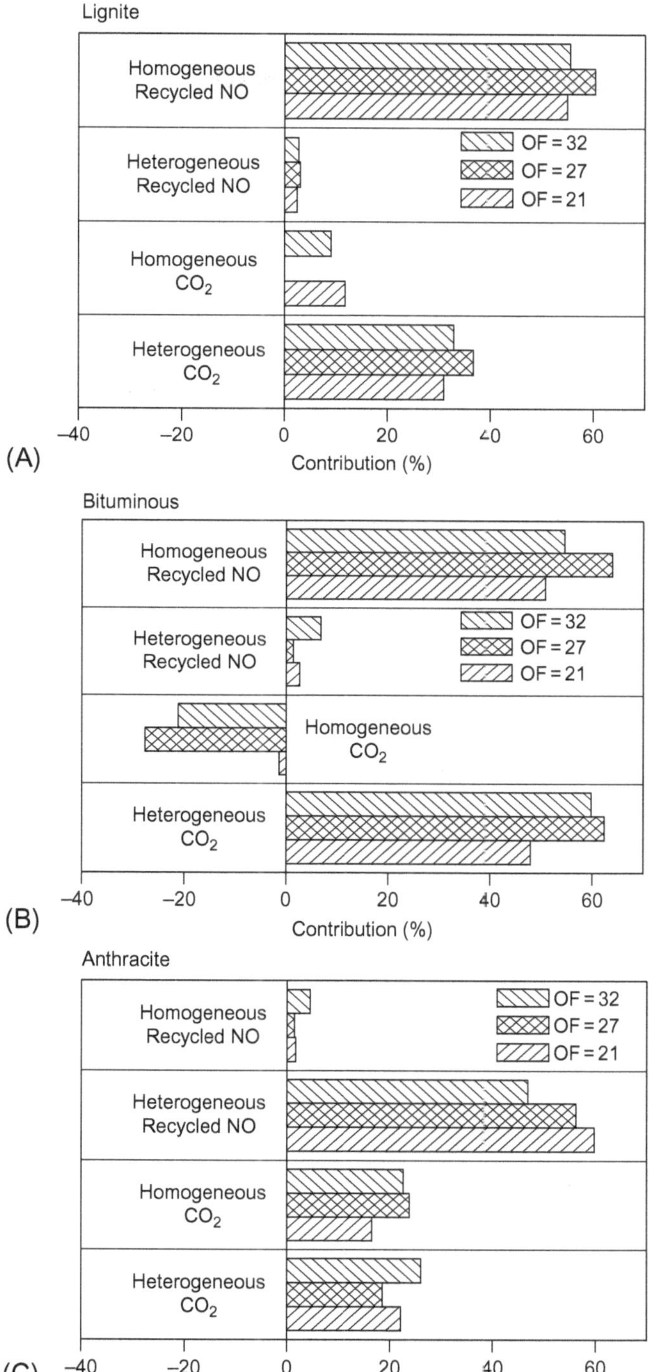

FIG. 5.5 Homogeneous and heterogeneous contributions of CO_2 and recycled NO to the overall NO decrements from O_2/N_2 to O_2/RFG for the three coals. *(From Hu X, Li X, Luo G, Yao H. Homogeneous and heterogeneous contributions of CO2 and recycled NO to NO emission difference between air and oxy-coal combustion. Fuel 2016;163:1–7.)*

The contribution of recycled NO ranged from 55% to 65% in the different coals, which was the main reason for the NO decrement. The influence of highly concentrated CO_2 contributed about 40% to the lower NO emission in oxy-coal combustion. As shown in Fig. 5.5, the heterogeneous effect of CO_2 contributed >20% to the lower NO emission, which indicated that NO formation from char-N decreased for all three coals in an oxy-coal condition [9]. However, the influence of the CO_2 on homogeneous reactions was totally different. The contribution of the CO_2 homogeneous effect was minor to lower NO emissions during the lignite combustion or even negative during the bituminous coal combustion, while it accounted for about 20% of NO decrement in the case of anthracite combustion. More NO was formed in O_2/CO_2 than in O_2/N_2 by volatile-N during the bituminous coal combustion at the O_2 fraction (OF)=27% and 32%. The nitrogen-containing intermediate could be oxidized by the higher concentration of oxygen-containing species in a high CO_2 condition through the Fenimore mechanism [10]. During the lignite or bituminous volatile matter combustion, the change of NO emissions owing to the increase of volatile-N may have been equal to or larger than that caused by the CO_2 suppression effect. During the anthracite volatile matter combustion, the effect of CO_2 suppressing homogeneous NO formation became more important than the promotion effect caused by the increasing volatile-N. The different CO_2 homogeneous impacts may be attributed to the volatile matter composition. The content and composition of volatile-N were decisive factors affecting the homogeneous CO_2 effect [9].

The proportions of NO homogeneous and heterogeneous removal were also different for the three coals. Homogeneous reductions of recycled NO were more important than the reactions between NO and char during lignite and bituminous coal combustion, while the reactions between recycled NO and char contributed more to NO reduction than homogeneous NO removal in the anthracite combustion. Because of the abundant NO from the RFG in the O_2/RFG atmosphere, the key factor for NO removal was the amount of the reducing agents in the gas and solid samples. The amount of the reducing agents were further determined by the volatile released and the carbon remaining in the char. Higher volatile content coal can reduce more NO through homogeneous reactions [11]. In addition, only a small amount of carbon remained in the lignite and bituminous coal chars. Therefore the homogeneous NO reduction reactions played a more important part in reducing NO compared to the reactions between NO and char. Most of the carbon remained in the anthracite char after the devolatilization stage in oxy-coal conditions. The main NO removal route during the anthracite combustion in the O_2/RFG atmosphere was through the reaction between NO and char [9].

5.3 MODELING OF NO_X CONVERSION USING DETAILED CHEMICAL KINETICS

NO_x conversion chemistry in combustion is a very complex chemical process, controlled by the formation and consumption of hydrocarbon radicals and nitrogen species. It is important to understand the NO_x combustion chemistry under an O_2/CO_2 atmosphere, which is done by conducting detailed chemical kinetic modeling. The Pisupati and Bhalla mechanism [12] is proper for NO_x combustion chemistry but relatively short on describing the hydrocarbon combustion process; therefore a combination of the NIST mechanism's hydrocarbon chemistry [12] with the Pisupati and Bhalla mechanism's NO_x chemistry was applied as an N-Pisupati mechanism to enhance its ability [12]. This mechanism was used to study nitrogen oxides reaction behaviors at different conditions under an air atmosphere and an O_2/CO_2 atmosphere [12].

Fig. 5.6 shows the rate of production analysis of NO elemental reactions at fuel-rich conditions, and the dominant reactions for NO production are highlighted. NO formation and reduction process under O_2/CO_2 and air atmosphere shared common issues [12]. The same productive elemental reactions are shown as follows:

$$\text{R1040} \quad NO + O_2 = NO_2 + O$$

$$\text{R1033} \quad NO + N = O + N_2$$

$$\text{R1070} \quad CO_2 + N = NO + CO$$

$$\text{R1085} \quad HCCO + NO = HCNO + CO$$

The reaction HCCO + NO = HCNO + CO was one of the most dominant elemental reactions. The reaction $NO + N = O + N_2$ was not able to consume NO under an O_2/CO_2 atmosphere at a fuel-rich condition. However, it was one of the most productive reactions contributing to NO reduction under an air atmosphere [12].

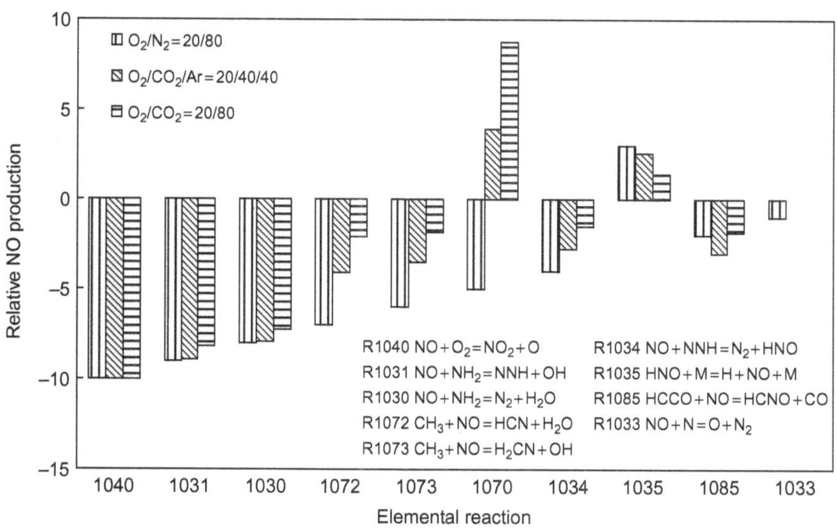

FIG. 5.6 Rate of production analysis of NO elemental reactions at fuel-rich condition. *(From Zhao R, Liu H, Zhong X, Wang Z, Hu H, Qiu J. Modeling of NOx conversion during combustion under high CO2 concentration using detailed chemical kinetics. Fuel Process Technol 2011;92(5):939–45.)*

High CO_2 concentration can affect some of the most productive elementary reactions. To study the reaction mechanism responsible for the chemical effects of CO_2 concentration, the elementary reactions that were potentially affected by the change in CO_2 concentration were identified as follows:

$$\text{R1029} \quad NO + HO_2 = NO_2 + OH$$

$$\text{R1033} \quad NO + N = O + N_2$$

$$\text{R1035} \quad HNO + M = H + NO + M$$

$$\text{R1039} \quad NO_2 + H = NO + OH$$

$$\text{R1070} \quad NO + CO = CO_2 + N$$

The sensitivity analysis was further carried out to identify the important reactions that contribute to NO conversion, and then the NO reaction path diagram under O_2/CO_2 was derived. Figs. 5.7 and 5.8 summarize the results of the cases under an O_2/CO_2 atmosphere at fuel-rich and fuel-lean conditions, illustrating the most important reactions routes for the formation and destruction of the nitrogenous reactants [12]. The reaction paths at fuel-rich conditions were more complicated compared to those of fuel-lean conditions, mainly because of the high concentration of CH_4 activation of various reaction channels and the increased number of species allowed in the diagram at fuel-rich conditions.

The similarities of N-oxidation pathways in different atmospheres broke down in the NO_x precursor's reaction pathways. Some alternative reaction pathways for HCN and NH_3 were competitive. The main NO_x precursor's pathways under O_2/CO_2 atmosphere were $HCN \rightarrow NCO \rightarrow N_2O \rightarrow NO$ and

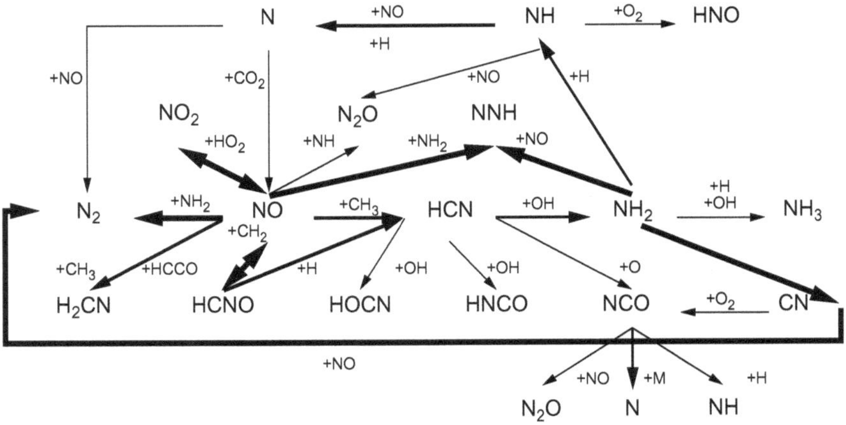

FIG. 5.7 Flow diagrams of the reactions of NO under O_2/CO_2 atmosphere at fuel-rich condition. *(From Zhao R, Liu H, Zhong X, Wang Z, Hu H, Qiu J. Modeling of NOx conversion during combustion under high CO2 concentration using detailed chemical kinetics. Fuel Process Technol 2011;92(5):939–45.)*

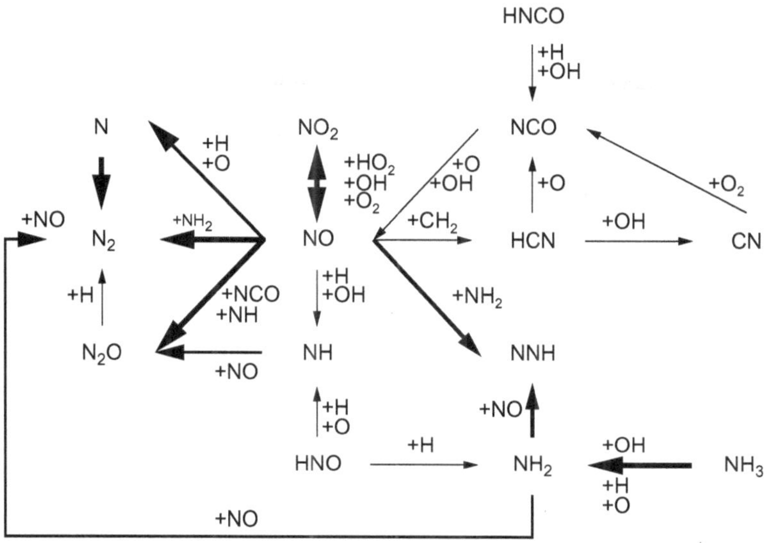

FIG. 5.8 Flow diagrams of the reactions of NO under O_2/CO_2 atmosphere at fuel-lean condition. *(From Zhao R, Liu H, Zhong X, Wang Z, Hu H, Qiu J. Modeling of NOx conversion during combustion under high CO2 concentration using detailed chemical kinetics. Fuel Process Technol 2011;92(5):939–45.)*

$NH_3 \rightarrow NH_2 \rightarrow (NNH,N_2) \rightarrow NO$. The main NO_x precursor's pathways were $NH_3 \rightarrow NH_2 \rightarrow NNH \rightarrow N_2$ and $HCN \rightarrow CN \rightarrow NCO \rightarrow N_2$ under an air atmosphere [12].

5.4 SO_2 FORMATION AND EMISSION DURING OXY-FUEL COMBUSTION

Tan et al. [13] found that the concentration of SO_2 in the furnace for the O_2/RFG case was three to four times higher than the corresponding air case because of the accumulated effects of flue gas recycling and the reduced volume of flue gas. Croiset et al. [14] and Kiga et al. [15] observed that the conversion of sulfur in coal to SO_2 decreased in O_2/CO_2 combustion compared to traditional air combustion. At temperatures below 500°C, the SO_3 formed from SO_2 reacting with water vapor (H_2O) in the flue gas, forming gaseous sulfuric acid (H_2SO_4), which would cause low-temperature corrosion if it were to condense on a metal surface. Knowledge about sulfur transfer characteristics during an oxy-fuel combustion process is necessary in order to better control SO_2 emission during its formation process.

Sulfur exists in coal in two major forms simultaneously, that is, organic and inorganic sulfur. Both might influence each other because of the interaction or competition reaction with minerals during pyrolysis/combustion [16,17]. Additionally, sulfur reacting with the mineral matter in coal, particularly with alkali and alkaline earth metals, may cause slagging, fouling, agglomeration, and defluidization during the use of coal. It is necessary to study the effects of minerals in coal on the transformation of organic and inorganic sulfur under oxy-fuel conditions. Density fractionation techniques were used to get coal fractions with various mineral contents from the same coal. The float-sink method [18] was widely used in the studies for determining major and trace element affinities in coal, particulate matter emission, char structure, and burnout. Few studies had been reported about the sulfur transformation in the case of density-separated coal fractions. The main objective was to investigate the behaviors and mechanisms of organic and inorganic sulfur transformations under high CO_2 conditions based on systematically comparing the sulfur transformations of the density-separated coal fractions. Furthermore, the roles of indigenous and added minerals on the sulfur transformation were also investigated by adding sodium and kaolinite.

5.4.1 The Behaviors of Organic and Inorganic Sulfur Transformation

A Chinese pulverized bituminous coal was fractionated via the float-sink method into three density fractions: light (<1.4 g/cm^3), medium (1.4–2.0 g/cm^3), and heavy (>2.0 g/cm^3), which were denoted as $C_{<1.4}$, $C_{1.4\text{–}2.0}$, and $C_{>2.0}$, respectively. The weight percentages of each fraction ($C_{<1.4}$, $C_{1.4\text{–}2.0}$ and $C_{>2.0}$) were 30%, 59%, and 11%, respectively. $C_{<1.4}$ consisted almost entirely of organic sulfur that presented as sulfidic, disulfidic, and thiophenic. $C_{<1.4}$ was almost free of indigenous minerals (nearly 4% ash content and 70% fixed carbon). $C_{1.4\text{–}2.0}$ contained a maximum amount of sulfates in the three coal fractions. It contained much more indigenous minerals than $C_{<1.4}$ (nearly 30% ash content and 50% fixed carbon). $C_{>2.0}$ contained mainly inorganic sulfur (pyrite), and it contained about 80% indigenous mineral.

To understand the effect of organic and inorganic sulfur on sulfur transformation, the volatile sulfur (primarily as H_2S and SO_2) between the three density fractions was compared. The evolution of hydrogen sulfide (H_2S) increased with the increasing temperature. H_2S emitting from $C_{<1.4}$ and $C_{1.4\text{–}2.0}$ appeared at low temperatures of about 400°C and rapidly increased above the temperature. But H_2S releasing from $C_{>2.0}$ appeared at about 500°C; moreover, its value was the smallest of the three coal samples. For $C_{<1.4}$, the amount of H_2S at 400°C was 4.2% and increased to 12.5% at 800°C, 3.4% at 400°C, and 10.6% at 800°C for $C_{1.4\text{–}2.0}$, while for $C_{>2.0}$, the amount of H_2S at 400°C never appeared and had a minimum value of about 3.5% at 800°C, respectively. As discussed previously, the more labile the organic sulfur, the lower the decomposition temperature. H_2S was the dominant gaseous sulfur during pyrolysis of $C_{<1.4}$ because of the SH* radicals from the decomposition of organic sulfur reacting with hydrogen provided by the surroundings.

However, sulfur dioxide (SO_2) was the major sulfurous gas in $C_{1.4\text{–}2.0}$ and $C_{>2.0}$. The SO_2 increased as the temperature increased. Although the amount of SO_2 in $C_{>2.0}$ was the lowest of the three coal fractions at a low temperature of 400°C, above this temperature, the SO_2 increased drastically. Even at the end temperature of 800°C, the SO_2 value reached the highest of the three coal fractions and was 12% higher than that in $C_{<1.4}$, in part because of the high oxygenated functional groups present in $C_{>2.0}$. During pyrolysis, the pyrite decomposes and then reacts with oxygen provided by the surroundings to produce SO_2 [19,20]. Another reason relates to the reactions between CO_2 and pyrite and/or decomposition of sulfates. Under a pure CO_2 atmosphere, pyrite first decomposes to pyrrothite; then pyrrothite reacts with CO_2 to produce magnetite and S_2, or iron sulfates, and the sulfates decompose and generate SO_2 [16,21].

5.4.2 The Effect of Indigenous Minerals on Transformation of Organic and Inorganic Sulfur

The total sulfur retention decreased with the increase of the pyrolysis temperature from 400°C to 800°C, the retention especially decreased drastically from 400°C to 600°C. The sulfur in $C_{<1.4}$ decomposed at low temperatures because of the release of the labile organic sulfurs. Then the total sulfur retention further lowered to slightly above 600°C, which contributed to the decomposition of some stable organic sulfur (such as cyclic sulfide and aryl sulfide). The decomposition of total sulfur in $C_{1.4\text{–}2.0}$ decreased because of the significant decomposition of sulfates and pyrite. The decrease of total sulfur retention in $C_{>2.0}$ was dominated by pyrite decomposition. The pyrite mineral (S/Fe = 2) decomposed to pyrrothite present in a mixed form of minerals (FeS_x with $1 < x < 2$) before complete decomposition. With increased temperatures, some of the troilite further decomposed to form elemental iron (S/Fe < 1) [19].

The total sulfur retention increased as the coal density increased. The amount of sulfur retention at 800°C was 49% for $C_{<1.4}$, 57% for $C_{1.4\text{–}2.0}$, and 70% for $C_{>2.0}$. Organic sulfur content decreased, while inorganic sulfur content clearly increased with the increase of coal density. During pyrolysis, nascent organic sulfur converted from the decomposition of pyrite was very stable. It existed in thiophenic or condensed thiophenic sulfur forms. Moreover, the indigenous minerals contained in coal increased drastically with the increase of coal density. The minerals reacted with sulfur to form new, stable sulfur-containing compounds, leading to a retention of sulfur in coal char.

The organic sulfur retention of $C_{<1.4}$ decreased with increasing temperatures, whereas those of $C_{1.4\text{–}2.0}$ and $C_{>2.0}$ clearly increased at temperatures of 400–600°C and stayed almost constant above 600°C. For $C_{1.4\text{–}2.0}$, the organic sulfur retention at 400°C was 43% and increased to 58% at 700°C, while for $C_{>2.0}$, the organic sulfur retention was 43% at 400°C and 69% at 700°C, which is explained by the fact that the pyrite in the coal char converted into nascent organic sulfur. Pyrite (FeS_2) was transformed into a more reactive ferrous sulfide and a mixture of allotropic species. S_2 was the dominant gaseous sulfur for pyrite decomposition. During pyrolysis, the active S_2 reacted with active sites of the organic matrix in a char structure to form new organic sulfur [22,23].

The inorganic sulfur retention decreased with an increase of temperatures from 400°C to 800°C and decreased especially rapidly at 400–600°C. The inorganic sulfur contents in the three density coal fractions were significantly different: 7.3% for $C_{<1.4}$, 52% for $C_{1.4\text{–}2.0}$, and 67% for $C_{>2.0}$. But after pyrolysis, the inorganic sulfur retention in coal char was very low in all cases: 0.57% for $C_{<1.4}$, 1.34% for $C_{1.4\text{–}2.0}$, and 3.21% for $C_{>2.0}$ at a temperature of 800°C. The results indicated that the inorganic sulfur almost completely converted into nascent organic sulfur, which was more stable than the inorganic sulfur at high temperatures. Also, the large amount of indigenous minerals in $C_{1.4\text{–}2.0}$ and $C_{>2.0}$ slightly suppressed the decomposition of inorganic sulfur.

5.4.3 The Effect of Added Minerals on the Transformation of Organic and Inorganic Sulfur

Added minerals affected the total sulfur retention during pyrolysis. The total sulfur retention increased with the addition of NaCl and kaolin. NaCl had a remarkable effect on sulfur retention at low temperatures of 400–700°C. The SH* free radicals from the decomposition of organic sulfur were more reactive toward Na to form nascent sulfur and to be retained in coal char. But the sulfur retentions decreased rapidly at temperatures above 700°C because the volatilization of Na increased quickly with the rising temperature, with almost 80% of Na volatilizing above 700°C. As for kaolin, the reason might be due to physical adsorption leading to a small increase of the sulfur retention at low temperatures. With temperatures above 700°C, kaolin reacted with sulfur to form more stable sulfates, thus presenting an improved ability to retain sulfur at high temperatures. The inorganic sulfur retentions of $C_{<1.4}$+NaCl, and $C_{<1.4}$+kaolin were both higher than $C_{<1.4}$ at low temperatures, but drastically decreased to above 700°C because the indigenous minerals had little effect on inorganic sulfur retention.

The sulfur retentions among $C_{<1.4}$+pyrite, $C_{<1.4}$+pyrite+kaolin, $C_{<1.4}$+pyrite+NaCl, and $C_{>2.0}$ were compared. The total sulfur retention in $C_{>2.0}$ char was higher than $C_{<1.4}$+pyrite at 800°C, even higher than $C_{<1.4}$+pyrite+kaolin and $C_{<1.4}$+pyrite+NaCl, which also contained a high content of added minerals. The total sulfur retentions were about 52% for $C_{<1.4}$+7%pyrite, 55% for $C_{<1.4}$+7%pyrite+kaolin, and 60% for $C_{<1.4}$+7%pyrite+NaCl; but the highest amount was 70% for $C_{>2.0}$. The indigenous minerals played a more significant role on sulfur retention than the other added minerals because the indigenous minerals were inherently embedded in the organic matrix of coal, whereas the added minerals were physically dispersed in coal and were independent of the coal matrix.

5.4.4 SO_2 Emission During Oxy-fuel Combustion

The SO_2 concentration in the furnace was more than three times higher during oxy-fuel combustion than in air combustion. Pilot-scale experiments [24] revealed that SO_2 in the flue gas was directly proportional to the fuel sulfur content, and the concentration did not change significantly with an increasing recycle ratio in oxy-fuel combustion. The SO_2 emission (mass per unit of energy released from the coal) was not affected significantly by the variations of O_2 or CO_2 concentrations [25]. The decrease in SO_2 emission during oxy-fuel combustion was due to SO_3 formation and subsequent sulfur retention. The higher SO_2 concentration in oxy-fuel combustion was accompanied by higher sulfur self-retention by ash. SO_2 was captured during char combustion by alkaline earth and alkali metals (Mg, Ca, Na, and K) to form sulfates. Some sulfur remained in the ash without being released. More sulfur self-retention by ash occurred in oxy-fuel than in air-fired conditions [26]. The high SO_2 concentration favored sulfation and stabilized the sulfates formed [27–29]. Calcium had a dominant role in sulfur self-retention by ash, reacting with SO_2 into $CaSO_4$. The Ca/S molar ratio in the coal was an important factor, along with the reactivity of Ca in the coal.

The conversion of fuel S to SO_2 is lower in oxy-fuel combustion than in air-fuel combustion, which led to about 35% lower SO_2 emission (mg/MJ fuel input) under oxy-fuel conditions. About 67% of fuel S was converted to SO_2 under air-fired conditions, while only about 43% was converted during oxy-fuel combustion. The main reason was the higher SO_2 concentration, which favored the formation of sulfates. Croiset et al. [14] and Kiga et al. [15] demonstrated the low SO_2 release in O_2/CO_2 combustion because a considerable amount of sulfur was absorbed in ash in the system. The formation of sulfates in oxy-fuel cases was also improved by the longer gas residence time in the furnace compared to air-fired case. Another reason for the lower SO_2 emission in oxy-fuel combustion is the possibility of improved sulfation of calcium in a high CO_2 atmosphere, which will be further discussed in Section 5.6.

5.4.5 Summary

The sulfur retention and the impact of minerals on sulfur transformation and organic/inorganic retention in coal char were studied in this section by using density-separated coal fractions. The comparison of sulfur transformation was made between indigenous minerals originally existing in coal and added minerals with NaCl and kaolin. In comparison to added minerals, indigenous minerals have more significant effects on sulfur transformation because they are inherently embedded in the organic matrix of coal, while added minerals are physically dispersed within coal. Moreover, the emission of SO_2 under oxy-fuel combustion was discussed. SO_2 emission is decreased during oxy-fuel combustion because of SO_3 formation and subsequent sulfur retention. The high SO_2 concentration in furnace favors sulfur self-retention by ash in an oxy-fuel condition.

5.5 SO_3 FORMATION AND EMISSION DURING OXY-FUEL COMBUSTION

It is well known that SO_3 in flue gas causes serious low-temperature (<200°C) corrosion for coal-fired boiler systems. SO_3 mainly formed during the cooling process of the flue gas in the post-flame region through the oxidation of SO_2 by homogenous and heterogeneous mechanisms. It has been confirmed that the SO_3 concentration in oxy-firing flue gas is several times higher than air-fuel combustion system [30–32]. This in turn enhances the tube corrosion during an oxy-fuel combustion process.

One of commonly used methods to suppress the low-temperature corrosion is to raise the exhaust flue gas temperature above its sulfuric acid dew point temperature (T_{adp}). The T_{adp} of flue gas is mainly determined by the concentrations of SO_3 and steam in flue gas. There are several methods for estimating T_{adp}. The former Soviet Union's empirical formula and its modified formulas [33,34] have been widely used. However, these formulas were found unsuitable for the flue gas with low SO_3 concentration. Several other formulas [35–37] were also proposed based on the SO_3 concentration in flue gas. However, the SO_3 concentration is not easily measured or estimated, since it is rather low and is affected interrelatedly by a large number of factors, including coal type, concentrations of SO_2, O_2, steam in flue gas, and the concentration/composition of fly ash [38,39]. Moreover, the conventional methods were developed based on the air-firing process, which should not be extended to the oxy-firing systems without validation and/or modification.

5.5.1 Influence of Temperature on SO_3 Formation

With increasing temperature, SO_3 concentrations for different initial SO_2 concentrations had the same variation tendency of increasing first and then decreasing, achieving a maximum value at around 950°C. Below 750°C, SO_3 concentration was very small and almost beyond the level of detection in some cases. At the elevated temperatures, SO_3 concentrations were relatively sensitive to the temperature and the maximum concentration reached about 950°C. The formation and decomposition mechanism of SO_3 involved a great many interactions with the radical pool compositions. Obviously, the O-cycle was of great significance for the formation and consumption of SO_3 based Eqs. (5.7), (5.8):

$$SO_2 + O(+M) \rightleftharpoons SO_3(+M) \tag{5.7}$$

$$SO_3 + O(+M) \rightleftharpoons SO_2 + O_2 \tag{5.8}$$

The O radical was derived mostly from the decomposition of oxygen. The SO_3 formation absolutely depended on the temperature and O radical concentration, and the temperature was able to affect the SO_2 oxidation indirectly by influencing the amount of the O radical [40]. The rate constants of Eq. (5.7) didn't change significantly with an increase in temperature [41]. However, the rate constants of Eq. (5.8) were apparently sensitive to variations in temperature [42]. The consumption of SO_3 was mainly realized via Eq. (5.8), and its reaction rate escalated as the temperature increased, which meant a faster rate converting SO_3 back to SO_2 at a higher temperature. Because the O radical concentration was very low before 750°C, the rate and the oxidation extent of SO_2 were consequently very small. When the temperature increased to 950°C, the O radical concentration increased drastically, and the SO_2 oxidation rate became faster and formed more SO_3. Conversely, when the temperature was above 950°C, SO_3 decomposition accelerated roughly, which led to a decreased SO_3 concentration. The discrepancy of SO_3 formation between Eqs. (5.7), (5.8) gradually became smaller, but the former always had a relative advantage, so the concentration of formed SO_3 was still smaller compared to its concentration at 950°C.

5.5.2 Homogenous Effects on the Conversion of SO_2 to SO_3

The formed SO_3 concentration improved with an increasing initial SO_2 concentration. SO_3 concentration decreased significantly with the injection of steam, which indicated steam was able to inhibit SO_3 formation. When steam accounted for 15% of the total simulated flue gas, the inhibiting effects on SO_3 formation were very obvious. Moreover, for three cases with different concentrations of steam from 15% to 35%, the difference on SO_3 concentrations with the same initial SO_2 concentration was relatively small. The high concentration of steam helped to sharply increase the OH radical, but the OH cycle did not increase the SO_3 concentration. On the contrary, the formation of the OH radical was mainly due to steam competing O radical with SO_2, which prevented SO_2 from being oxidized directly. Moreover, the contribution of steam to the formation of SO_3 was particularly small, so steam in flue gas can significantly decrease the formation of SO_3. In addition, as for the wet flue gas recycling, steam influenced the reactivity of gasification, and the gasification rate of coal char with steam was several times higher than that with CO_2 [43]. Consequently, CO concentration was much higher in the wet recycling than in the dry recycling during oxy-fuel combustion. Because SO_2 has a weaker competitive ability for OH radicals than CO does, the recombination reaction between the SO_2 and OH radical hardly happened, which may have indirectly facilitated the inhibition of CO on SO_3 formation [44].

An empirical formula was proposed between the conversion rate of SO_2 to SO_3 (α_{SO_2}) and the initial SO_2 concentration ($C^*_{SO_2}$) at a fixed temperature, as shown in Eq. (5.9). The values of f and g were obtained by regression analysis. Their weighted average values were obtained by Eq. (5.10). As a result, the weighted average value of SO_2 conversion rate under homogenous effect was obtained, as shown in Eq. (5.11).

$$\alpha_{\text{hmSO}_2,\text{temperature}} = f \times C_{\text{SO}_2}^{*\ g} \tag{5.9}$$

Here $\alpha_{\text{hmSO}_2,\text{temperature}}$ represents SO_2 conversion rate under homogenous effect (%).

$$W_k = \frac{\sum_{i=1}^{n} \frac{k_i}{T_i + 273.15}}{\sum_{i=1}^{n} \frac{1}{T_i + 273.15}} \tag{5.10}$$

Here W_k is the weighted average of k values obtained at different temperatures. T_i denotes the temperature (°C). k represents f or g.

$$\alpha_{\text{hmSO}_2} = 127.65 C_{\text{SO}_2}^{*\ -0.585} \tag{5.11}$$

5.5.3 Heterogeneous Effects on the Conversion of SO_2 to SO_3

In addition to homogenous oxidation, the heterogeneous effect of fly ash also influenced the conversion of SO_2 to SO_3, which includes the catalysis effect of alumina/iron oxides and the adsorption effect by calcium/magnesium oxides [39,45,46]. The optimum temperature for the catalytic influence of pure iron oxides (α-Fe_2O_3, γ-Fe_2O_3, Fe_3O_4) and Al_2O_3 was around 700°C and 850°C, respectively. The α_{SO_2} under the effects of fly ashes in the simulated flue gas (O_2: 5%, CO_2: 95%, SO_2: 3000 ppm) at different temperatures were investigated. Overall, the fly ashes improved the conversion of SO_2 to SO_3. The high α_{SO_2} were obtained by the ashes with high Fe and Al contents. On the contrary, the low α_{SO_2} were obtained by ashes with a high Ca and Mg content. To illustrate the coupled effects of Fe, Al, Ca, and Mg, ternary phase diagrams of Fe, *Al*, and Ca and Mg mass proportions were plotted versus α_{SO_2} at 750°C and 900°C, as shown in Fig. 5.9 The SO_2 conversion rate ranged from −0.1% to 0.7% at 750°C. The low SO_2 conversion rate at 750°C was in the region with high Ca & Mg and low Fe and Al contents. The highest SO_2 conversion rate at 750°C was found in the region with low Ca & Mg and a Fe/Al mass ratio of 2:3. The SO_2 conversion rate at 900°C ranged from −0.6% to 0.9%, greater than that at 750°C. The SO_2 conversion rate distribution clearly proved the catalytic effect of Fe/Al and the adsorption effect of Ca/Mg for SO_3. The former effect enhanced the formation of SO_3, whereas the latter one negated it. The different α_{SO2} distribution at 750°C and 900°C implied the different optimum temperatures for the effects of the four elements.

An empirical formula shown in Eq. (5.12) was established to correlate the α_{SO2} and Fe, Al, Ca, and Mg oxide contents in fly ash. The coefficient values (*a*, *b*, *c*, *d*, and *e*) at each temperature are shown in Table 5.2, were obtained by linear regression fitting of the experiment data by Eq. (5.12). The table showed

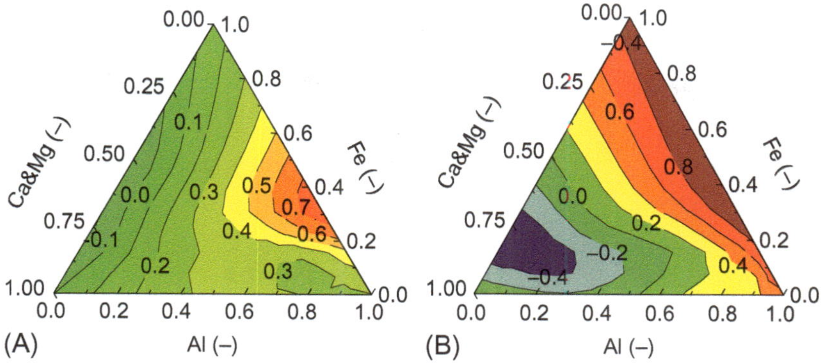

FIG. 5.9 Ternary phase diagrams of mass proportions of Fe, Al, and Ca & Mg versus SO_2 conversion rates at (A) 750°C and (B) 900°C.

TABLE 5.2 Values of *a*, *b*, *c*, *d*, and e

Temperature (°C)	*a* ($\times10^{-2}$)	*b* ($\times10^{-2}$)	c ($\times10^{-2}$)	*d* ($\times10^{-2}$)	e ($\times10^{-2}$)
650	0.014	*N*	−0.005	−0.084	20.830
750	0.050	*N*	−0.019	−0.147	29.411
900	0.169	0.151	−0.171	−0.123	−8.280
W	0.071	0.044	−0.058	−0.117	15.191

N, negligible; *W*, weighted average.

that the influence of the four oxides increased with the temperature increase. At 650°C and 750°C, the catalytic effect was mainly attributed to iron oxide, rather than alumina. In contrary, at 900°C, the catalytic effect of alumina was more significant.

$$\alpha_{htSO_2,\mathrm{temperature}} = a \times \mathrm{Fe} + b \times \mathrm{Al} + c \times \mathrm{Ca} + d \times \mathrm{Mg} + e \qquad (5.12)$$

where $\alpha_{htSO_2,\mathrm{temperature}}$ denotes the α_{SO_2} under heterogeneous effect at each temperature, %; Fe, Al, Ca, Mg: Fe, Al, Ca, Mg contents in ash, g/kg.

The weighted average values of the coefficients, shown in Table 5.2, were obtained by the same method as shown in Eq. (5.10). Furthermore, the ash content of coal (A_{ar}, %, d.b.); the fly ash proportion of the inherent coal ash (α_{fly}), which was fixed at 0.95; and the initial SO_2 concentration ($C^*_{SO_2}$, ppm) were also included for the α_{SO_2} calculation. These factors were reflected by the parameter *n*, shown in Eq. (5.13). The SO_2 conversion rate under heterogeneous effect was finally summarized as Eq. (5.14),

$$n = \lg\left(\frac{A_{ar} \times \alpha_{fly}}{C^*_{SO_2}}\right)^{\frac{1}{2}} \tag{5.13}$$

where A_{ar} represents the ash content of coal (% on dry basis), α_{fly} represents the fly ash proportion of the inherent coal ash, and $C^*_{SO_2}$ represents the initial SO_2 concentration (ppm).

$$\alpha_{htSO_2} = \lg\left(\frac{A_{ar} \times \alpha_{fly}}{C^*_{SO_2}}\right)^{\frac{1}{2}} \times \begin{pmatrix} 15.191 + 0.071\text{Fe} + 0.044\text{Al} \\ -0.058\text{Ca} - 0.117\text{Mg} \end{pmatrix} \times 10^{-2} \tag{5.14}$$

Here $\alpha_{htSO_2,\text{temperature}}$ represents SO_2 conversion rate under heterogeneous effect (%).

5.5.4 SO_3 Concentration and Acid Dew Point Temperature Calculation Formula Developments

The total SO_3 concentration (C_{SO_3}) in flue gas contributed by homogenous and heterogeneous effects can be calculated by combining Eqs. (5.11), (5.14), and $C^*_{SO_2}$ together, as shown in Eq. (5.15).

$$C_{SO_3} = C^*_{SO_2}\left(\begin{array}{l} 127.65C^{*\ -0.585}_{SO_2} + \lg\left(\dfrac{A_{ar} \times \alpha_{fly}}{C^*_{SO_2}}\right)^{\frac{1}{2}} \\ \times\begin{pmatrix} 15.191 + 0.071\text{Fe} + 0.044\text{Al} \\ -0.058\text{Ca} - 0.117\text{Mg} \end{pmatrix} \times 10^{-2} \end{array}\right) \tag{5.15}$$

The И. А. Bapahoba formula, based on the steam concentration (C_{H_2O}, ppm) and C_{SO_3} (ppm) in flue gas, was used widely for the T_{adp} estimation of flue gas [34,47,48], as shown in Eq. (5.16). An updated T_{adp} calculation formula, combining the И. А. Bapahoba formula and the SO_3 concentration calculation Eq. (5.15), was proposed as shown in Eq. (5.17). The T_{adp} calculation formula included the influential factors of steam and the $C^*_{SO_2}$, ash content of raw coal, the fly ash proportion of coal ash, and the Fe, Al, Ca, and Mg contents in coal ash.

$$T_{adp} = 186 + 20\lg C_{H_2O} + 26\lg C_{SO_3} \tag{5.16}$$

$$T_{adp} = 186 + 20\lg C_{H_2O} + 26\lg\left(C^*_{SO_2}\left(\begin{array}{l} 127.65C^{*\ -0.585}_{SO_2} + \lg\left(\dfrac{A_{ar} \times \alpha_{fly}}{C^*_{SO_2}}\right)^{\frac{1}{2}} \\ \times\begin{pmatrix} 15.191 + 0.071\text{Fe} + 0.044\text{Al} \\ -0.058\text{Ca} - 0.117\text{Mg} \end{pmatrix} \times 10^{-2} \end{array}\right)\right) \tag{5.17}$$

In order to check the accuracy of the proposed Eqs. (5.11), (5.14), (5.15), (5.17), extra experimental data obtained from laboratory-scale experiments, pilot-scale experiments, and industrial-scale coal-fired power plants of air- and oxy-fuel combustions were collected from the literature, and are shown in Table 5.3 [30,31,51,52].

The α_{SO_2} and C_{SO_3} contributed by the homogenous and heterogeneous effects, shown in Fig. 5.10 were calculated by Eqs. (5.11), (5.14), (5.15), respectively. It showed that α_{SO_2} ranged from 0.51% to 6.71%. It was over 4% for the coals labeled from c1, c2, d1, e1, I, and l. One reason may have been their relatively low $C^*_{SO_2}$. The α_{SO_2} of oxy-fuel cases were all significantly lower than their corresponding air-fired cases. Furthermore, the percentages contributed by the heterogeneous effect ranged from 1.33% to 20.86%. The low percentages contributed by the heterogeneous effect for c1, c2, and l may have been caused by the low ash content of the raw coals and the high Ca content in their ashes, as shown in Table 5.3. On the contrary, the largest percentages contributed by the heterogeneous effect for g and k may have been caused by the high Fe and Al content in their ashes. It is worth noting that the percentages contributed by the heterogeneous effect for oxy-fuel processes were 1.2 to 1.6 times their corresponding air-fired cases, which may have been caused by the higher $C^*_{SO_2}$ in the oxy-fuel flue gas, implying that the heterogeneous effect was more sensitive to the SO_2 concentration variation than to the homogenous effect. It was also found that the C_{SO_3} of the oxy-fuel processes were always higher than their corresponding air-fired cases, although the α_{SO_2} of the former were lower than that of the latter.

Two widely used calculation methods of T_{adp} for flue gases, P. A. Лероеян (Лes)empirical formula [49] and the modified A. Ф. Гаврцлов (modified Гa's) empirical formula [33,50] were employed for comparison purposes. The Лe's formula, shown in Eq. (5.18), involved the effects of the water dew point temperature (T_{dp}), the sulfur and ash contents, and the fly ash proportion in coal ash. The modified Гa's formula also involved the effects of alkali and alkaline earth metals in coal ash, shown in Eq. (5.19).

$$T_{adp} = T_{dp} + \frac{125\sqrt[3]{S_{SP}}}{1.05^{\alpha_{fly} A_{sp}}} \tag{5.18}$$

$$T_{adp} = T_{dp} + \frac{201\sqrt[3]{S_{SP}}}{f(R)} \tag{5.19}$$

$$R = \frac{91.5 S_{ar} \left(Q_{net}/1000\right)^{2/3}}{\sqrt{\alpha_{fly}\left(A_{ar} + \overline{CaO}_{ar}\right)}} \tag{5.20}$$

$$\frac{1}{f(R)} = 0.92 e^{-2.433(2.75 - \lg R)^{3.95}} \text{ for } \lg R \geq 1.9 \tag{5.21}$$

$$\frac{1}{f(R)} = 3.524 \times 10^{-3} e^{3.34(2.75 - \lg R)^{-1.5}} \text{ for } \lg R \leq 1.9 \tag{5.22}$$

TABLE 5.3 Data Collected From Literature

		A_{ar}	Ash Composition (g/kg)				C_{SO2}	S_{ar}	Q_{net}	A_{sp}	S_{sp}	H_2O	Measured	$\overline{CaO}_{ar}$	
Coal	Coal Type	%	Fe_2O_3	Al_2O_3	CaO	MgO	ppm	%	MJ/kg	%	%	%	T_{adp} (°C)	%	Boiler
a	HVM[a]-bituminous coal	8.0	145.7	176.6	18.7	9.8	3500[b]/18000[c]	3.98	10.11	0.79	0.39	10	146.2[b]/156.1[c]	3.0	1.5 MW pilot-scale combustor, O%: 27% for oxy-firing [30]
b	Subbituminous coal	8.8	50.9	145.2	61.1	13.9	500/2000	0.53	11.31	0.78	0.05	10	128.0/128.0	7.4	
c	LVM[d]-bituminous coal	4.9	52.0	147.8	221.9	51.7	200/300	0.23	7.09	0.70	0.03	10	122.2/125.5	27.4	
d	LVM-bituminous coal	19.3	160.0	286.0	16.4	11.2	178/552	0.24	22.81	0.85	0.01	19	115.8/137.0	2.8	60–120 kg/h pilot-scale combustor, O%: 28% for oxy- firing [31]
e	Lignite	18.2	10.5	285.0	3.7	3.4	387/1296	0.57	26.57	0.69	0.02	20	133.4/147.5	0.7	
f	HVM-bituminous coal	6.9	194.0	220.0	37.9	20.9	508/1612	0.88	28.62	0.24	0.03	20	120.2/141.7	6.0	

g	Lean coal	30.2	25.3	384.8	6.1	29.1	3206	2.32	19.53	1.55	0.12	7	136.3	4.0	600 MW power plant [49]
h	Bituminous coal	14.1	15.8	265.5	8.2	9.7	1088	0.78	22.81	0.62	0.03	10	109.4	2.0	
i	Lignite	10.8	76.0	166.5	88.3	19.3	290	0.15	11.60	0.93	0.01	18	103.7	8.0	
j	Bituminous coal	14.1	176.0	221.0	37.0	12.0	3000	3.74	13.63	4.33	1.15	10	137.4	5.0	Full-scale utility boilers [50]
k	Bituminous coal	11.9	243.0	206.0	29.0	8.0	2600	3.57	13.73	3.64	1.09	10	138.0	3.5	
l	Subbituminous coal	7.5	52.7	177.0	254.0	47.0	250	0.41	12.30	2.54	0.14	10	102.0	30.0	

A_{sp}, equivalent value of coal ash content; $A_{sp}=4182\times A_{ar}/Q_{net}$; $\overline{CaO}_{ar}$, Ca and Mg content in ash; Q_{net}, lower heating value; S_{ar}, S content in coal on received basis; S_{sp}, equivalent value of coal S content; $S_{sp}=4182\times S_{ar}/Q_{net}$.
a High-volatile matter.
b Air-fired case.
c Oxy-fuel case.
d Low-volatile matter.

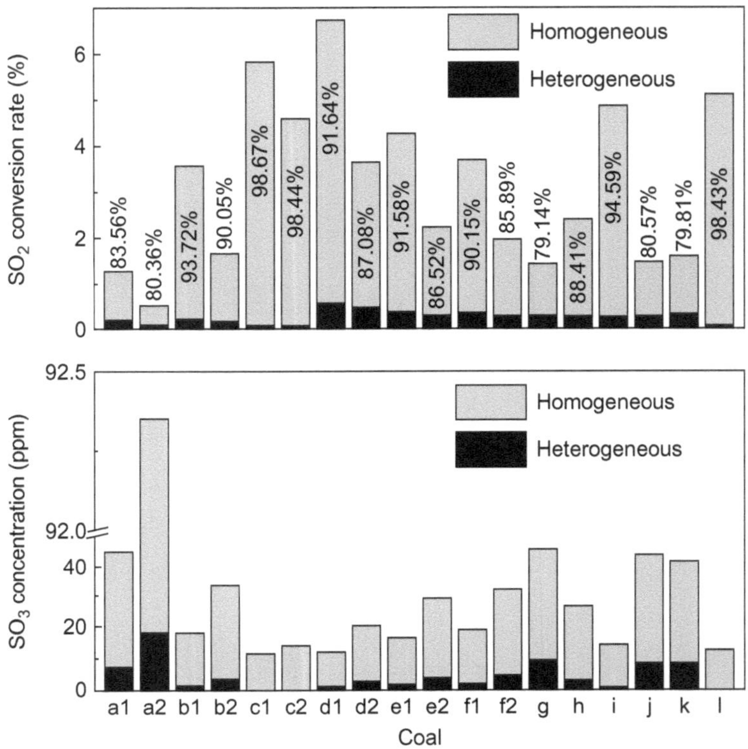

FIG. 5.10 SO_2 conversion rates and concentrations contributed by homogenous and heterogeneous effects. Symbol a1 to f1 and a2 to f2 were the results of air- and oxy-fuel combustion of coal a to f, respectively. The numbers on the columns represent the contribution percentage of homogenous effect.

The calculated results of T_{adp} by the updated formula, the Лe's and modified Гa's formulas, and the measured T_{adp} are shown in Table 5.4. The relative deviations of the three calculation methods are shown in Fig. 5.11. The T_{adp} calculations for oxy-fuel cases by Лe's and modified Гa's formulas were not performed because both formulas were obtained according to the air-fired combustion process. The T_{adp} ranges of air- and oxy-fuel cases obtained by the updated formula were rather similar to and much closer to the T_{adp} range of measured values, respectively, compared to the results obtained by Лe's and modified Гa's formulas. In particular, the T_{adp} obtained by the updated formula of oxy-fuel processes were always higher than their corresponding air-fuel cases, which was in agreement with the measured results. Fig. 5.11 shows that the updated formula presented the smallest relative deviation among the three calculation methods, which were 1.7% –27.9% for air-fired cases and 0.5%–7.8% for oxy-fuel cases. Thus the updated formula proposed was an effective method for acid dew point temperature estimation. Furthermore, it was suitable for both air- and oxy-fuel processes. The updated formula proposed in future work; for instance,

TABLE 5.4 Calculated and Measured Acid Dew Point Temperature (°C)

	This Method	Ле's Method	Гa's Method	Measured T_{adp}
a1	141.9	140.0	182.0	146.2
a2	150.5	–	–	156.1
b1	130.5	95.0	57.1	128.0
b2	138.0	–	–	128.0
c1	125.1	90.2	49.4	122.2
c2	127.2	–	–	125.5
d1	131.3	88.9	50.3	115.8
d2	137.7	–	–	137.0
e1	135.6	96.2	76.5	133.4
e2	142.5	–	–	147.5
f1	137.2	101.2	102.0	120.2
f2	143.7	–	–	141.7
g	139.6	104.8	136 2	136.3
h	135.6	91.5	95 6	109.4
i	132.6	88.7	61.7	103.7
j	141.7	159.7	171.6	137.4
k	141.1	161.1	169.5	138.0
l	126.3	109.5	54.3	102.0

–, not calculated.

the effect of steam for SO_3 formation will be involved in the SO_3 concentration calculation Eq. (5.15).

The calculation results show that the accuracy of the updated formula was significantly higher than that of the two conventional empirical formulas. The reason may be that the updated formula was developed based on systematic study of the heterogeneous effect, such as the effects of special minerals (Fe, Al, Ca, and Mg) on the conversion of SO_2 to SO_3. These minerals were considered separately during the formulas development. In addition, the coupling effects between the heterogeneous and homogenous factors were considered by involving initial SO_2 concentration in the SO_2 conversion rate calculation under the heterogeneous effect shown in Eq. (5.14). To our knowledge, the catalytic effects of Fe and Al were rarely considered by the conventional formulas for

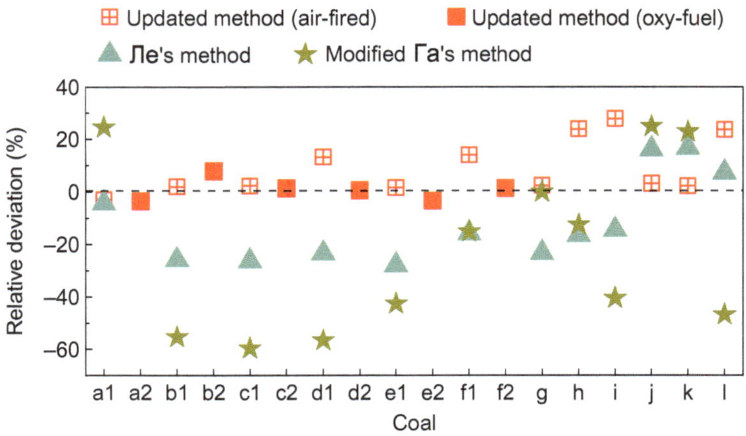

FIG. 5.11 Relative deviations of the T_{adp} calculation by the updated formula, and the Ле's and modified Га's formulas.

the acid dew point calculation of flue gas. The updated formula will be further modified by involving the effect of steam and other factors, after these factors are studied in the future research.

5.5.5 Summary

The homogeneous and heterogeneous effects for the conversion of SO_2 to SO_3 in flue gas were discussed in detail. An updated formula for the acid dew point temperature estimation of flue gas was proposed as follows:

$$T_{adp} = 186 + 20\lg C_{H_2O} + 26\lg\left(C^*_{SO_2}\left(\begin{array}{l} 127.65 C^{*\ -0.585}_{SO_2} + \lg\left(\dfrac{A_{ar}\times\alpha_{fly}}{C^*_{SO_2}}\right)^{\frac{1}{2}} \\ \left(\begin{array}{l} 15.191 + 0.071\text{Fe} \\ +0.044\text{Al} - 0.058\text{Ca} \\ -0.117\text{Mg} \end{array}\right)\times 10^{-2} \end{array}\right)\right)$$

The formula was further validated by various data obtained from laboratory-scale experiments, pilot-scale experiments, and industrial-scale coal-fired power plants of air- and oxy-fuel combustions, and was also compared to the conventional empirical formulas to confirm its accuracy. The contribution percentage of the heterogeneous effect for the conversion of SO_2 to SO_3 ranged from 1.33% to 20.86%, which was smaller than the homogenous effect. Furthermore, the contribution percentage of the heterogeneous effect for the oxy-fuel process was 1.2 to 1.6 times that for the air-fired process. The relative deviation of the calculation by the updated formula was significantly smaller than the conventional empirical formulas, which was due to the introduction of the detailed research results of the effect of fly ash on the conversion of SO_2 to SO_3 to the formula. The formula was suitable for both oxy-fuel and air-fired combustion processes.

5.6 EFFICIENT DESULFURIZATION DURING OXY-FUEL COMBUSTION

In conventional air combustion, the removal of sulfur with limestone involves two consecutive steps: calcination of limestone and sulfation of lime (calcium oxide); see Eqs. (5.23), (5.24). The efficiency of in-furnace desulfurization in conventional pulverized coal combustion is usually low, typically around 20%, according to the operational condition of the boiler [53]. The drastic decomposition of $CaSO_4$ formed from the desulfurization reaction owing to the high temperatures in boilers is responsible for the very low desulfurization efficiency. However, in O_2/CO_2 pulverized coal combustion, the SO_2 concentration inside the furnace is very high owing to the recirculation of flue gas, and consequently an ideal desulfurization condition is provided. Moreover, the $CaSO_4$ decomposition is significantly inhibited because of a high SO_2 concentration inside the furnace. The high CO_2 concentration makes the calcination reaction of limestone slow, and the pore structure of calcined limestone is more suitable for high-temperature desulfurization in an O_2/CO_2 atmosphere. When the CO_2 partial pressure is higher than equilibrium, the calcination of limestone is restrained, and limestone is subject to direct sulfation; see Eq. (5.25). The rate of direct sulfation does not decrease as much with sulfation degree as with CaO-SO_2 sulfation, because sintering is quite mitigated during direct sulfation of limestone. The diffusivity in the product layer demonstrates the high-temperature dependence and hardly changes with sulfation degree, which is associated with the fact that the $CaSO_4$ product layer is porous owing to CO_2 formation. The different behaviors of calcination and sintering may account for efficient in-furnace desulfurization during O_2/CO_2 coal combustion. Capturing SO_2 through in-furnace injection of limestone under oxy-fuel combustion is one promising option for cleaning flue gas, as it is cost-effective and easily implemented, which could offset the decrease of efficiency during production.

$$\text{Calcination}: \quad CaCO_3 \rightarrow CaO - CO_2 \tag{5.23}$$

$$\text{Indirect sulphation}: \quad CaO + SO_2 + 1/2O_2 \rightarrow CaSO_4 \tag{5.24}$$

$$\text{Direct sulphation}: \quad CaCO_3 + SO_2 + 0.5O_2 \rightarrow CaSO_4 + CO_2 \tag{5.25}$$

Both direct and indirect sulfation mechanisms govern the efficiency of desulfurization by limestone during oxy-fuel combustion. However, the quantified contribution of direct and indirect sulfation on desulfurization efficiency under O_2/CO_2 combustion is unclear. So far investigations concerning desulfurization in O_2/CO_2 combustion have been limited, and many unknowns about in-furnace desulfurization in O_2/CO_2 pulverized coal combustion need to be studied.

5.6.1 Calcination and Desulfurization Mechanism of Limestone Under Oxy-fuel Combustion

The degree of calcination of pure limestone and 18% limestone mixed with coal in a drop-tube furnace under 1273 K was investigated. The case of 18% limestone addition is discussed because the interference of the inherent sulfur and ash in coal was quite small and even negligible in this case. For the pure $CaCO_3$ injected into an O_2/CO_2 atmosphere, undecomposed $CaCO_3$ at a reactor length of 1200 mm still accounted for 55.5%, compared to an undecomposed fraction of 18.9% in air atmosphere. The high CO_2 partial pressure in oxy-fuel atmosphere was the sole explanation for this difference. Once limestone together with coal was injected into the furnace, as occurs in a real boiler burning coal either in air or in the oxy-fuel mode, the extent of limestone undecomposed in O_2/CO_2 at a reactor length of 1200 mm (13.9%) decreased to a level very close to that in air (12.3%). Because the interference of coal impurities had been ruled out in this case, the only explanation was an increase of the limestone particle temperature to a higher level overriding the inhibitory effect of the CO_2 partial pressure. The source of the extra heat must have been the flame derived from coal particles. Experimental observation at a reactor length of 600 mm further dictated a kinetically slow process for the calcination of limestone during oxy-coal combustion, leading to 20.5% undecomposed limestone relative to 13.6% in air. The slower decomposition of limestone in O_2/CO_2 was due to a slow and incomplete oxidation of coal in abundant CO_2 so that less heat was generated to heat the limestone and flue gas. At the very first step of coal combustion—volatile release and ignition—the large partial pressure of CO_2 in the bulk gas remained a large barrier, inhibiting the calcination reaction; see Eq. (5.23).It was confirmed that the mass ratio of limestone to coal played a noticeable role in determining the extent of the limestone calcination. As shown in Fig. 5.12A, for either air or O_2/CO_2, the fraction of the unreacted $CaCO_3$ was reduced remarkably by decreasing the amount of limestone added, possibly because of the higher local particle temperature for limestone at a lower fraction in the mixture. The endothermic characteristics of limestone calcination dictated a higher local particle temperature for the lower limestone fraction. The results of adding 1000 ppm SO_2 to the flue gas, shown in Fig. 5.12B, demonstrate the importance of the SO_2 concentration for calcination of limestone via direct sulfation; see Eq. (5.25). Irrespective of the bulk gas composition (air versus O_2/CO_2) and the limestone percentage, the unreacted limestone fraction dropped to a level of approximately 7% for all the conditions tested, which is also the same as that achieved when adding 1% limestone to coal and no extra SO_2 to the flue gas in panel (a). Looking at the dash-dot curves for the reduction in the unreacted limestone fraction quantified by subtracting the results in panel (a) from the corresponding results in panel (b), the largest improvement in limestone calcination (at least 8%) upon

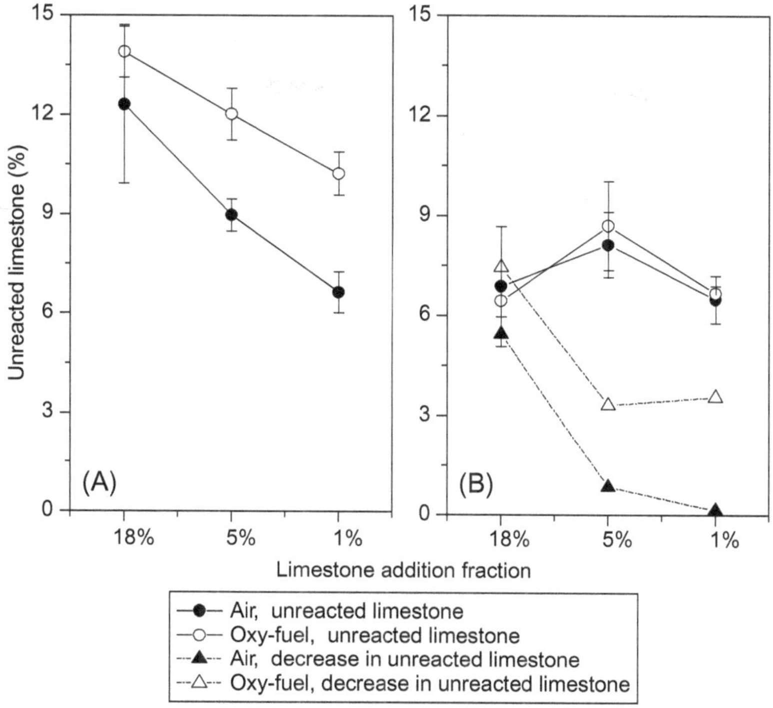

FIG. 5.12 The fractions of unreacted limestone as a function of the percentage of limestone addition to coal during combustion in drop-tube furnace. Panels (A) and (B) depict the gas added without and with extra SO_2 (1000 ppmV), respectively.

the addition of SO_2 was achieved for 18% limestone. The direct sulfation was much more pronounced with the addition of 18% limestone to coal in either air or O_2/CO_2. This prominence of direct sulfation is reasonable because ash interference is minimal and the direct injection of extra SO_2 with limestone requires no calcination step, particularly in the oxy-fuel combustion mode. For the other two limestone addition percentages, the direct sulfation was clearly less pronounced, contributing to a maximum of 3% of the total calcium under the oxy-fuel mode.

Table 5.5 indicates the quantified contribution of direct and indirect sulfation on desulfurization efficiency of limestone under O_2/CO_2 combustion. Clearly, the direct sulfation mechanism accounts for nearly all the desulfurization efficiency of limestone when Ca/S = 16 under O_2/CO_2 combustion. Moreover, the contribution of direct sulfation is reduced dramatically upon decreasing the Ca/S ratio. For the medium and low Ca/S ratio of Ca/S = 4.4 and Ca/S = 0.9, indirect sulfation is also the principal route for desulfurization during oxy-fuel combustion.

TABLE 5.5 The Contribution of Direct and Indirect Sulfation on Desulfurization Efficiency of Limestone Under O_2/CO_2 Combustion

Limestone Addition Percentage	Ca/Fuel-S Molar Ratio	System Desulfurization Efficiency, η(%)	Desulfurization Efficiency of Different Mechanism, η(%)	
			Direct Sulfation	Indirect Sulfation
18%	16	68.1 ± 4.7	68.1 ± 4.7	0
5%	4.4	36.2 ± 3.0	13.3 ± 1.1	22.9 ± 1.9
1%	0.9	12.6 ± 0.5	2.4 ± 0.1	10.2 ± 0.4

5.6.2 Efficient In-Furnace Desulfurization and Influence of Combustion Conditions and Sorbent Properties in O_2/CO_2 Combustion

System Approach Using Theoretical Analysis on Desulfurization

The modeling of desulfurization in O_2/CO_2 pulverized coal combustion was based on the global balance of all material, as shown in Fig. 5.13.

If the feed rate of coal was G_C (g/s), then the stoichiometric amount of O_2 necessary for burnout of the coal was

$$R_{O2} = \frac{G_C}{100}\left(\frac{W_C}{12} + \frac{W_H}{4} + \frac{W_S}{32} - \frac{W_O}{32}\right)(\text{mol/s}) \tag{5.26}$$

where W_C, W_H, W_S, and W_O represented the carbon, hydrogen, sulfur, and oxygen contents of coal (wt%), respectively.

In a wet recycle, the moisture was recycled with the flue gas. We supposed that the ratio of concentrations among CO_2, H_2O, N_2, and O_2 was CO_2: H_2O:N_2: $O_2 = x:y:u:z$, where $x+y+u+z=1$; the flux of recycled gas was R_E (mol/s), and the oxygen-fuel stoichiometric ratio was λ. Considering that the feeding oxygen may contain some nitrogen because of the cost of oxygen production, the amount of each gas after combustion was

$$CO_2 : \frac{G_C}{100} \cdot \frac{W_C}{12} + R_E \cdot x \tag{5.27}$$

$$H_2O : \frac{G_C}{100} \cdot \frac{W_H}{2} + R_E \cdot y \tag{5.28}$$

$$O_2 : R_{O_2}(\lambda - 1) \tag{5.29}$$

$$N_2 : R_E u + \left(\lambda R_{O_2} - R_E \cdot z\right) \cdot e \tag{5.30}$$

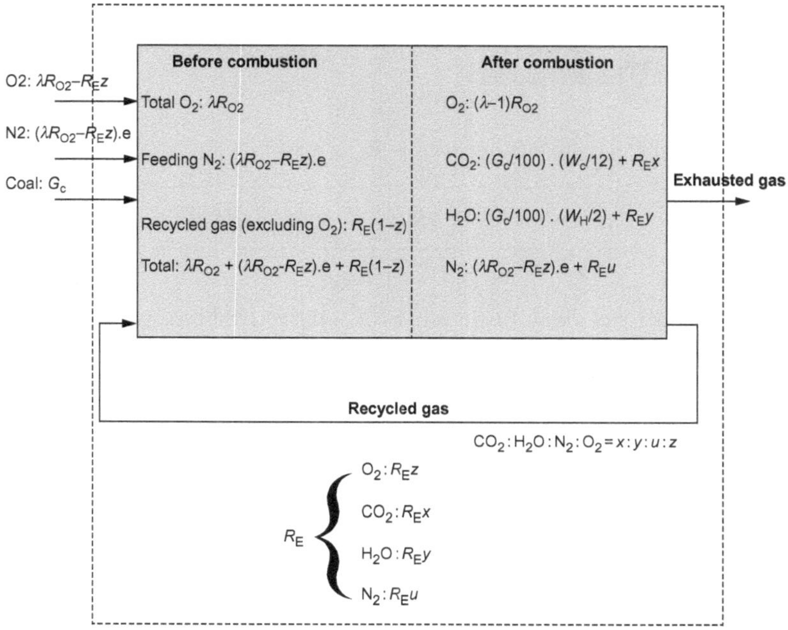

FIG. 5.13 Global balance of all material in O_2/CO_2 combustion with a wet recycle. *(From Hao L, Hong Y, Xing Y, Xiaojie X, Yibo F, Takashi A, et al. Efficient desulfurization in O2/CO2 combustion: dependence on combustion conditions and sorbent properties. Chem Eng Commun 2012;199:991-1011.)*

where e represented the concentration ratio between oxygen and nitrogen in the feeding oxygen. To satisfy the requirement of the initial O_2 concentration at the entrance of the furnace prior to combustion, the flux of recycled gas was derived as

$$R_E = \frac{\lambda R_{O_2}\left(100 - C_{O_2} - e\cdot C_{O_2}\right)}{C_{O_2}\left(1 - z - z\cdot e\right)} \tag{5.31}$$

Considering CO_2: H_2O: N_2: $O_2=x$: y: u: z, the O_2 concentration after combustion could be derived as

$$z = \frac{\left[a+b+c+R_E+\left(\lambda R_{O_2}-R_E\cdot z\right)\cdot e\right] \pm \sqrt{\left[a+b+c+R_E+\left(\lambda R_{O_2}-R_E\cdot z\right)\cdot e\right]^2 - 4\cdot R_E\cdot c}}{2\cdot R_E} \tag{5.32}$$

where $a=G_CW_C/1200$, $b=G_CW_H/200$, and $c=(\lambda-1)\cdot R_{O_2}$, respectively.

In a dry recycle, the moisture was removed and not recycled with the flue gas. Similarly to a wet recycle, for a dry recycle, the O_2 concentration after combustion could be derived as

$$z = \frac{\left[a + c + R_E + \left(\lambda R_{O_2} - R_E \cdot z\right) \cdot e\right] \pm \sqrt{\left[a + c + R_E + \left(\lambda R_{O_2} - R_E \cdot z\right) \cdot e\right]^2 - 4 \cdot R_E \cdot c}}{2 \cdot R_E} \tag{5.33}$$

The mass balance of sulfur in an O_2/CO_2 pulverized coal combustion system, shown in Fig. 5.14, was given by

$$\left(S_F + S_R\right)\left(1 - \eta_{loc}\right)\alpha = S_R \tag{5.34}$$

yielding the recycled sulfur

$$S_R = \frac{S_F\left(1 - \eta_{loc}\right)\alpha}{1 - \left(1 - \eta_{loc}\right)\alpha} \tag{5.35}$$

where S_F = flux of fuel-S (mol/s); S_R = flux of recycled-S (mol/s); gas recirculation ratio α = (recycled gas amount)/(total gas amount); η_{loc} was the local desulfurization efficiency inside the furnace during O_2/CO_2 pulverized coal combustion defined as

$$\eta_{loc} = \frac{(\text{removed} - \text{S})}{(\text{the sum of fuel} - \text{S and recycled} - \text{S})} = R^*_{Ca/S} \cdot X \tag{5.36}$$

where X was the molar fractional conversion of sorbent and $R^*_{Ca/S}$ was the local Ca/S molar ratio inside the furnace during O_2/CO_2 pulverized coal combustion:

$$R^*_{Ca/S} = \frac{(\text{Ca in sorbent})}{(\text{the sum of fuel} - \text{S and recycled} - \text{S})} \tag{5.37}$$

Correspondingly, the system Ca/S molar ratio was defined as $R_{Ca/S}$ = (Ca in sorbent)/(fuel-S). The SO_2 concentration in the bulk gas (C_0) was related to $R^*_{Ca/S}$ and X as

$$C_0 = C_{O_2/CO_2}\left(1 - \eta_{loc}\right) = C_{O_2/CO_2}\left[1 - R^*_{Ca/S} \cdot X\right] \tag{5.38}$$

where $C_{O2/CO2} = (S_F + S_R)/M$ was the SO_2 concentration at the furnace entrance during O_2/CO_2 pulverized coal combustion; M = flue gas amount in O_2/CO_2 pulverized coal combustion (mol/s).

As a system, the desulfurization efficiency in O_2/CO_2 pulverized coal combustion was defined as

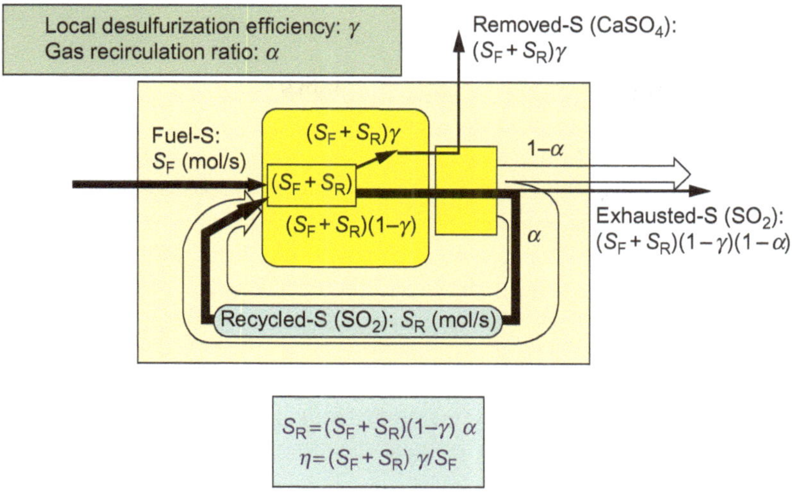

FIG. 5.14 Material balance of sulfur in O_2/CO_2 combustion. *(From Hao L, Hong Y, Xing Y, Xiaojie X, Yibo F, Takashi A, et al. Efficient desulfurization in O2/CO2 combustion: dependence on combustion conditions and sorbent properties. Chem Eng Commun 2012;199:991-1011.)*

$$\eta=\frac{(\text{removed}-\text{S})}{(\text{fuel}-\text{S})}=1-\frac{(\text{exhausted}-\text{S})}{(\text{fuel}-\text{S})}=\frac{(S_F+S_R)\eta_{loc}}{S_F} \tag{5.39}$$

Deriving Eqs. (5.27), (5.28), (5.31) yielded

$$\eta=\frac{\eta_{loc}}{1-(1-\eta_{loc})\alpha}=\frac{R^*_{Ca/S}\cdot X}{1-\left[1-R^*_{Ca/S}\cdot X\right]\alpha} \tag{5.40}$$

where X, the molar fractional conversion of sorbent, was a very important item to derive the global desulfurization efficiency in O_2/CO_2 coal combustion. To obtain X, a single particle pore diffusion model was proposed. This model assumed that the desulfurization reaction was the first order with respect to the SO_2 concentration at the surface of unreacted core and that the desulfurization reaction $CaO+SO_2+1/2\,O_2\rightarrow CaSO_4$ took place at the surface of a shrinking core of unreacted CaO within the particle.

The unsteady mass conservation equation was given by (in particle)

$$\frac{1}{r^2}\frac{\partial}{\partial r}\left(r^2 D_e\frac{\partial C}{\partial r}\right)=\frac{\partial C}{\partial t} \tag{5.41}$$

with boundary conditions given by

$$D_e\frac{\partial C}{\partial r}=h_D\left(C_0-C_s\right)\text{ at } r=r_0 \tag{5.42}$$

$$D_e \frac{\partial C}{\partial r} = k_s C \quad \text{at} \quad r = r_f \tag{5.43}$$

where D_e = effective diffusivity in particle; the subscripts s and f referred to the particle surface and interface between product layer and unreacted core, respectively; $C = SO_2$ molar concentration (mol/m^3); r = particle radius (m); h_D = mass transfer coefficient of SO_2 in gas film (m/s). Moreover, h_D was obtained from Sherwood number $Sh = 2r_0 h_D / D = 2.0 + 0.6Re^{1/2}Sc^{1/3}$; Reynolds number $Re = 2r_0 U / v$; Schmidt number $Sc = v/D$; U = linear velocity of the gas stream flowing past the particle (m/s); ν = kinematic viscosity (m^2/s); D = diffusivity of SO_2 in gas film (m^2/s). Considering the low Reynolds number, $Sh = 2.0$ was taken to derive h_D. The initial condition was given by $C = 0$ at $t = 0$. The molar fractional conversion of sorbent from CaO to $CaSO_4$, X, was calculated as $X = 1 - (r - r_0)^3$. The previously mentioned single particle model and system approach were combined to investigate the desulfurization in O_2/CO_2 coal combustion.

Eqs. (5.41)–(5.43) in the single particle model were mass conservation equations and boundary conditions for SO_2. Solving these equations gave the SO_2 concentration at the surface of unreacted CaO core and the radius of the unreacted shrinking core. Furthermore, the molar fractional conversion of sorbent was obtained from $X = 1 - (r - r_0)^3$. The decomposition of $CaSO_4$ was considered by subtracting it from X. The local Ca/S molar ratio inside the furnace during O_2/CO_2 pulverized coal combustion, $R^*_{Ca/S}$, was obtained according to the sorbent amount fed into the combustor and the sum of fuel-S and recycled-S. Finally the system desulfurization efficiency was determined through Eq. (5.40).

Influence of O_2 Purity on Desulfurization Efficiency

The global desulfurization efficiency increased with O_2 purity in O_2/CO_2 combustion with a wet recycle. At O_2 purity of 1.0, that is, when pure O_2 was fed into the combustor, the global desulfurization efficiency was able to reach 54.6% in O_2/CO_2 combustion with a wet recycle.

For O_2/CO_2 combustion with a dry recycle, similarly to the case of a wet recycle, the global desulfurization efficiency increased with O_2 purity and reached 63% at O_2 purity of 1.0. Moreover, the global desulfurization efficiency in a dry recycle was higher than that in a wet recycle with the same O_2 purity. This was mainly because of the higher gas recirculation ratio for a dry recycle.

Desulfurization Efficiency at Various Initial O_2 Concentrations

The global efficiency of in-furnace desulfurization decreased with the increase of the initial O_2 concentration. The initial O_2 concentration referred to the O_2 concentration in a mixture of newly fed pure O_2 and the RFG. In the case of a higher initial O_2 concentration, less RFG was needed to dilute the newly fed pure O_2, provided the amount of the newly fed pure O_2 was the same. Therefore

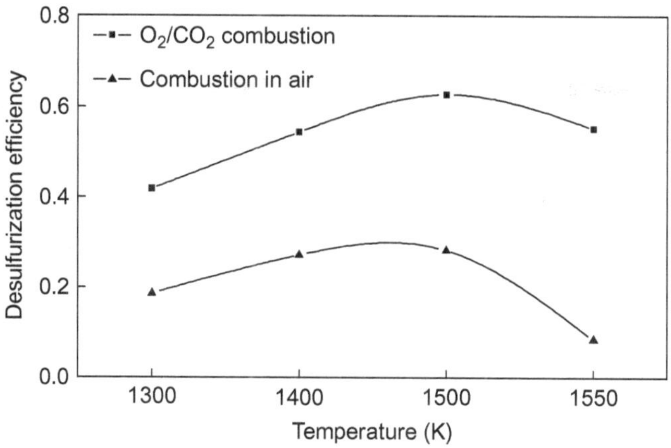

FIG. 5.15 Global desulfurization efficiency at various temperatures (system Ca/S ratio = 5). *(From Hao L, Hong Y, Xing Y, Xiaojie X, Yibo F, Takashi A, et al. Efficient desulfurization in O2/CO2 combustion: dependence on combustion conditions and sorbent properties. Chem Eng Commun 2012;199:991-1011.)*

the gas recirculation ratio decreased when the initial O_2 concentration increased, and consequently, the global efficiency of in-furnace desulfurization decreased.

Desulfurization Efficiency at Various Temperatures

Fig. 5.15 shows the global desulfurization efficiency at various temperatures. As the temperature increased, the global desulfurization efficiency increased first and then decreased due to decomposition of $CaSO_4$. For comparison, the desulfurization efficiency for combustion in air is also plotted in Fig. 5.15. It can be seen in the temperature range investigated that the global desulfurization efficiency in O_2/CO_2 coal combustion was much higher (about six times) than conventional coal combustion in air.

Influence of Ca/S Ratio

A higher Ca/S ratio resulted in a lower SO_2 concentration because more SO_2 was captured through the desulfurization reaction when the Ca/S ratio increased. When the Ca/S ratio increased from 1 to 5 (radius of sorbent particles $=1.415\times10^{-5}$ m), the SO_2 concentration decreased from 9000 ppm to about 5000 ppm, nearly half. However, the global desulfurization efficiency increased from 20% to 62.7%, that is, almost tripled when the Ca/S ratio increased from 1 to 5. This nonlinear increase was due to the decrease in the SO_2 concentration in the presence of more sorbent particles.

Influence of Sorbent Size on Desulfurization Efficiency

The global desulfurization efficiency increased with sorbent size decrease (radius of sorbent particles). When the particle radius decreased to one-quarter (radius of sorbent particles $=7.07\times10^{-6}$ m), the global desulfurization efficiency doubled, becoming as high as 80%. This result implied that employing fine sorbent particles can enhance the global desulfurization efficiency effectively.

Influence of Sorbent Property

Fig. 5.16 shows the global desulfurization efficiency when using various sorbents. The global desulfurization efficiency was very different among the three sorbents, whether in O_2/CO_2 combustion or in conventional air combustion. The global desulfurization efficiency increased in the order of $Ca(OH)_2$, scallop, and limestone in O_2/CO_2 combustion, but in the order of scallop, $Ca(OH)_2$, and limestone in conventional air combustion. These results were attributed to the different desulfurization kinetics of various sorbents in different atmospheres. Nevertheless, all three sorbents demonstrated much higher desulfurization efficiency in O_2/CO_2 combustion than in conventional air combustion. The high SO_2 concentration facilitated the desulfurization reaction and led to effective utilization of the sorbent.

5.6.3 Summary

Based on experiments on desulfurization and the system approach using theoretical analysis, the desulfurization mechanism and the influence of combustion conditions and sorbent properties on the system desulfurization efficiency during O_2/CO_2 combustion were investigated. In the temperature range

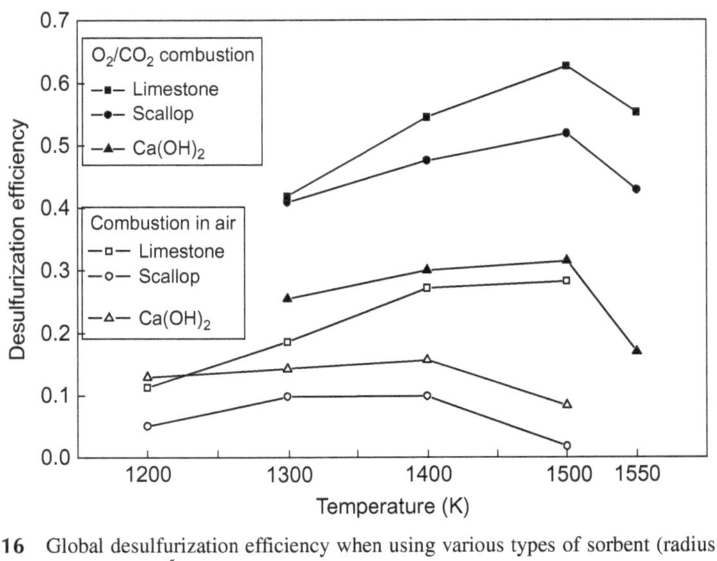

FIG. 5.16 Global desulfurization efficiency when using various types of sorbent (radius of sorbent particles $=1.415\times10^{-5}$ m). *(From Hao L, Hong Y, Xing Y, Xiaojie X, Yibo F, Takashi A, et al. Efficient desulfurization in O2/CO2 combustion: dependence on combustion conditions and sorbent properties. Chem Eng Commun 2012;199:991-1011.)*

investigated, the system desulfurization efficiency in O_2/CO_2 coal combustion was much higher than conventional coal combustion in air. The direct sulfation mechanism accounted for nearly all the desulfurization efficiency of limestone when Ca/S = 16 under O_2/CO_2 combustion. Upon decreasing the Ca/S ratio, the contribution of direct sulfation was reduced dramatically, and the role of indirect sulfation increased for desulfurization during oxy-fuel combustion.

The global desulfurization efficiency increased with O_2 purity. The global desulfurization efficiency in a dry recycle was higher than that in a wet recycle. As the initial O_2 concentration increased, the global efficiency of in-furnace desulfurization decreased. As the temperature increased, the global desulfurization efficiency increased first and then decreased because of the decomposition of $CaSO_4$.

The global desulfurization efficiency decreased with sorbent size. When the particle radius decreased to one-quarter, the global desulfurization efficiency doubled, becoming as high as 80%. The global desulfurization efficiency was very different among the different sorbents investigated, whether in O_2/CO_2 combustion or in conventional air combustion. The global desulfurization efficiency increased in the order of $Ca(OH)_2$, scallop, and limestone in O_2/CO_2 combustion, but in the order of scallop, $Ca(OH)_2$, and limestone in conventional air combustion. Nevertheless, all three sorbents demonstrated much higher desulfurization efficiency in O_2/CO_2 combustion than in conventional air combustion.

REFERENCES

[1] Toftegaard MB, Brix J, Jensen PA, et al. Oxy-fuel combustion of solid fuels. Prog Energy Combust Sci 2010;36(5):581–625.

[2] Cao H, Sun S, Liu Y, et al. Computational fluid dynamics modeling of NOx reduction mechanism in oxy-fuel combustion. Energy Fuel 2009;24(1):131–5.

[3] Fujimori T, Yamada T. Realization of oxyfuel combustion for near zero emission power generation. Proc Combust Inst 2013;34(2):2111–30.

[4] Normann F, Andersson K, Leckner B, et al. High-temperature reduction of nitrogen oxides in oxy-fuel combustion. Fuel 2008;87(17–18):3579–85.

[5] Normann F, Andersson K, Leckner B, et al. Emission control of nitrogen oxides in the oxy-fuel process. Prog Energy Combust Sci 2009;35(5):385–97

[6] Jia L, Tan Y, Anthony E. Emissions of SO2 and NO x during oxy-fuel cfb combustion tests in a mini-circulating fluidized bed combustion reactor. Energy Fuel 2009;24(2):910–5.

[7] Shoji M, Yamamoto T, Tanno S, et al. Modeling study of homogeneous NO and N2O formation from oxidation of HCN in a flow reactor. Energy 2005,30(2–4):337–45.

[8] Liu H, Yuan Y, Yao H, et al. Factors affecting NO reduction during O2/CO2 combustion. Energy Fuel 2011;25(6):2487–92.

[9] Hu X, Li X, Luo G, et al. Homogeneous and heterogeneous contributions of CO2 and recycled NO to NO emission difference between air and oxy-coal combustion. Fuel 2016;163:1–7.

[10] Okazaki K, Ando T. NOx reduction mechanism in coal combustion with recycled CO2. Energy 1997;22(2):207–15.

[11] Álvarez L, Gharebaghi M, Jones JM, et al. Numerical investigation of NO emissions from an entrained flow reactor under oxy-coal conditions. Fuel Process Technol 2012;93(1):53–64.

[12] Zhao R, Liu H, Zhong X, et al. Modeling of NOx conversion during combustion under high CO2 concentration using detailed chemical kinetics. Fuel Process Technol 2011;92(5):939–45.
[13] Tan YW, Croiset E, Douglas MA, Thambimuthu KV. Combustion characteristics of coal in a mixture of oxygen and recycled flue gas. Fuel 2006;85:507–12.
[14] Croiset E, Thambimuthu KV, Palmer A. Coal combustion in O2/CO2 mixtures compared with air. Can J Chem Eng 2000;78:402–7.
[15] Kiga T, Takano S, Kimura N, Omata K, Okawa M, Mori T, et al. Characteristics of pulverized-coal combustion in the system of oxygen/recycled flue gas combustion. Energy Convers Manag 1997;38:S129–34.
[16] Yan J, Yang J, Liu Z. SH radical: the key intermediate in sulfur transformation during thermal processing of coal. Environ Sci Technol 2005;39(13):5043–51.
[17] Telfer M, Zhang DK. The influence of water-soluble and acid-soluble inorganic matter on sulfur transformations during pyrolysis of low-rank coals. Fuel 2001;80(14):2085–98.
[18] Querol X, Klika Z, Weiss Z. Determination of element affinities by density fractionation of bulk coal samples. Fuel 2001;80(1):83–96.
[19] Hu GL, Johansen KD, Wedel S, Hansen J. Decomposition and oxidation of pyrite. Prog Energy Combust Sci 2006;32(3):295–314.
[20] Joseph MLJ, George S, Philip LWJ. Production of pyrrhotites by pyrite reduction. Fuel 1980;59(10):687–90.
[21] Duan LB, Zhao CS, Zhou W, Qu CR, Chen XP. Investigation on coal pyrolysis in CO2 atmosphere. Energy Fuel 2009;23(23):3826–30.
[22] Chen H, Li B, Zhang B. Decomposition of pyrite and the interaction of pyrite with coal organic matrix in pyrolysis and hydropyrolysis. Fuel 2000;79(13):1627–31.
[23] Gryglewicz G. Sulfur transformations during pyrolysis of a high sulfur Polish coking coal. Fuel 1995;74(3):356–61.
[24] Wall T, Liu YH, Spero S, Rathnam R, Zeenathal F, Moghtaderi B, et al. An overview on oxyfuel coal combustion—state of the art research and technology development. Chem Eng Res Des 2009;8(7):1003–16.
[25] Croiset E, Thambimuthu KV. NOx and SO2 emission from O2/CO2 recycled coal combustion. Fuel 2001;80:2117–21.
[26] Fleig D, Andersson K, Johnsson F, Leckner B. Conversion of sulfur during pulverized oxy-coal combustion. Energy Fuel 2011;25:647–55.
[27] Cheng J, Zhou J, Liu J, Zhou Z, Huang Z, Cao X, et al. Sulfur removal at high temperature during coal combustion in furnaces: a review. Prog Energy Combust Sci 2003;29:381–405.
[28] Liu H, Okazaki K. Simultaneous easy CO2 recovery and drastic reduction of SOx and NOx in O2/CO2 coal combustion with heat recirculation. Fuel 2003;82:1427–36.
[29] Liu H, Katagiri S, Okazaki K. Drastic SOx removal and influences of various factors in O2/CO2 pulverized coal combustion system. Energy Fuel 2001;15:403–12.
[30] Ahn J, Okerlund R, Fry A, Eddings EG. Sulfur trioxide formation during oxy-coal combustion. Int J Greenhouse Gas Control 2011;Supplement 1:S127–35.
[31] Stanger R, Wall T. Sulfur impacts during pulverized coal combustion in oxy-fuel technology for carbon capture and storage. Prog Energy Combust Sci 2011;37:69–88.
[32] Fleig D, Andersson K, Normann F, Johnsson F. SO3 formation under oxyfuel combustion conditions. Ind Eng Chem Res 2011;50:8505–14.
[33] Zhang J. Discriminate and discuss on some mistake region and doubtful point where the gas acid dew point calculating methods had be studies. Boiler Technol 2013;44:10–6.
[34] Ceng K, Fan J, Chi Z, Shen L. The prevention principle and calculation of fouling, slagging, wear and corrosion of boiler and heat exchanger. Beijing: Science Press; 1994; p. 381–2.

[35] Muller PA. Contribution to the problem of the action of sulfuric acid on the dew point temperature of flue gases. Chem Eng Technol 1959;31:345–50.
[36] Verhoff F, Banchero J. Predicting dew points of flue gases. Chem Eng Prog 1974;70:71–2.
[37] Okkes AG, Badger BV. Get acid dew point of flue gas. Hydrocarb Process 1987;66:53–5.
[38] Fleig D, Alzueta MU, Normann F, Abián M, Andersson K, Johnsson F. Measurement and modeling of sulfur trioxide formation in a flow reactor under post-flame conditions. Combust Flame 2013;160:1142–51.
[39] Jørgensen TL, Livbjerg H, Glarborg P. Homogeneous and heterogeneously catalyzed oxidation of SO2. Chem Eng Sci 2007;62:4496–9.
[40] Merryman EL, Levy A. Enhanced SO3 emissions from staged combustion. Symp Int Combust 1979;727–36.
[41] Mueller MA, Yetter RA, Dryer FL. Kinetic modeling of the CO/H2O/O2/NO/SO2 system: implications for high-pressure fall-off in the SO2 + O (+M) = SO3 (+M) reaction. In J Chem Kinet 2000;32:317–39.
[42] Alzueta MU, Bilbao R, Glarborg P. Inhibition and sensitization of fuel oxidation by SO2. Combust Flame 2001;127:2234–51.
[43] Liu H, Zhu H, Kaneko M, Kato S, Kojima T. High-temperature gasification reactivity with steam of coal chars derived under various pyrolysis conditions in a fluidized bed. Energy Fuel 2009;24:68–75.
[44] Gleason JF, Sinha A, Howard CJ. Kinetics of the gas-phase reaction HOSO4 + O2 → HO2 + SO2. J Phys Chem 1987;91:719–24.
[45] Belo LP, Elliott LK, Stanger RJ, Spörl R, Shah KV, Maier J, et al. High-temperature conversion of SO2 to SO3: homogeneous experiments and catalytic effect of fly ash from air and axy-fuel firing. Energy Fuel 2014;28:7243–51.
[46] Marier P, Dibbs HP. The catalytic conversion of SO2 to SO3 by fly ash and the capture of SO2 and SO3 by CaO and MgO. Thermochim Acta 1974;8:155–65.
[47] Li P, Tong H. Comparison and analysis on the calculation methods of acid dew point of flue gas. Boiler Technol 2009;40:5–8.
[48] Tang Z, Jin B, Sun K, Zhong Z. The contrast and evaluation of the calculation formulas of acid dew point temperature of the flue gas after the WFGD. J Power Eng 2005;25:18–21.
[49] Halstead W, Talbot J. Sulphuric acid dewpoint in power station flue gases. J Inst Energy 1980;53:142–5.
[50] Zhang J. Review on current calculating method of gas acid dewpoint temperature for coal fired boilers and inquire into its optimization and improvement. Total Corros Control 2012;26:1–7.
[51] Xie T, Li Y, Zhou Y, Jiang H. A SO2 concentration based calculation method for acid dew point of flue gas. Thermal Power Gener 2012;42:118–21.
[52] Cao Y, Zhou H, Jiang W, Chen C-W, Pan W-P. Studies of the fate of sulfur trioxide in coal-fired utility boilers based on modified selected condensation methods. Environ Sci Technol 2010;44:3429–34.
[53] Falkenberry HL, Slack AV. SO2 removal by limestone injection. Chem Eng Prog 1969;65:61–6.

Chapter 6

Mineral Behavior During Oxy-fuel Combustion

Liqi Zhang, Yongchun Zhao, Hong Yao
Huazhong University of Science and Technology, Wuhan, PR China

6.1 INTRODUCTION

Mineral behavior and ash-related problems are recognized as major sources of uncertainty in oxy-fuel combustion technology. According to the studies of many researchers, the temperature and the gas phase composition within and around the coal particles significantly affect the transformations of coal mineral matter during coal combustion [1]. Because of the higher heat capacity of CO_2 in oxy-fuel combustion, more heat needs to be supplied to burn the particles from the gas around the particles. In addition, the endothermic reaction of C and CO_2 decrease the particles burning temperature with the conditions of high CO_2 and temperature. Thus the formation of ash, the migration and transformation of iron-containing minerals, the characteristics of ash melting, and the slag in oxy-fuel conditions are different from those in air combustion.

6.2 ASH FORMATION DURING O_2/CO_2 COMBUSTION OF PULVERIZED COALS

Generally, it is thought that ash particles are produced by two mechanisms in coal combustion. The composition and the size distribution of inorganic particles depend on the content and distribution of inorganic minerals in coal, the composition and size distribution of coal particles, the combustion conditions, the system geometry, and so on [2,3]. Fig. 6.1 shows the pathways of mineral transformation and particle formation during coal combustion. The vaporization-nucleation-condensation mechanism of volatile metals and suboxides of nonvolatile metals is the pathway to the formation of submicrometer particles. The other nonvolatile metals which are not released as suboxides are released from the coarse particles. The differences in size, elements, and morphology distribution of submicrometer particles and the fly ash between oxy-fuel combustion and air combustion have been studied worldwide by researchers.

Oxy-fuel Combustion. http://dx.doi.org/10.1016/B978-0-12-812145-0.00006-2

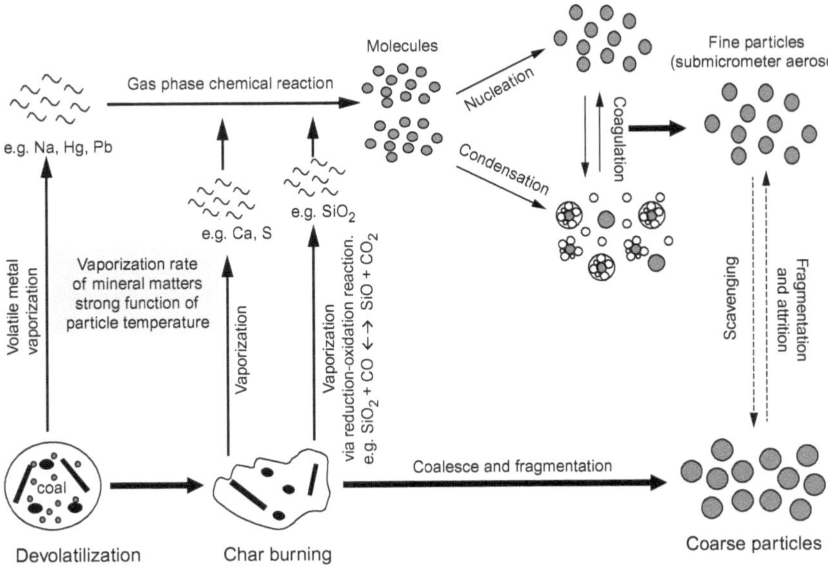

FIG. 6.1 Pathways of mineral transformation and particle formation during coal combustion.

6.2.1 Size Distribution of Submicrometer Particles

Under O_2/CO_2 combustion, the burning temperature is reduced, and CO_2 can inhibit the redox reaction ($MO_n + CO \rightarrow MO_{n-1} + CO_2$) [3]. Compared to air combustion, O_2/CO_2 combustion decreases refractory oxide vaporization and the quality of submicrometer particles. On the other hand, the mass diffusivity of O_2 in CO_2 is lower than in N_2, and O_2 diffusivity is decreased on the surface of char particles, so the evaporation and nucleation rate is reduced under O_2/CO_2 atm. The total residence time is constant. These factors result in the decrease of the particles growth rate and decrease the size of submicrometer particles under O_2/CO_2 combustion.

Under different atmospheres, Sheng et al. [1,4,5] studied particle-generating characteristics in a coal with high alumina and in three different Chinese coals using a drop-tube furnace. They concluded that when the concentration of oxygen was the same, the submicrometer particles size was smaller under O_2/CO_2 combustion than under air combustion. The mass of two kinds of modal particles was reduced. Under the O_2/N_2 atm, the changes in oxygen concentration had little effect on the generating characteristics of submicrometer particles. However, under the O_2/CO_2 atmosphere and an increasing oxygen concentration, the size of the submicrometer particles increased, and the mass of two kinds of modal particles increased. Suriyawong et al. [6] did similar experiments with a subbituminous coal and concluded that for O_2/CO_2 combustion, compared to a conventional air combustion atmosphere, the total mass of submicrometer modal particles (<0.5 μm) reduced and that the geometric average size of submicrometer aerosols reduced about 28%. With increasing concentrations of O_2, the fine particle concentration changed little, and the geometric average particle size increased, because with increasing concentrations of O_2, when the char combustion temperature and the evaporation rate of refractory oxide increased, the fine particle size increased. The coarse particles average particle size also decreased because the higher particle temperatures contributed to a decomposition effect.

6.2.2 Element Distribution of Submicrometer Particles

It can be seen from the studies of Sheng [1,4,5] that compared with air conditions, that the content of Na_2O, K_2O, SO_3 increased and the content of SiO_2, Fe_2O_3, CaO decreased in 20% O_2/80% CO_2 combustion. The change of O_2 concentration had little effect on the elements distribution in air combustion. However, increasing the O_2 concentration in O_2/CO_2 combustion can increase the content of refractory metal oxides in submicrometer particles. Suriyawong et al. [6] observed that Si, Al, Mg, Fe, and Ca were released much less in the gas phase than in air.

A modeling study by Krishnamoorthy and Veranth [7] investigated the effect of the gas composition of flue gas on the ratio of CO/CO_2 through a detailed combustion model of a single char particle. The studys results indicated that CO_2 concentration had a great effect on the ratio of CO/CO_2, as well as on the evaporation of refractory metal oxide and the formation of submicrometer particles.

6.2.3 Composition and Size Distribution of Fly Ash

Many researchers have focused on the composition and size distribution of fly ash under oxy-fuel combustion [1,4,5,8–12]. Their studies showed little difference between air and oxy-fuel combustion in terms of the composition and size distribution of fly ash, but there were differences between air and oxy-fuel combustion in terms of the number of mineral phases and the slagging characteristics of fly ash.

In addition to the preceding conclusions, Yu et al. [8] suggested that the high ash deposition rate under oxy-fuel conditions was determined by the lower gas flow rate, the higher combustion temperature, and the compositional changes of the flue gas on a 100kW oxy-fuel combustion facility. Zhan et al. [9] found that the compositions of deposits were different between samples from oxy-firing and those from air firing simply by using the CCSEM technique. Zhang et al. [10] investigated the ash formation characteristics of a high-calcium coal under air and three oxy-fuel combustion conditions. Fig. 6.2 shows the type and content of ash components under air-firing and oxy-fuel combustion as identified by the CCSEM technique, and the minerals were divided into three types according to the melting temperature of the mineral: high melting temperature (HMT: >1773 K), moderate melting temperature (MMT: 1573–1773 K), and low melting temperature (LMT: <1573 K).

It can be found from Fig. 6.2 that, compared to air combustion, in oxy-fuel combustion, the transformations of coal minerals were different and that the

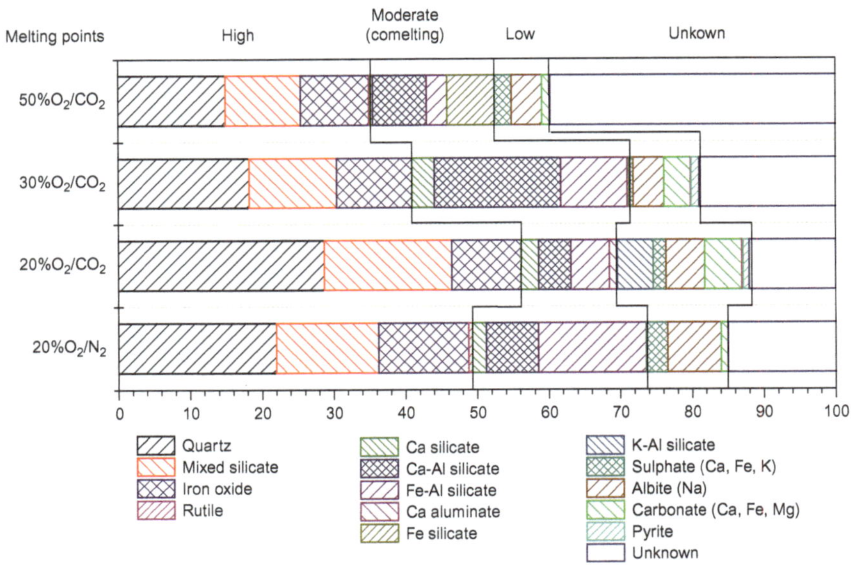

FIG. 6.2 Type and content of ash components under air-firing and oxy-fuel combustion as identified by CCSEM.

decomposition process of carbonate minerals was delayed. Therefore the increasing content of unclassified minerals with the increasing O_2 concentrations also indicated an enhanced interaction among different minerals and increased the slagging tendency.

Thermodynamic equilibrium calculation is also an important method for studying the evaporation characteristics of elements under different conditions [11,12]. Zheng and Furimsky [11] studied the evaporation characteristics of elements under oxy-fuel combustion by using the FACT software. They suggested that compared to air combustion, in the O_2/CO_2 condition, the combustion medium had little effect on the chemical composition of ash and the number and species of trace elements. However, with the effects of carbonate and sulfate on the ash composition of low S and low Cl, high alkaline coal cannot be ignored. Zhao et al. [12] found that atmosphere and temperature had a significant effect on the evaporation type and on the content of mineral elements. For example, under reducing conditions with a flue gas temperature of 2400 K, the total evaporation content of mineral elements was 4.46% during O_2/CO_2 conditions and could be as high as 9.65% in air conditions.

6.3 TRANSFORMATION BEHAVIOR OF IRON-CONTAINING MINERALS DURING OXY-FUEL COMBUSTION

Iron-containing minerals are considered as one of the major factors in fouling and slagging during pulverized coal combustion [13]. The most common iron-containing minerals in coal are pyrite, siderite, dolomite, and iron-bearing clay minerals [14]. The migration and transformation behaviors of iron-bearing minerals in the combustion process play an important role in the formation of slag.

6.3.1 Pyrite Transformation in High CO_2 Concentration

Pyrite (FeS_2) is a naturally occurring material that can be found as impurities in coal, oil shale, and other fossil resources [15,16] and is one of the main sources of sulfur dioxide (SO_2) emissions in combustion processes [17,18]. To utilize pulverized coal, slagging, corrosion, and fouling in the furnace must be managed and mitigated in commercial power plants [19,20]. Among the inorganic constituents of coal, FeS_2 is the primary contributor to slagging on fireside heat transfer surfaces, leading to operational problems and reducing the lifetime of facilities. Previous investigations suggested that the formation of low-eutectic-point FeO-FeS during the FeS_2 transformation present in coal combustion was an important factor in promoting slag fluxing [21–23]. McLennan et al. [22] suggested that excluded FeS_2 was decomposed into pyrrhotite (FeS_x, $1 \leq x \leq 2$) and then oxidized to form a FeO-FeS melt phase, which was stable under reducing conditions. The FeO-FeS phase was able to control the melting behavior, partly because of the large amount of FeS_x produced under reducing conditions [21]. Because of the exothermic oxidation stage, the particle temperature exceeded the eutectic temperature of the FeO-FeS phase (940°C), which resulted in melt formation during FeS_2 combustion in a drop-tube furnace under oxidation conditions [24]. The burning FeS_2 particles in a drop-tube furnace can form Fe-S-O droplets that have a low melting point, suggesting that it is possible for FeS_2 particles to completely melt because of the low eutectic point of Fe-S-O droplets, which can reach 915°C [23]. Therefore insight into the FeS_2 transition mechanism is necessary in order to explain the slag or agglomerate formation.

The FeS_2 transformation process is complicated and can be significantly influenced by particle size, oxygen concentration, temperature, combustion atmosphere, and so on. In many thermochemical processes of FeS_2, thermal decomposition is one of the first steps and significantly influences the subsequent oxidation process [25]. Hu et al. [15] reviewed FeS_2 decomposition under various nonoxidizing environments, such as nitrogen, argon (Ar), vacuum, and carbon dioxide (CO_2), and suggested that the transformation mechanisms depend upon the reaction conditions; especially in a CO_2 atmosphere, FeS_2 or its decomposition products may react with CO_2 to form iron oxides [16,26].

Experimental studies show that [27] an amazingly small amount of SO_2 is produced at 550°C, which can be attributed to the evolution of sulfur on the pyrite's surface. When the temperature reaches 600°C, the release rate of SO_2 dramatically increases, which indicates that FeS_2 is being decomposed. At temperatures below 950°C, the cumulative production and the cumulative production rate increase with rising temperatures. Considerable sulfur will be trapped in the melt at temperatures above 950°C, which results in a decrease in the cumulative production of SO_2. In addition, elemental sulfur (e.g., S_x, primarily S_2)

condensation begins to be observed at the outlet of the reactor at 550°C, at which point carbonyl sulfide (COS) also begins to be detected by gas chromatography.

As Fig. 6.3A [27] shows, FeS_2 is the only phase detected in the ashes formed at 500°C and 550°C, while the intensities of the peaks of the FeS_2 decrease with rising temperatures, which implies that FeS_2 is not being decomposed at temperatures below 550°C in a CO_2 atmosphere and that small amounts of sulfur on the pyrite's surface are released. At 600°C, FeS_2 disappears completely, and FeS_x is found to be the dominant phase, indicating that FeS_2 is being decomposed. The intensities of the peaks of the FeS_x increase with rising temperatures.

As the reactor temperature reaches 750°C, Fe_3O_4 begins to appear in the X-ray diffraction (XRD) spectrogram of the residual ash, and the intensities of the peaks are low, as shown in Fig. 6.3B. However, the intensities of the Fe_3O_4 in the XRD spectrogram of the residual ashes increase, and the intensities of the FeS_x decrease with rising temperatures. These results suggest that the decomposition product FeS_x from FeS_2 can be oxidized into Fe_3O_4, which is consistent with the observation that FeS_2 thermal decomposition occurs under CO_2 conditions. At higher temperatures (e.g., 1000°C or 1050°C), the detected mineralogical phases include Fe_2O_3, Fe_3O_4, and FeS_x. An amorphous molten phase is also present, which implies that further oxidation of Fe_3O_4 did occur, and due to the sample particles' temperature exceeding the melting point of the eutectic FeO-FeS system, the amounts of the Fe_2O_3 are small.

As shown in Fig. 6.3, when the temperature is below 600°C, the S/Fe molar ratio dramatically decreases with rising temperatures. At temperatures above 600°C, the S/Fe molar ratio decreases at a slower rate with rising temperatures. However, at 950°C, the S/Fe molar ratio begins to increase as the temperature rises, which is most likely because of the eutectic material retaining a large amount of sulfur in the solid phase.

As shown in Fig. 6.4, the FTIR spectra of the raw FeS_2 and those of the solid products formed at 500°C and 550°C are very similar, and the strongest peak at 410–430 cm^{-1} matches the peak of FeS_2, indicating that FeS_2 is not being decomposed at temperatures below 550°C in a CO_2 atmosphere. This finding agrees with the XRD results; that is, the residual ashes formed below 550°C do not experience significant changes. Then the peak at 430–410 cm^{-1} disappears completely from the spectra at 650°C and 700°C, suggesting that FeS_2 has completely decomposed. The FTIR spectra of the solid products formed at 750°C, 850°C, and 950°C are also quite similar; the 590–560 cm^{-1} peak can be attributed to Fe_3O_4. By raising the reaction temperature, the residue ashes will contain a considerable amount of iron oxide during pyrite's thermal decomposition in a CO_2 atmosphere.

6.3.2 Effect of H_2O on Pyrite Thermal Decomposition During Oxy-fuel Combustion

H_2O concentration in an oxy-fuel combustion process is much higher than that in conventional air combustion because of the large amount of flue gas circulation [28]. When burning high H-containing fuel or lignite, the proportion of

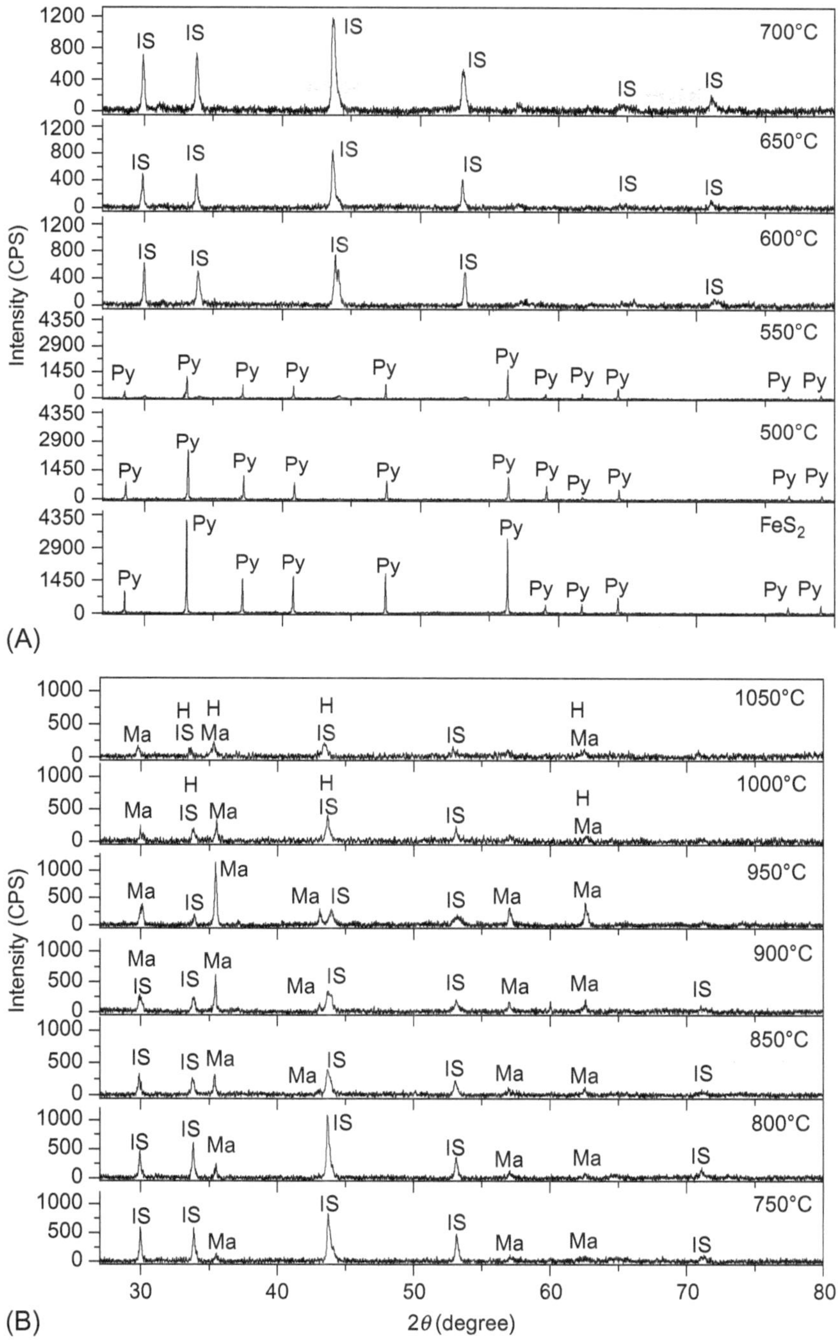

FIG. 6.3 Evolution of FeS_2 in CO_2 atmosphere at reactor temperatures of 500–1050°C: (A) XRD patterns of the solid product at temperatures ≤700°C, (B) XRD patterns of the solid product at temperatures >700°C,

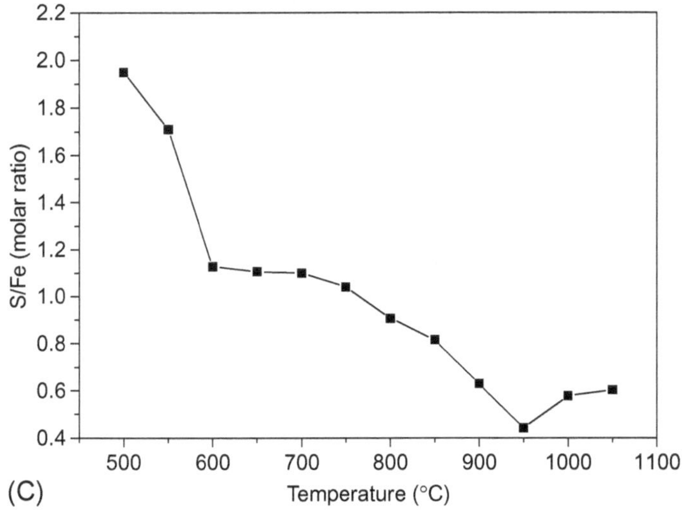

FIG. 6.3, CONT'D (C) S/Fe (i.e., molar ratio) of the solid product at different temperatures. Peaks in the XRD patterns are shown for *Py*, pyrite (FeS_2); *IS*, pyrrhotite (FeS_x); *Ma*, magnetite (Fe_3O_4); *H*, hematite (Fe_2O_3).

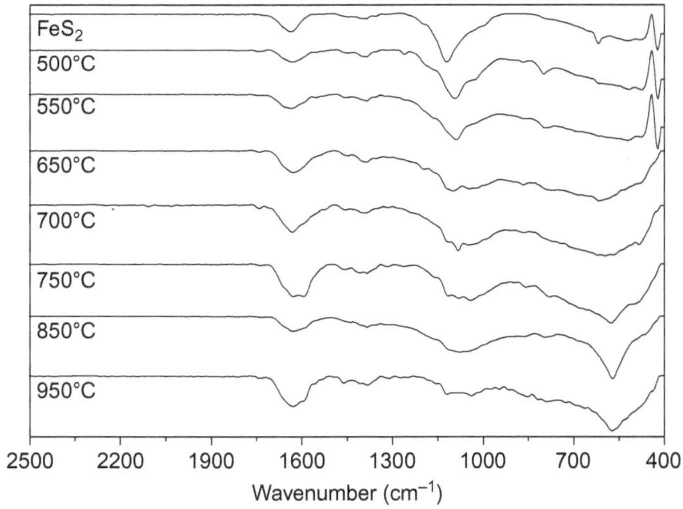

FIG. 6.4 FTIR spectra of FeS_2 samples at different temperatures in CO_2.*(From Huang F, Zhang LQ, Yi BJ, et al. Transformation pathway of excluded mineral pyrite decomposition in CO2, atmosphere. Fuel Process Technol 2015;138:814–24.)*

H_2O in the flue gas can reach 45% in the oxy-fuel combustion with wet flue gas recycle [29]. The high content of H_2O not only affects the combustion of pulverized coal but also the transformation of pyrite.

In H_2O atmospheres, pyrite and pyrrhotite are the dominant mineral phases at 500°C. Magnetite also appears in the XRD diffractogram, although the intensity

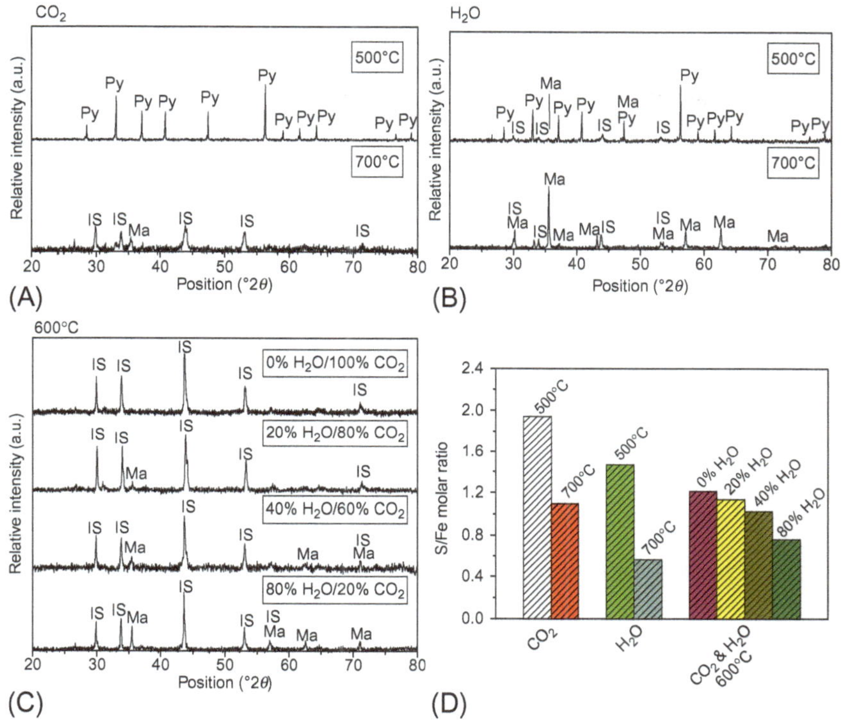

FIG. 6.5 Solid products characteristic of pyrite decomposition: (A) XRD patterns of products in CO_2, (B) XRD patterns of products in H_2O, (C) XRD patterns of products in CO_2 mixed with H_2O at 600°C, and (D) S/Fe molar ratio of solid products: *Py*, pyrite (FeS_2); *IS*, pyrrhotite (FeS_x); *Ma*, magnetite (Fe_3O_4).*(From Huang F, Zhang LQ, Yi BJ, et al. Transformation pathway of excluded mineral pyrite decomposition in CO2, atmosphere. Fuel Process Technol 2015;138:814–24.)*

is low, as shown in Fig. 6.5B [30]. When the temperature increases to 700°C, pyrite disappears completely, and the main mineral phases are pyrrhotite and magnetite in the residual ash. The S/Fe molar ratios are compared in Fig. 6.5D; significantly, the S/Fe ratio is lower in H_2O than in CO_2 at the same temperature, which suggests that pyrite can decompose at a lower temperature and that the release of sulfur is greatly increased in a H_2O atmosphere. H_2S and SO_2 are the main composition of the produced gases during the pyrite heating in a H_2O atmosphere. Previous literature reported that pyrite may be oxidized by water, which occurs on the pyrite surface as the pyrite electrode leaches in acidic solutions [31,32]. The observations show that pyrite could first decomposes to form pyrrhotite, which is then converted to magnetite in a H_2O atmosphere.

At 600°C, pyrrhotite is the only mineral phase in ash particles under pure CO_2. As 20% CO_2 is replaced by H_2O, magnetite starts to appear in the XRD pattern of the residual ash. Fig. 6.5C shows that the mineral phases in the residual ashes are similar as the H_2O concentration increases. However, the peak intensities of the magnetite phase are the most obvious in H_2O concentrations of up to 80%. The S/Fe molar ratio decreases (Fig. 6.5D), and the release of

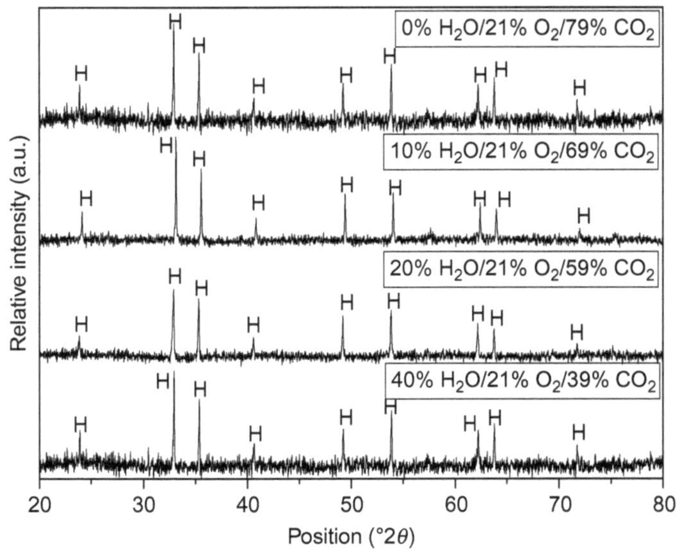

FIG. 6.6 The XRD patterns of pyrite oxidation in O_2-containing atmosphere: *H*, hematite (Fe_2O_3).

H_2S and SO_2 increases as H_2O partial pressure increases, indicating that H_2O promoted the release of sulfur in the experimental conditions.

The presence of H_2O in the atmosphere improves the pyrite combustion process, which may be attributed to the bigger mass diffusivity of H_2O than that of CO_2. However, the final product of pyrite oxidation is not affected by the presence of water vapor. Fig. 6.6 [30] shows that the main species phase is hematite in all 21% O_2-containing atmospheres. The final product remains unchanged by adding H_2O.

6.3.3 Transformation Pathway of Pyrite Thermal Decomposition in Oxy-fuel Combustion

The transformation of pyrite thermal decomposition in oxy-fuel combustion is summarized in Fig. 6.7 [30]. In an O_2-free atmosphere without H_2O, pyrite is first desulfurized to form Fe-rich pyrrhotite (FeS_x) at the particle surface layer, and the core is still unreacted pyrite. Then the pyrrhotite reacts with CO_2 or H_2O to form magnetite (Fe_3O_4). The unreacted pyrite decomposes to form Fe-rich pyrrhotite. However, in an O_2-free atmosphere with H_2O, the magnetite further reacts with CO_2 to form hematite. In an O_2-containing atmosphere without H_2O, pyrite is directly oxidized to form hematite and ferrous sulfate, with the release of SO_2. At the same time, the desulfurization of the pyrite core, which does not contact with oxygen, occurs and forms pyrrhotite. Finally, the ferrous sulfate decomposes, and pyrrhotite is oxidized to form hematite. In an O_2-containing atmosphere with H_2O, pyrite is first rapidly oxidized to form hematite and ferrous sulfate at the particle surface layer, with the core still being unreacted pyrite.

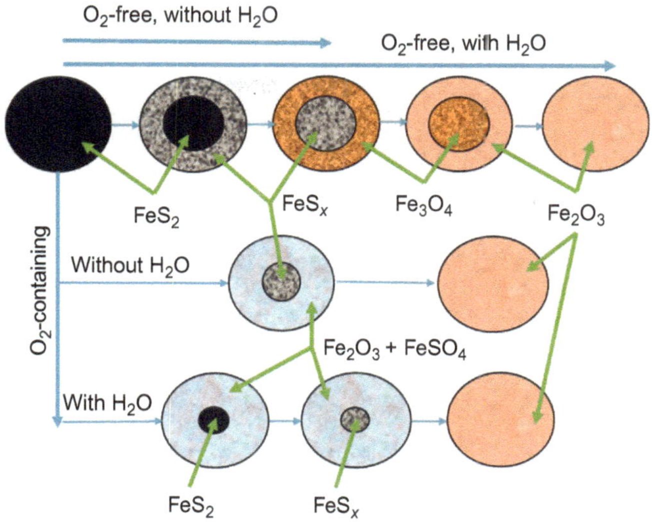

FIG. 6.7 The schematic diagram of pyrite conversion in oxy-fuel combustion atmosphere.

Then the unreacted pyrite decomposes to form pyrrhotite. Finally, ferrous sulfate decomposes, and pyrrhotite is oxidized to form hematite. Note that Pyrite would experience three significant steps in an O_2-containing atmosphere with H_2O.

6.4 CHARACTERISTICS OF ASH MELTING AND SLAG IN OXY-FUEL CONDITIONS

The burning behavior of coal under the atmosphere of O_2/CO_2 has been briefly presented [33–36]. However, the behavior partitioning, volatilization, condensation, capture, and retention of different minerals and trace elements in O_2/CO_2 recycle combustion are still uncertain or are incompletely understood, and additional investigations are still required, especially on the evaporation of mineral elements, which is much different from that in conventional coal combustion. The evaporation of mineral elements not only responds to the characteristics of ash deposits but also is the only source of fine particular matter. So it is important to understand the partitioning of minerals during coal combustion in an O_2/CO_2 mixture atmosphere. The present work provides a systematic investigation on the thermal behavior of minerals in O_2/CO_2 recycle combustion, analyzes the factors that affect mineral vaporization, and provides a basic method for industrial application of O_2/CO_2 recycle combustion.

6.4.1 Samples and Experiment

The coal samples used in experiments were collected from Xiaolongtan (XLT), Shenfu (SF), and Yangzonghai1 (YZH1) power plants in China. All the coal

TABLE 6.1 Proximate Analysis and Ultimate Analysis of Coals (wt%)

	Proximate (ad)				Ultimate (ad)			
Coal Sample	M	V	A	FC	C	H	N	S
Xiaolongtan	15.0	39.8	17.9	27.4	45.4	4.3	1.2	1.9
Yangzonghai1	11.1	37.1	30.4	21.4	37.5	3.4	1.0	2.0
Shenfu	10.2	31.6	5.8	52.4	80.5	4.8	0. 9	0.4

samples were pulverized to <74 μm in the laboratory. The proximate and ultimate analyses of these coals are listed in Table 6.1.

Low-temperature ash (LTA) was prepared at low temperature for about 12–24 h in an EMITECH K1050X Plasma Asher. The plasma process was accomplished through the use of a low-pressure, radio frequency-induced, and gaseous discharge. With high purity oxygen as the process gas, the molecules dissociated into chemically active atoms and molecules. The unique property of the process was the relatively low temperature at which it occurred when ashing typically organic material and avoiding the transform of inorganic mineral during the raw coal ashing [37].

XRD spectrometry was used to characterize the minerals in the coal and their LTAs. XRD studies were carried out on a χ'Pert PRO diffractometer equipped with a graphite diffracted-beam monochromator. The accelerating voltage was 40 kV, and the current was 40 mA. Diffraction patterns were collected at 5°2θ to 70°2θ using Cu-Kα radiation. The thermal analysis (TG-DTG-DTA) of these samples was conducted by using a NETZSCH STA 409 thermogravimetric analyzer. These experiments were carried out for 16–20 mg of coals and their LTAs. The samples were heated from ambient temperature to 1450°C in corundum crucibles under a static air atmosphere at a heating rate of 20°C min^{-1} and 50°C min^{-1}, respectively.

6.4.2 Mineral in Coal and LTA

The mineral composition of these coals and their LTAs were identified by XRD, as shown in Table 6.2. The mineral content range in these coals was 10.2%–41.0%. The YZH1 sample had the highest mineral content. The mineral composition varied in different coals. The principal minerals found in a coal and LTA were divided into six categories [38]. The mineral identified in the three coals and LTAs included (1) silicates—quartz, clay minerals, kaolinite, illite, montmorillonite; (2) sulfides—pyrite; (3) carbonates—calcite, dolomite; (4) sulfates—gypsum, bassanite; and (5) others—anatase. All of the coals had a high content of calcium-bearing minerals. After the low-temperature ashing, gypsum in coal lost water and transformed to bassanite.

TABLE 6.2 Mineral Composition of Three Typical High Calcium Coals and Their LTAs (wt%)

	Xiaolongtan		Shenfu		Yangzonghai1	
	Coal	LTA	Coal	LTA	Coal	LTA
Organic	82.1		89.8		59.0	
Minerals	17.9		10.2		41.0	
Kaolinite	2.5	8.8	4.5	12.7	6.2	16.9
Montmorillonite					3.4	
Illite	5.2	21.0			7.6	25.0
Quartz	4.3	19.1	3.1	30.6	6.7	18.6
Calcite	1.4	15.4	2.6	40.6		
Gypsum	4.4				9.1	
Bassanite		26.3		10.1		15.9
Dolomite						10.6
Pyrite		2.8		5.2	2.2	5.3
Anatase		4.0			5.9	7.7

6.4.3 Thermal Analysis of Coal and LTA

The thermal curves (TG-DTG-DTA) of coals and their LTAs in both air and O_2/CO_2 atmospheres are shown in Figs. 6.8–6.11. Usually, the coal combustion process was performed in three stages that were classified based on the weight change and heat change, such as preheating (first stage), volatile matter releasing (second stage), and coal char combustion (third stage) [39]. The DTG curve of the XLT coal sample in an air atmosphere (Fig. 6.8) exhibited four obvious weight loss peaks.

The first peak showed an endothermic trend from room temperature to around 200°C with a peak at 105°C accompanied by a weight loss of around 4%, as calculated by the TG curve. The weight loss was attributed to the elimination of physically adsorbed water [40]. The second weight loss peaked at around 400°C, ranging from 280°C–600°C. With the temperature increasing, the volatile matter and carbon started to combust step by step. From the TG-DTA curves of XLT coal, it was found that the ignition point of the coal was 291.9°C, after which point, the reaction turned too acutely with an obvious increase in the weight loss, as shown in TG curve in Fig. 6.11. The DTA curve then exhibited an exotherm up to 600°C. The DTG curve exhibited an overlap peak at 400°C, which was attributed to the release of volatile matter and the combustion of the carbonaceous

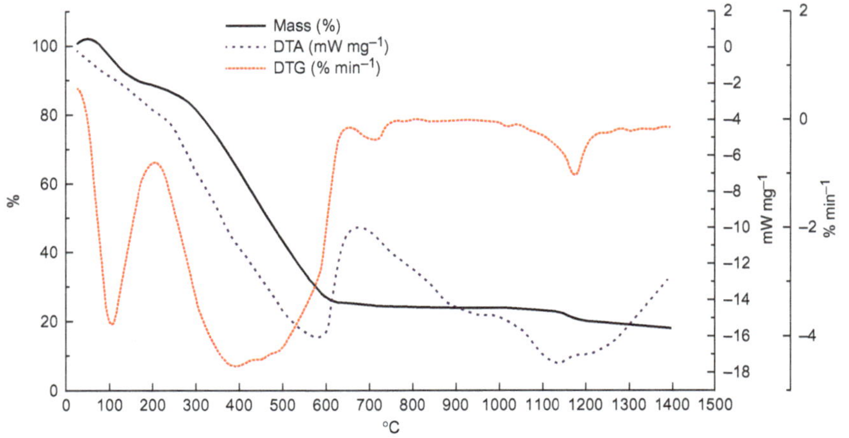

FIG. 6.8 Thermal curves (TG-DTA-DTG) of XLT coal in O_2/N_2 condition.

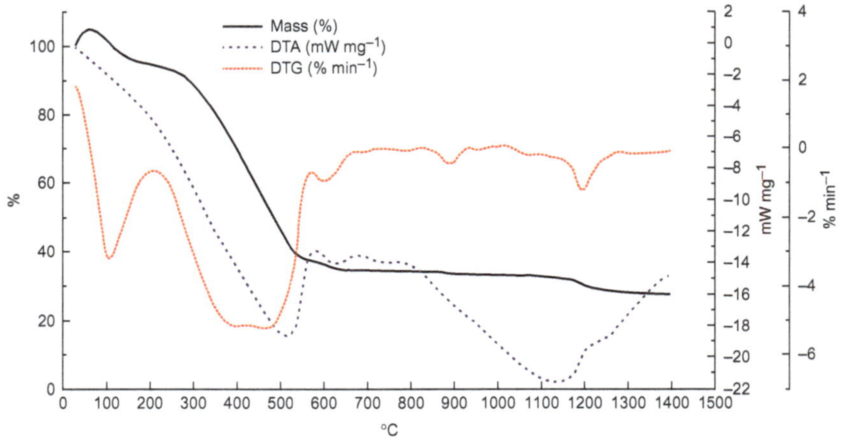

FIG. 6.9 Thermal curves (TG-DTA-DTG) of XLT coal in O_2/CO_2 condition.

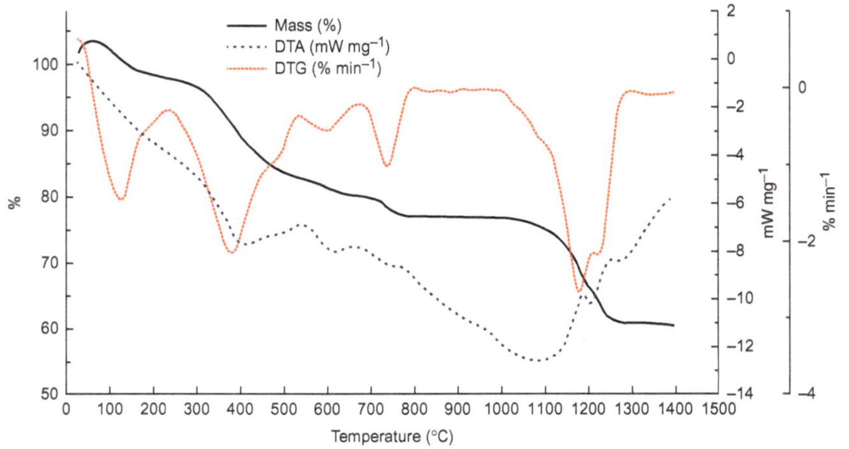

FIG. 6.10 Thermal curves (TG-DTA-DTG) of XLT low-temperature ash in O_2/N_2 condition.

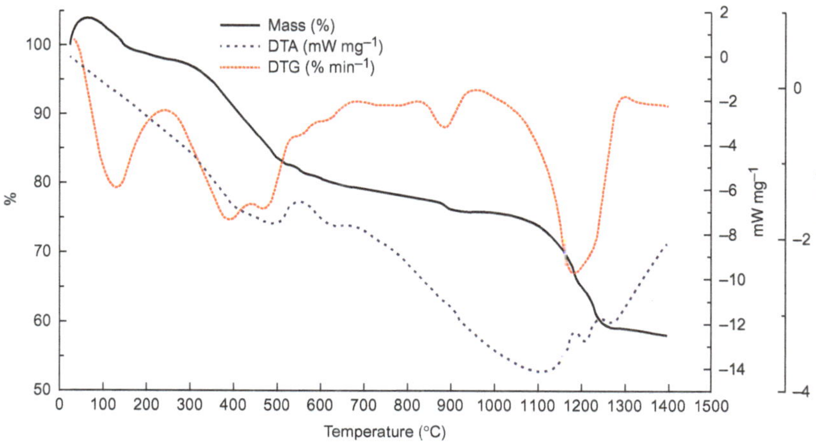

FIG. 6.11 Thermal curves (TG-DTA-DTG) of XLT low-temperature ash in O_2/CO_2 condition.

materials. The third peak, ranging from 600°C to 1000°C, occurred in the burnout stage. The burnout temperature was 910.3°C. An acromion was formed on the DTA curve; the acromion represented the resistant burn performance. The coal burned out completely after the acromion disappeared. The TG curve began a slow-moving descent from 600°C, and then continued after burnout. During this stage, some weight loss peaks appeared, which may have been due to the transformation of minerals. The last peak appeared at 1000°C, with an obvious weight loss peak at 1200°C on the DTG curve. Based on mineral transformation processes discussed in several references, the last weight loss peak may have been due to the vaporization of volatile mineral elements.

As shown in Table 6.2, the minerals in LTA mainly included clay minerals, quartz, carbonates, sulfides, and sulfates, and so on. The TGA curves of LTA were similar with the coal; however, the ash sample exhibited a prominent endotherm from about room temperature to around 200°C, with a peak at around 140°C. All these changes were accompanied by weight loss. The total weight loss up to 200°C was around 3% and was attributed to the loss of adsorbed moisture and crystallized water. From the XRD analysis results, it also found that the gypsum in coal transformed to asanite after ashing. For the second stage, several weight loss peaks occurred at 300–1100°C, which were credited to the minerals decomposition and transformation. After the transformation of minerals, a typical weight loss occurred after 1100°C, and peaked at around 1200°C, which may have been caused by the mineral melt and vaporization of the volatile mineral elements.

6.4.4 Thermal Behavior in Different Atmospheres

The characteristic temperature of coal and LTA in different conditions are listed in Table 6.3. The characteristic temperature of coals varied in different conditions. Compared with air atmosphere, the firing temperature (T_s), the peak temperature at maximum weight loss rate (T_m), and the burnout temperature (T_h)

TABLE 6.3 Characteristic Temperature of Coal and Its Low-Temperature Ash in Different Conditions

		50°C min^{-1}			20°C min^{-1}		
Sample	**Atmosphere**	T_s	T_h	T_m	T_s	T_h	T_m
YZH1-coal	O_2/N_2	283.5	877.7	415.1	305.2	585.9	502.2
	O_2/CO_2	275.8	852	407.5	278.5	536.4	497
YZH1-LTA	O_2/N_2	336	1214.7	457.9	327.9	1185.5	367.6
	O_2/CO_2	306.5	1200.8	436.6	–	–	–
SF-coal	O_2/N_2	380.6	1155.6	957.7	367.7	869.3	494.5
	O_2/CO_2	379.1	1141.2	1062.2	365.8	595.3	489.5
SF-LTA	O_2/N_2	758.6	1235.8	813.5	488	1215.7	833.8
	O_2/CO_2	864	1299.2	882.4	416.9	1229.6	894
XLT-coal	O_2/N_2	321.9	910.3	404.1	283	606.6	487.1
	O_2/CO_2	307.6	907	394.6	277.8	544.4	470.1
XLT-LTA	O_2/N_2	315.1	1265.1	450.3	330.9	1250	1185
	O_2/CO_2	306.1	1274.1	370.4	–	–	1184.6

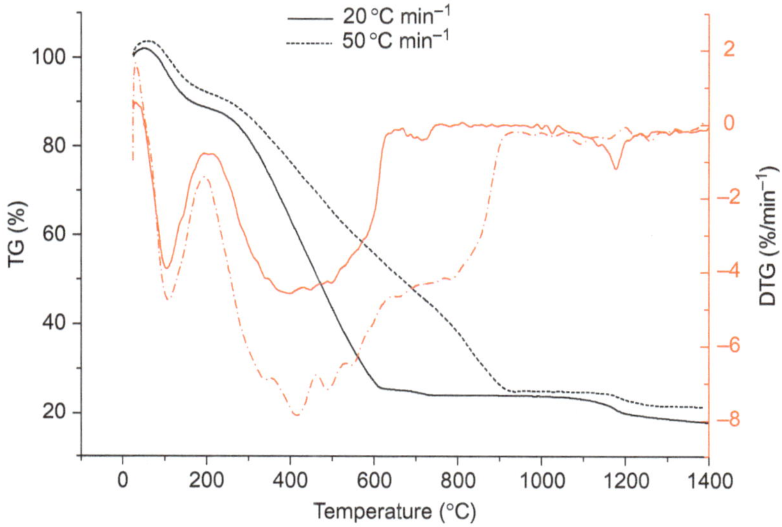

FIG. 6.12 TG-DTA curves of XLT coal in different heating rates.

were lower in O_2/CO_2 conditions. The character temperatures of LTAs showed less regularity.

The heating rates had a big impact on the coals thermal behavior. The TG-DTA curves of XLT coal in different heating-up rates are shown in Fig. 6.12. The TG curve turned toward the high temperature area with increasing temperatures, and also for the DTG curve. The maximum weight loss rate also increased with increasing temperatures [41].

As listed in Table 6.3, the peak temperature at maximum weight loss rate (T_m) within 50°C min^{-1} heating rate occurred at about 1000°C, higher than coal samples at 360–500°C., which may have resulted from the high fixed-carbon content in Shenbei coal, which was up to 52.35%. In addition, the weight loss rate increased as the heating rate increased, also for the maximal weight loss rate. The higher heating rate strongly affected the coal structure, which increased the cracking of the coal macromolecular side chains and aromatic fused rings. Furthermore, the high heating rate enhanced the deepness of the rupture, thereby producing more gas and liquid.

The state of plastomer before the formation of semicoke was the important factor in defining coal cohesiveness [42]. Thus the pyrolysis temperature transferred to the high temperature side because of the increasing heating rate. The temperature area of high-intensity pyrolysis was shortened, resulting in a more liquid consistency in plastomer and improvement of cohesiveness process.

6.4.5 Vaporization of Mineral Elements

After 1100°C, the carbon in coal have been burnout completely, and the water also volatilization, thus the temperature at 1100–1400°C was chosen to analysis the mineral vaporization. Assume that all of the mass loss after 1100°C are result of the mineral elements vaporization. YZH1 coal sample which has the highest mineral content was chosen to calculate the vaporization amount. Then the vaporization amounts of YZH1 coal and LTA in the typical temperature are listed in Table 6.4.

All of the vaporization amounts were based on the mass of coal. The mineral element vaporization amount increased with increasing temperatures (Table 6.4). There was an aberration peak at approximately 1150–1200°C for mineral vaporization, which may have resulted from the vaporization of most alkali volatile elements. The total mineral element vaporization, which amounted to 1100–1400°C, was also calculated (Table 6.4).

The organic carbonaceous in coal promoted the formation of a local reducing atmosphere around the coal particles, especially in the O_2/CO_2 mixture atmosphere. Therefore the amount of mineral vaporization was greater during the coal heating process than during the LTA heating process in thermal gravimeter. However, the total amount of vaporization for coal mineral elements was almost equal to that of the LTA in the O_2/N_2 combustion atmosphere, which indicated that the carbon in coal had little effect on the vaporization of the mineral elements, compared to LTAs.

However, the total amount of mineral element vaporization was higher in the O_2/N_2 atmosphere than in the O_2/CO_2 atmosphere, which was consistent with the calculation [12]. In the O_2/CO_2 atmosphere, the CO_2 in the main gas did not shorten the burning time of the coal or lower the combustion temperature of coal. The observed suppression of evaporation in the O_2/CO_2 atmosphere qualitatively supported the mechanism [43]

$$MO_n(s) + CO(g) \rightarrow MO_{n-1}(g) + CO_2(g)$$

In addition, with the increase of the heating rate from 20°C min^{-1} to 50°C min^{-1}, the total vaporization amount also exhibited an obvious increase. Under the high heating rate, the coal structures received a strong heat impact, which accelerated the separation of the mineral elements from the organic carbonaceous materials.

The mineral content range in these coals was 10.2%–41.0%. The YZH1 sample had the highest mineral content. All of the coals had a high content of calcium-bearing minerals. The TG-DTA curves of the coal samples exhibited the four typical weight loss stages, including the elimination of physically adsorbed water, the release of volatile matter, the combustion of the carbonaceous materials, the transformation of minerals, and the vaporization of mineral elements. Compared with an air atmosphere, the characteristic temperatures (T_s, T_m, T_h) were lower in the O_2/CO_2 combustion atmosphere. The heating rates had a great affinity for the coals thermal behavior. The TG-DTA curves turned toward the high temperature area with increasing temperatures. The weight loss rate increased with as the heating rate increased, also for the maximal weight loss rate. The amount of mineral element vaporization increased with increasing

TABLE 6.4 Mineral Elements Vaporization of YZH Coal and Its LTA (wt%) ($m_{LTA}/m_{coal} = 0.1786$)

	50 K min^{-1}				20 K min^{-1}			
	Coal		LTA		Coal		LTA	
Temperature (°C)	O_2/N_2	O_2/CO_2	O_2/N_2	O_2/CO_2	O_2/N_2	O_2/CO_2	O_2/N_2	O_2/CO_2
1100–1150	0.11	0.30	0.43	0.32	0.17	0.23	0.39	0.17
1150–1200	0.42	0.43	0.51	0.51	0.42	0.42	0.58	0.39
1200–1250	0.30	0.20	0.27	0.26	0.24	0.08	0.21	0.11
1250–1300	0.18	0.16	0.06	0.03	0.15	0.07	0.05	0.11
1300–1350	0.13	0.05	0.08	0.02	0.13	0.07	0.03	0.08
1350–1400	0.21	0.05	0.03	0.06	0.15	0.20	0.01	0.15
Total vaporization amount	1.35	1.19	1.38	1.20	1.25	1.07	1.27	1.01

temperatures. The organic carbonaceous material in coal promoted the vaporization of mineral elements, especially in the O_2/CO_2 mixture atmosphere. The total amount of mineral element vaporization was higher in the O_2/N_2 atmosphere than in the O_2/CO_2 atmosphere, which may have been due to the increase of the CO_2 partial pressure in the O_2/CO_2 mixture atmosphere. With an increased heating rate from $20°C\,min^{-1}$ to $50°C\,min^{-1}$, the total vaporization amount also exhibited an obvious increase.

REFERENCES

[1] Sheng C, Li Y. Experimental study of ash formation during pulverized coal combustion in O2/CO2 mixtures. Fuel 2008;87(7):1297–305.

[2] Smoot LD. Fundamentals of coal combustion for clean and efficient use. Coal Science and Technology. 20. Amsterdam: Elsevier; 1993, p. 299–313.

[3] Toftegaard MB, Brix J, Jensen PA, Glarborg P, Jensen AD. Oxy-fuel combustion of solid fuels. Prog Energy Combust Sci 2010;36(5):581–625.

[4] Sheng C, Lu Y, Gao X, Yao H. Fine ash formation during pulverized coal combustion—a comparison of O2/CO2 combustion versus air combustion. Energy Fuel 2007;21(2):435–40.

[5] Sheng C, Li Y, Liu X, Yao H, Xu M. Ash particle formation during O2/CO2 combustion of pulverized coals. Fuel Process Technol 2007;88(11):1021–8.

[6] Suriyawong A, Gamble M, Lee M, Axelbaum R, Biswas P. Submicrometer particle formation and mercury speciation under O2–CO2 coal combustion. Energy Fuel 2006;20(6):2357–63.

[7] Krishnamoorthy G, Veranth JM. Computational modeling of CO/CO2 ratio inside single char particles during pulverized coal combustion. Energy Fuel 2003;17(5):1367–71.

[8] Yu D, Morris WJ, Erickson R, Wendt JOL, Fry A, Senior CL. Ash and deposit formation from oxy-coal combustion in a 100 kW test furnace. Int J Greenhouse Gas Control 2011;5(1):S159–67.

[9] Zhan Z, Bool LE, Fry A, Fan W, Xu M, Yu D, et al. Novel temperature-controlled ash deposition probe system and its application to oxy-coal combustion with 50% inlet O2. Energy Fuel 2014;28(1):146–54.

[10] Zhang T, Liu Z, Huang X, Sun Q, Liu C, Li J, et al. Computer-controlled scanning electron microscopy investigation on ash formation characteristics of a calcium-rich coal under O2/CO2 environments. Energy Fuel 2017;31:319–27.

[11] Zheng L, Furimsky E. Assessment of coal combustion in O2+CO2 by equilibrium calculations. Fuel Process Technol 2003;81(1):23–34.

[12] Zhao YC, Zhang JY, Liu HT, Tian JL, Yang LI, Zheng CG. Thermodynamic equilibrium study of mineral elements evaporation in O2/CO2 recycle combustion. J Fuel Chem Technol 2006;34(6):641–50.

[13] Sheng C, Lin J, Li Y, Wang C. Transformation behaviors of excluded pyrite during O2/CO2 combustion of pulverized coal. Asia Pac J Chem Eng 2010;5(2):304–9.

[14] Taneja SP, Jones CHW. Mössbauer studies of iron-bearing minerals in coal and coal ash. Fuel 1984;63(5):695–701.

[15] Hu G, Dam-Johansen K, Wedel S, Hansen JP. Decomposition and oxidation of pyrite. Prog Energy Combust Sci 2006;32(3):295–314.

[16] Bhargava SK, Garg A, Subasinghe ND. In situ high-temperature phase transformation studies on pyrite. Fuel 2009;88(6):988–93.

[17] Liu Y, Che D, Xu T. Catalytic reduction of SO2 during combustion of typical Chinese coals. Fuel Process Technol 2002;79(2):157–69.

[18] Zhang D, Yani S. Sulphur transformation during pyrolysis of an Australian lignite. Proc Combust Inst 2011;33(2):1747–53.

[19] Tomeczek J, Palugniok H. Kinetics of mineral matter transformation during coal combustion. Fuel 2002;81(10):1251–8.

[20] Harding NS, O'Connor DC. Ash deposition impacts in the power industry. Fuel Process Technol 2007;88(11−12):1082–93.

[21] Huffman GP, Huggins FE, Dunmyre GR. Investigation of the high-temperature behaviour of coal ash in reducing and oxidizing atmospheres. Fuel 1981;60(7):585–97.

[22] McLennan AR, Bryant GW, Stanmore BR, Wall TF. Ash formation mechanisms during pf combustion in reducing conditions. Energy Fuel 2000;14(1):150–9.

[23] Li Y, Wu H. Ash cenosphere from solid fuels combustion. Part 1: an investigation into its formation mechanism using pyrite as a model fuel. Energy Fuel 2014;26(1):130–7.

[24] Srinivasachar S, Helble JJ, Boni AA. Mineral behavior during coal combustion 1. Pyrite transformations. Prog Energy Combust Sci 1990;16(4):281–92.

[25] Lv W, Yu D, Wu J, Zhang L, Xu M. The chemical role of CO2 in pyrite thermal decomposition. Proc Combust Inst 2014;35(3):3637–44.

[26] Aylmore MG, Lincoln FJ. Mechanochemical milling-induced reactions between gases and sulfide minerals: I. Reactions of SO 2 with arsenopyrite, pyrrhotite and pyrite. J Alloys Compd 2000;309(1):61–74.

[27] Huang F, Zhang LQ, Yi BJ, Xia ZJ, Zheng CG. Transformation pathway of excluded mineral pyrite decomposition in CO2 atmosphere. Fuel Process Technol 2015;138:814–24.

[28] Yi B, Zhang L, Huang F, Mao Z, Zheng C. Effect of H2O on the combustion characteristics of pulverized coal in O2/CO2 atmosphere. Appl Energy 2014;132(11):349–57.

[29] Jiang Z, Duan L, Chen X, Zhao C. Effect of water vapor on indirect Sulfation during oxy-fuel combustion. Energy Fuel 2013;27(3):1506–12.

[30] Fang H, Zhang L, Yi B, Xia Z, Zheng C. Effect of H2O on pyrite transformation behavior during oxy-fuel combustion. Fuel Process Technol 2015;131:458–65.

[31] Chandra AP, Gerson AR. The mechanisms of pyrite oxidation and leaching: a fundamental perspective. Surf Sci Rep 2010;65(9):293–315.

[32] Holmes PR, Crundwell FK. The kinetics of the oxidation of pyrite by ferric ions and dissolved oxygen: an electrochemical study. Geochim Cosmochim Acta 2000;64(2):263–74.

[33] Buhre BJP, Elliott LK, Sheng CD, Gupta RP, Wall TF. Oxy-fuel combustion technology for coal-fired power generation. Prog Energy Combust Sci 2005;31(4):283–307.

[34] Kiga T, Takano S, Kimura N, Omata K, Okawa M, Mori T, et al. Characteristics of pulverized-coal combustion in the system of oxygen/recycled flue gas combustion. Energy Convers Manag 1997;38(96):S129–34.

[35] Komi N. The characteristics of pulverized coal combustion in O2/CO2 mixtures for CO2 recovery. Fuel Energy Abstr 1995;36(6–9):805–8.

[36] Luo Z, Mao Y, Wu X. Test study on and analysis of burning behavior for coal under atmosphere of O2/CO2. Thermal Power Gener 2004;8(12):1657–71.

[37] Gluskoter HJ. Electronic low temperature ashing of bituminous coal. Fuel 1965;44:285–91.

[38] Ward CR. Analysis and significance of mineral matter in coal seams. Int J Coal Geol 2002;50(1–4):135–68.

[39] Chen Y, Mori S, Pan WP. Estimating the combustibility of various coals by TG-DTA. Energy Fuel 2002;9(1):71–4.

[40] And SM, Srivastava SK. Minerals transformations in northeastern region coals of india on heat treatment. Energy Fuel 2006;20(3):1089–96.

[41] Tai L, Chao W. Study on the effect of coal diameter and heating rate on the coal pyrolysis. Coal Conversion 2005;28:17–20.
[42] Han Y, Yao Z. The effect of heating-rate on coal pyrolysis kinetics. J East China Univ Metall 1999;16:318–22.
[43] Quann RJ, Sarofim AF. Vaporization of refractory oxides during pulverized coal combustion. Symp Combust 1982;19(1):1429–40.

Chapter 7

Mercury Behavior and Retention in Oxy-fuel Combustion

Jianping Yang, Yongchun Zhao, Junying Zhang, Chuguang Zheng
Huazhong University of Science and Technology, Wuhan, China

7.1 INTRODUCTION

Carbon dioxide (CO_2), as the main greenhouse gas emitted through human activities, has drawn worldwide attention in recent years because of its potential effects on climate change. Coal-fired power plants are considered as a primary anthropogenic source of CO_2. Oxy-fuel combustion is considered as a promising technology for capturing CO_2 from coal-fired power plants. However, the elemental mercury (Hg^0) present in the flue gas can damage the CO_2 compression and purification units through liquid metal embrittlement (LME) and amalgam corrosion (AMC) [1,2]. Moreover, mercury is a great environmental concern because of its volatility, persistence, and bio-accumulation [3,4]. Coal-fired power plants are one of the biggest anthropogenic emitters of mercury into the atmosphere [3]. The concentration of mercury is generally higher in oxy-fuel combustion flue gas compared to that in air combustion because of the recycled flue gas. Thus special attention needs to be paid to the removal of mercury in oxy-fuel combustion systems. In this chapter, the partition of mercury during oxy-fuel combustion is summarized. The distribution of mercury in combustion products and flue gas were investigated in different demonstration oxy-fuel furnaces. Different sorbents were used to capture mercury in an oxy-fuel flue gas atmosphere, and the removal efficiency was compared to an air combustion atmosphere. Finally, the effect of mercury on oxy-fuel CO_2 compression unit is discussed.

7.2 HG PARTITION DURING OXY-FUEL COMBUSTION

The fate of mercury is important for the operation of oxy-fuel combustion because of its environmental concerns [3,4] and because of the potential corrosion of CO_2 compression and purification units [1]. In the past few decades, extensive studies were carried out in order to better understand the fate of mercury in conventional air combustion [5–11]. Because it is the most volatile trace element in

Oxy-fuel Combustion. http://dx.doi.org/10.1016/B978-0-12-812145-0.00007-4

coal, most of the mercury is volatilized as gaseous elemental mercury (Hg^0) at a high-temperature combustion zone. Cooling the flue gas and interactions of Hg^0 with other flue gas components could result in the oxidation of Hg^0 to oxidized mercury (Hg^{2+}).Also, several parts of Hg^{2+} could be converted into particulate-bound mercury (Hg^p) with the participation of fly ash and flue gas.

Roy et al. [5] predicted the mercury speciation during oxy-fuel combustion of Victorian brown coal at different temperatures (800–1400°C) by thermodynamic equilibrium calculations. The results showed that Hg^0 was the primary species over a wide range of temperatures (800–1400°C) in air and oxy-fuel combustion, as shown in Fig. 7.1. The atmosphere and temperature had little effect on the distribution of mercury. When above 98.5%, mercury occurred as gas-phase Hg^0 under both air and oxy-fuel combustion conditions.

Fig. 7.2 shows the effect of temperature on the equilibrium distribution of mercury during oxy-fuel combustion of Yallourn coal. It was observed that gas-phase Hg^0 was the dominant species over the entire temperature range with a small amount of HgO (g) at low temperatures. Moreover, the emission of Hg increased slightly with increasing combustion temperatures.

A combined homogeneous-heterogeneous model developed by Gharebaghi et al. [12] showed that the mercury retention on fly ash particles depended on temperature, burnout, and particle sizes. Also, a higher amount of particulate

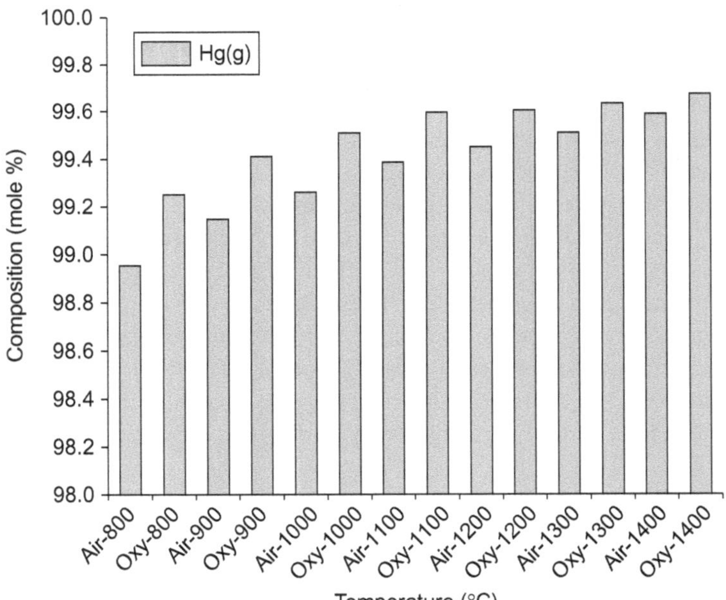

FIG. 7.1 Equilibrium composition of major toxic Hg species during air and oxy-fuel combustion. (Balance of 100% on the vertical axis are the other Hg-species.) *(From Roy B, Choo WL, Bhattacharya S. Prediction of distribution of trace elements under oxy-fuel combustion condition using Victorian brown coals. Fuel 2013;114:135–42.)*

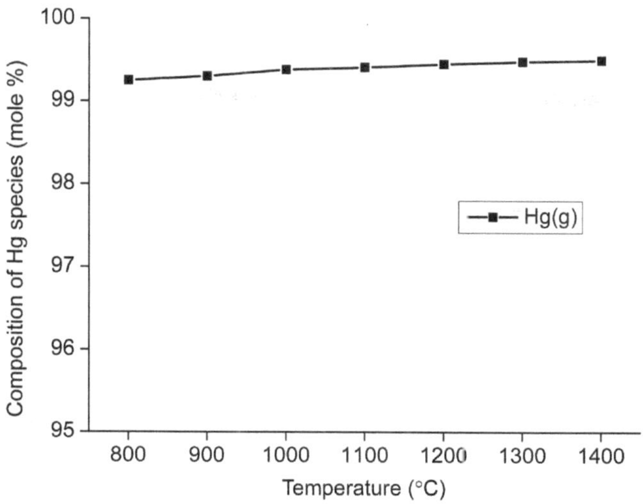

FIG. 7.2 Equilibrium distribution of mercury during oxy-fuel combustion using Yallourn coal. (Balance of 100% on the vertical axis are the other Hg-species.) *(From Roy B, Choo WL, Bhattacharya S. Prediction of distribution of trace elements under oxy-fuel combustion condition using Victorian brown coals. Fuel 2013;114:135–42.)*

mercury occurred in the oxy-fuel combustion conditions. However, the species and fate of mercury during coal combustion is complex and depends on many factors, such as the combustion temperature, the mass and heat transfer, the amount of Hg and Cl/Br in the coal, the flue gas composition, the fly ash components (i.e., unburned carbon (UBC), Fe_2O_3, CaO), the application of gas-cleaning equipment (SCR, ESP/FF, FGD), and so on. Moreover, in oxy-fuel combustion units, impurities in flue gas such as SO_2, SO_3, HCl, and H_2O are much higher than those in air combustion because of the flue gas recirculation [13–16]. This could result in the difference in the species and the fate of mercury between oxy-fuel and air combustion.

7.3 FIELD MEASUREMENT OF HG DISTRIBUTION DURING OXY-FUEL COMBUSTION

In the past 10 years, several pilot-scale results about the fate of mercury under oxy-fuel combustion conditions have been reported. Font et al. [17] studied the fate and abatement of mercury in a 90 kW_{th} bubbling fluidized bed (BFB) oxy-fuel combustion pilot plant fed with coal and limestone (bed material). The results showed that about 82% of the gaseous Hg occurred in Hg^0 with lower proportions of Hg^{2+} (18%). A similar percentage of Hg^0/Hg^{2+} in the exhaust gas from a 10 kW_{th} fluidized bed under oxy-fuel combustion conditions was also obtained by Roy et al. [18], where the percentage of Hg^0 in the total gaseous mercury was about 83%.

Zhuang et al. [19] investigated the fate of mercury during oxy-fuel combustion in a 173.5 kW_{th} experimental system with different flue gas recycle (FGR). The mercury speciation and distribution were studied via the EPA Method 29 at the inlet and outlet of the ESP and the scrubber outlet, respectively. Speciated sorbent traps were also used to verify the EPA Method 29 results and to provide gas-phase mercury speciation data. The mercury species partitioning during air and oxy-fuel combustion are summarized in Fig. 7.3.

The concentration of mercury in air combustion flue gas was 11.4 $\mu g/dNm^3$ with 97% of gas-phase mercury. The mercury capture efficiency of ESP was very low because most of the mercury was in the gas phase. About 52% of gaseous mercury occurred in a soluble oxidized form. Thus 40%–50% of the mercury was removed after passing through the wet scrubber. The mercury concentration in flue gas at the ESP inlet under oxy-fuel combustion with RFG was about 20 $\mu g/dNm^3$, which was almost double that under air combustion conditions, which may have been due to the cumulative effects of flue gas recycling and the reduced gas volume during oxy-fuel combustion. Moreover, the conversion of both Hg(g) to Hg(p) and Hg^0 to Hg^{2+} was promoted in oxy-fuel combustion, which was attributed to the continuous interactions of mercury with enriched HCl, UBC, and other reagents in the flue gas. Accordingly, the overall mercury capture efficiency was 80.9% under oxy-fuel combustion conditions, which was much higher than that under air combustion conditions (40.4%). However, different conclusions were obtained in a study by Obras-Loscertales et al. [20], in which the recirculation of SO_2 inhibited the Hg^0 oxidation and thus the mercury fixation as particle-bound mercury decreased.

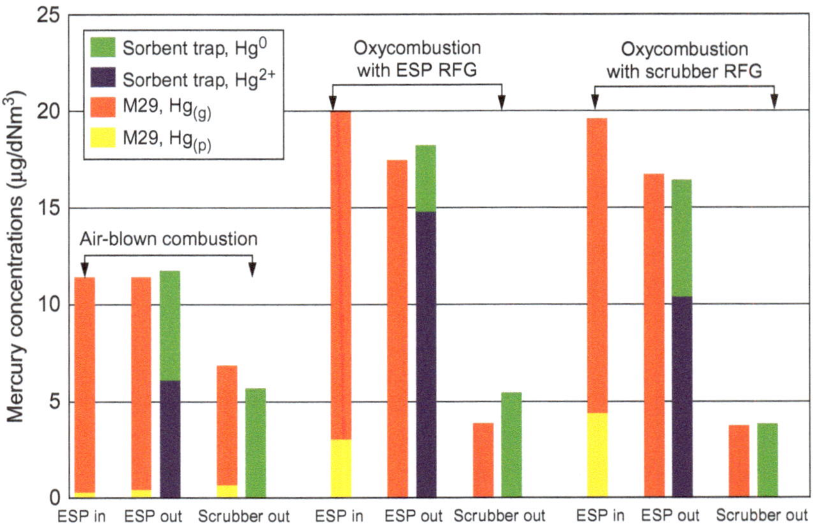

FIG. 7.3 Mercury species partitioning during air- and oxygen-blown combustion. *(From Zhuang Y, Pavlish JH. Fate of hazardous air pollutants in oxygen-fired coal combustion with different flue gas recycling. Environ Sci Technol 2012;46:4657–65.)*

The current data are based mostly on bench- and pilot-scale measurements, as information from industrial-scale facilities is limited. Córdoba et al. [6] reported the first study of Hg species and behavior in a 20 MW_{th} industrial-scale oxy-fuel combustion plant. As shown in Fig. 7.4, continuous CO_2-rich flue gas recirculation played a significant role in the partitioning and fate of the Hg. Calculations showed that 44 mg of Hg was recirculated to the boiler as a result of 2183 recirculations of CO_2-rich flue gas. Moreover, note especially the effect of the CO_2-rich exhaust flue gas recirculation and the boiler on the Hg enrichment in fly ashes.

The data originating from industrial-scale oxy-fuel combustion power plants are still very limited, especially on the behavior of mercury in oxy-fuel combustion with different FGR. The behavior of mercury during oxy-fuel combustion was investigated in a 35 MW_{th} industrial-scale oxy-fuel combustion power plant (HUST, Yingcheng, China) with dry and wet FGR.

The mercury concentrations in the gas and particle phases at the ESP inlet, ESP outlet, and FGD outlet under air and oxy-fuel combustion are shown in Table 7.1. Information about Hg_g concentration in oxy-fuel combustion with wet FGR is not included because the high SO_2 and moisture concentration in flue gas would interfere with the Hg adsorption by activated carbon traps [21,22]. The Hg_g concentration at the ESP inlet in air combustion was about 3.3 $\mu g/m^3$. For oxy-fuel combustion with dry FGR, the Hg_g concentration at the ESP inlet was 5.6 $\mu g/m^3$, which was nearly double that in air combustion flue gas. This was likely due to the cumulative effect of flue gas recirculation and the reduced gas volume in oxy-fuel combustion [19]. Across the pollution control devices, such as ESP and FGD, the emission of Hg_g reduced drastically. For oxy-fuel combustion with dry FGR, the Hg_g concentration reduced to 4.8

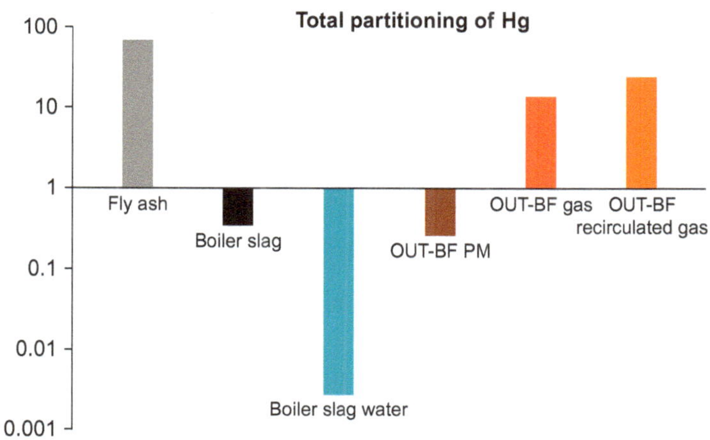

FIG. 7.4 Partitioning of Hg for the whole oxy-PCC installation. *(From Córdoba P, Maroto-Valer M, Delgado MA, Diego R, Font O, Querol X. Speciation, behaviour, and fate of mercury under oxy-fuel combustion conditions. Environ Res 2016;145:154–61.)*

TABLE 7.1 Mercury Concentration in Gas and Particle Phases

Combustion Conditions	Gas Phase (Hg^g)			Particle Phase (Hg^P)		
	ESP In	ESP Out	FGD Out	ESP In	ESP Out	FGD Out
Air combustion	3.3±0.4			98±13	nd	nd
Oxy-fuel combustion (dry FGR)	5.6±0.6	3.5±0.4	2.4±0.3	355±21	nd	nd
Oxy-fuel combustion (dry FGR)				300±15	nd	nd

nd, not detected.

and 2.2 μg/m^3 at the ESP outlet and the FGD outlet, respectively, which implied that about 14.2% of the Hg_g was adsorbed by fly ash and then captured by ESP. Across the FGD, about 46.4% of the Hg_g was further reduced, which may have been attributed to the capture of Hg^{2+} by FGD, since Hg^0 is insoluble and is difficult for the FGD system to capture.

The amount of Hg^P in fly ash from oxy-fuel combustion was much higher than that from air combustion. As shown in Table 7.1, the concentration of Hg^P was 355 and 300 ng/g at ESP inlet for oxy-fuel combustion with dry and wet FGR, respectively, while it was 98 ng/g for air combustion. Previous studies have demonstrated that the UBC content played an important role in the capture and oxidation of Hg. However, in this work, it was not the key factor affecting the enrichment of Hg in fly ash under oxy-fuel combustion, since the content of UBC (14.04%–17.27%) was similar under different combustion conditions. In such a case, the continuous recirculation of CO_2-rich flue gas with a high Hg concentration might be responsible for the enrichment of Hg in fly ash under oxy-fuel combustions [6]. At the ESP outlet and FGD outlet, the content of Hg^P in flue gas was very low because of the low content of fly ash in the flue gas, which implied that, across the ESP and FGD, the Hg^P was almost completely captured.

7.4 HG REMOVAL IN OXY-FUEL FLUE GAS

7.4.1 Mercury Oxidation Over Selective Reduction of NO_x (SCR) Catalysts

It has been demonstrated that SCR catalysts could enhance the oxidation of Hg^0 in conventional air combustion. However, the Hg^0 oxidation behaviors over SCR catalysts under oxy-fuel combustion atmosphere are not very clear. Wang et al. [23] studied the mercury removal performance over the vanadia-titania (V_2O_5-WO_3/TiO_2) catalyst in CO_2-enriched conditions. The results showed that Hg^0 and NO could be simultaneously removed over V_2O_5-WO_3/TiO_2, with a maximum Hg^0 and NO removal efficiency of 65% and 85% at 250–350°C, respectively. The Hg^0 removal capacity was enhanced in the CO_2-enriched (80 vol%) atmosphere, as shown in Fig. 7.5. The effects of NO on the Hg^0 removal performance depended on the NO concentrations (Fig. 7.6). The NH_3 presented in the SCR system played an inhibitive role in the removal of Hg^0 (Fig. 7.7). The oxidation of Hg^0 over V_2O_5-WO_3/TiO_2 catalysts at 250–350°C in the CO_2-enriched atmosphere was attributed to: (1) the adsorption of Hg^0 over vanadia oxide to form adsorbed mercuric oxide (HgO(ad)-V_2O_4), (2) the oxidation of Hg^0 by chemisorbed oxygen on the catalyst surface, and (3) the oxidation of Hg^0 by C=O function group (from CO_2) to form HgO and C–O–C.

Fernández-Miranda et al. [24] studied the Hg^0 oxidation capacity over several SCR catalysts under oxy-fuel combustion atmospheres, and the differences in conventional air coal combustion were compared as well. The results showed that a higher Hg^0 oxidation efficiency was obtained in the oxy-fuel combustion atmospheres, as shown in Figs. 7.8 and 7.9, because in the CO_2-enriched atmosphere, the conversion of NO_x was lower than in the N_2-riched atmosphere. Thus more NO and NO_2 were presented in flue gas, which were free to homogeneously oxidize mercury. The high amount of CO_2 and H_2O in oxy-fuel combustion flue gas decreased the number of active sites for mercury adsorption. Moreover, the effect of the gas composition eclipsed the catalysts different natures.

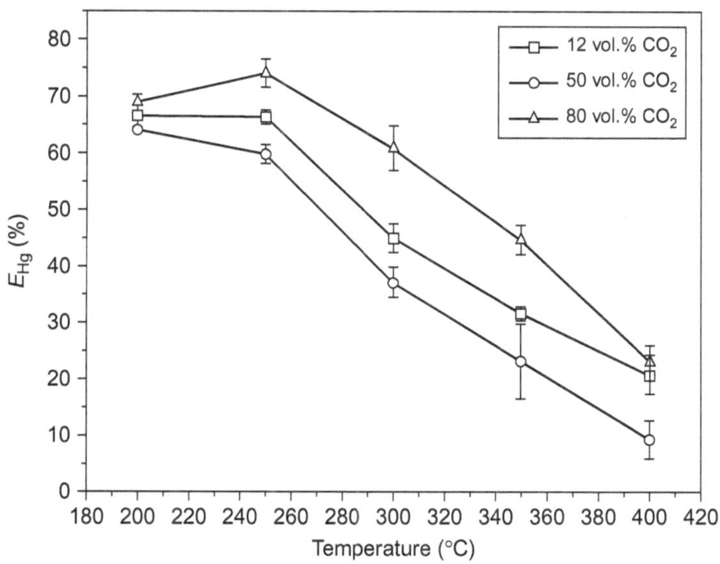

FIG. 7.5 Effects of CO_2 on the Hg^0 removal over V_2O_5-WO_3/TiO_2. Reaction conditions: 12–80 vol% CO_2, 5 vol% O_2, N_2 as balance gas, 35 μg/m^3 Hg^0, GHSV = 75,000 h^{-1}. *(From Wang F, Li G, Shen B, Wang Y, He C. Mercury removal over the vanadia-titania catalyst in CO2-enriched conditions. Chem Eng J 2015;263:356–63.)*

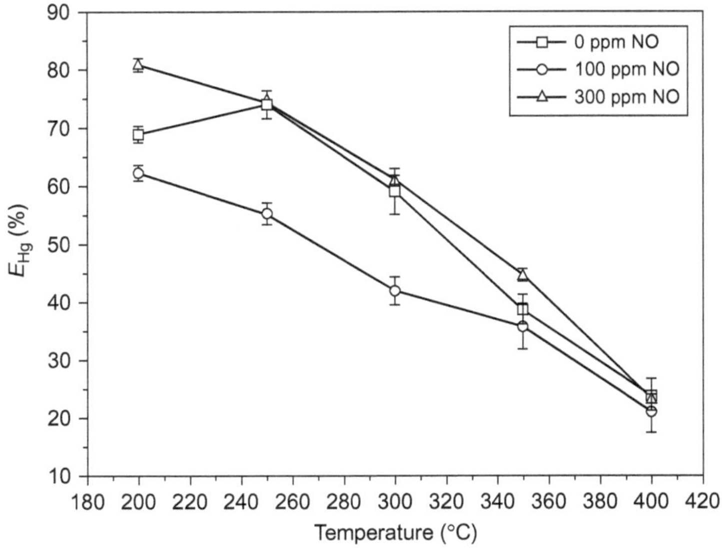

FIG. 7.6 Effects of NO on Hg^0 removal. Reaction conditions: 80% CO_2, 5% O_2, 35 μg/m^3 Hg^0, N_2 as balance gas, GHSV = 75,000 h^{-1}. *(From Wang F, Li G, Shen B, Wang Y, He C. Mercury removal over the vanadia-titania catalyst in CO2-enriched conditions. Chem Eng J 2015;263:356–63.)*

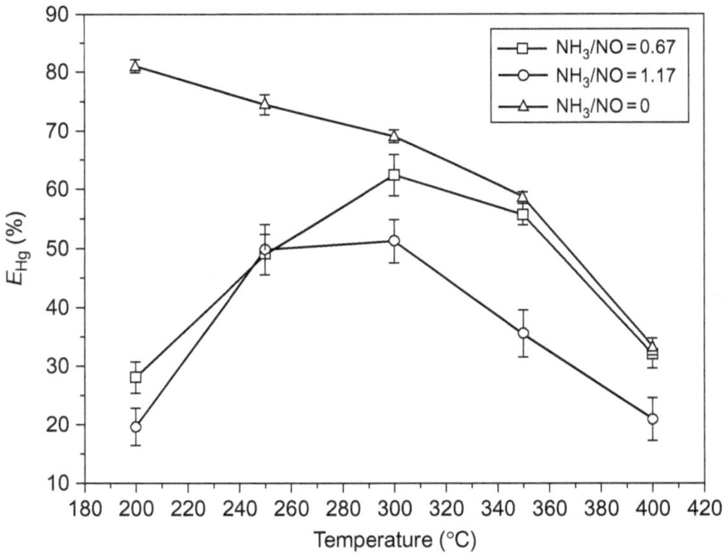

FIG. 7.7 Effects of NH_3 on the Hg^0 removal. Reaction conditions: 80 vol% CO_2, 5 vol% O_2, 300 ppm NO, 0–351 ppm NH_3, N_2 as balance gas, 35 μg/m^3 Hg^0, GHSV = 75,000 h^{-1}. *(From Wang F, Li G, Shen B, Wang Y, He C. Mercury removal over the vanadia-titania catalyst in CO2-enriched conditions. Chem Eng J 2015;263:356–63.)*

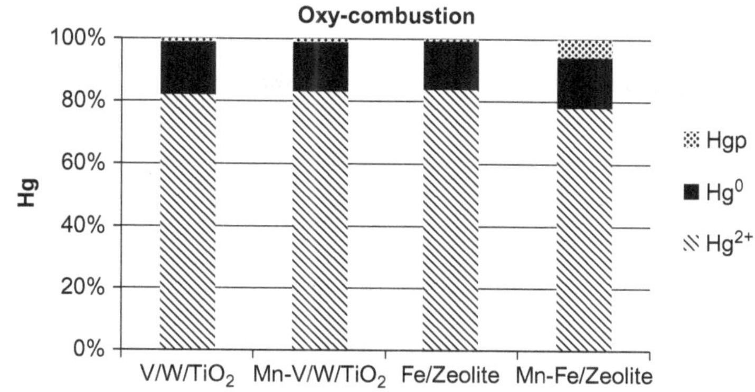

FIG. 7.8 Percentages of mercury retained in the catalysts (Hg^p), oxidized mercury (Hg^{2+}), and elemental mercury (Hg^0) in an oxy-fuel combustion atmosphere. *(From Fernández-Miranda N, Lopez-Anton MA, Díaz-Somoano M, Martínez-Tarazona MR. Mercury oxidation in catalysts used for selective reduction of NOx (SCR) in oxy-fuel combustion. Chem Eng J 2016;285:77–82.)*

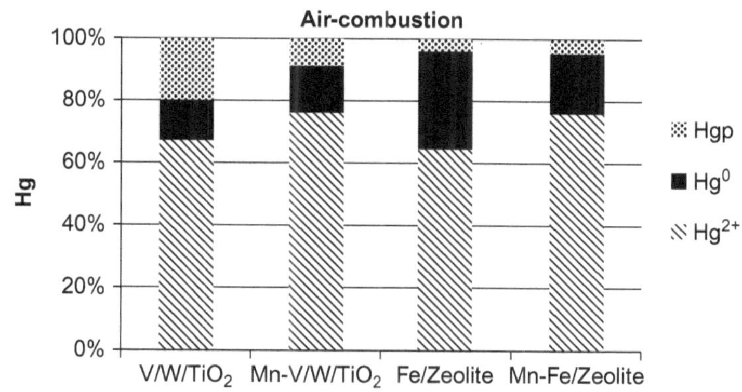

FIG. 7.9 Percentages of mercury retained in the catalysts (Hg^p), oxidized mercury (Hg^{2+}), and elemental mercury (Hg^0) in a coal air combustion atmosphere. *(From Fernández-Miranda N, Lopez-Anton MA, Díaz-Somoano M, Martínez-Tarazona MR. Mercury oxidation in catalysts used for selective reduction of NOx (SCR) in oxy-fuel combustion. Chem Eng J 2016;285:77–82.)*

7.4.2 Mercury Removal by Carbon-Based Sorbents

Different mercury sorbents have been developed for elemental mercury removal, such as activated carbon [25–28], noble metals [29–31], metal oxides [32–38], calcium-based sorbents [39,40], zeolite [41], fly ash [42–44], and so on. Among these sorbents, activated carbon is one of the most studied ones in terms of capturing elemental mercury from flue gas. In oxy-fuel combustion system, the flue gas components are significantly different those in air combustion systems. CO_2 was the main component in oxy-fuel combustion atmospheres, whereas N_2 was the main component in air combustion atmospheres. The concentration of water vapor (H_2O) and SO_2 was significantly higher in oxy-fuel combustion flue gas

[13,14], because the flue gas was recycled to the boiler. Moreover, different proportions of SO_2/SO_3 were present in the flue gas in the two atmospheres [16]. Thus the Hg^0 removal performance of activated carbon in oxy-fuel combustion and air combustion atmospheres may be different. Currently, the removal of mercury over activated carbon under oxy-fuel combustion atmospheres are limited.

Lopez-Anton et al. [45] studied the mercury adsorption capacity of a series of activated carbons obtained from the wastes in leather industries. The mercury removal behavior of these materials was studied under typical oxy-coal combustion atmosphere and $O_2 + N_2$ atmospheres. The activated carbon with the highest surface area and the greatest number of micropores and sulfur and acidic characteristics showed the optimal mercury adsorption capacity. Different mercury adsorption behaviors were observed in $O_2 + N_2$ and oxy-fuel combustion atmospheres, and the mercury adsorption capacity decreased in the presence of acid gases, as shown in Figs. 7.10 and 7.11. A high percentage of mercury oxidation was produced in oxy-fuel combustion atmospheres, which favored the removal of mercury by WFGD units. As a result, the risk of damage to the CO_2 compression and purification units was reduced, or perhaps even removed.

Lopez-Anton et al. [8] studied the impact of oxy-fuel combustion gases on mercury retention in activated carbons in macroalgae wastes. The results showed that mercury retention in overactive carbons was more inhibited in an oxy-fuel combustion atmosphere than in a 12.6% $O_2 + N_2$ atmosphere, as shown in Fig. 7.12. The presence of a high concentration of water vapor may have resulted because of mercury adsorption. However, the presence of reactive gases

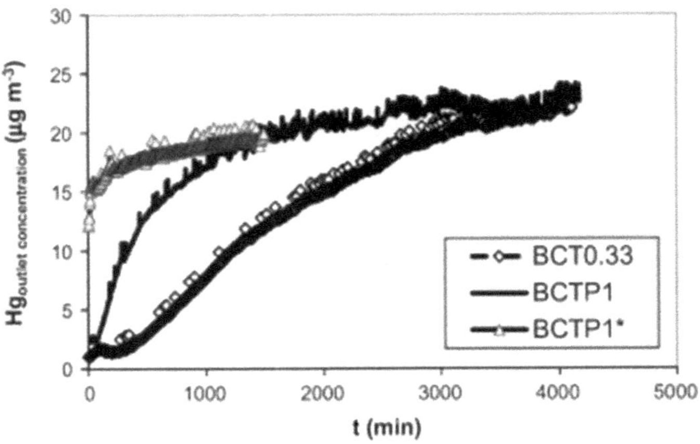

FIG. 7.10 Mercury adsorption curves obtained for the activated carbons in an $O_2 + N_2$ atmosphere. *(From Lopez-Anton MA, Gil RR, Fuente E, Díaz-Somoano M, Martínez-Tarazona MR, Ruiz B. Activated carbons from biocollagenic wastes of the leather industry for mercury capture in oxy-combustion. Fuel 2015;142:227–34.)*

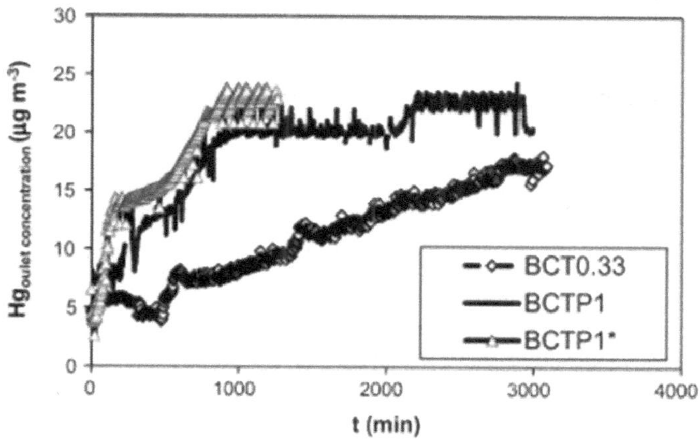

FIG. 7.11 Mercury adsorption curves obtained for the activated carbons in an oxy-fuel combustion atmosphere. *(From Lopez-Anton MA, Gil RR, Fuente E, Díaz-Somoano M, Martínez-Tarazona MR, Ruiz B. Activated carbons from biocollagenic wastes of the leather industry for mercury capture in oxy-combustion. Fuel 2015;142:227–34.)*

such as SO_2/SO_3 and vapor water enhanced Hg^0 oxidation, and the Hg^0 oxidation efficiency achieved 60%, which resulted in enhanced retention of mercury in the flue gas desulfurization (FGD) units.

The mercury removal capacity of commercial activated carbon with/without sulfur in oxy-fuel combustion conditions was also studied by Lopez-Anton et al. [46]. Fig. 7.13 shows that, under CO_2-enriched flue gas atmospheres, Hg^{2+} was the main species that resulted from the interactions between activated carbons and Hg^0. The carbonyl and quinone groups were responsible for mercury oxidation and retention in the carbons. The high retention capacities of overactivated carbon were obtained in an oxy-fuel combustion atmosphere. The high concentration of H_2O in an oxy-fuel combustion atmosphere inhibited mercury retention, possibly because the H_2O competed for the same active sites for mercury adsorption.

7.4.3 Mercury Removal by Metal Oxides [47]

Regarding activated carbon and other conventional sorbents, the common limitations are [48–51] (1) the recovery and recycle of spent sorbent, (2) the disposition of the mercury in the spent sorbent, and (3) the negative effect on the utilization of fly ash. Therefore developing recoverable and regenerable sorbents/catalysts to overcome these limitations is highly attractive. Yang et al. [47] developed a novel regenerable cobalt oxide loaded magnetosphere catalyst from fly ash. Magnetospheres were first separated from fly ash, and then the Co-MF catalyst (Co-MF) was synthesized by loading cobalt oxide. The mercury removal performances were studied under oxy-fuel combustion atmospheres.

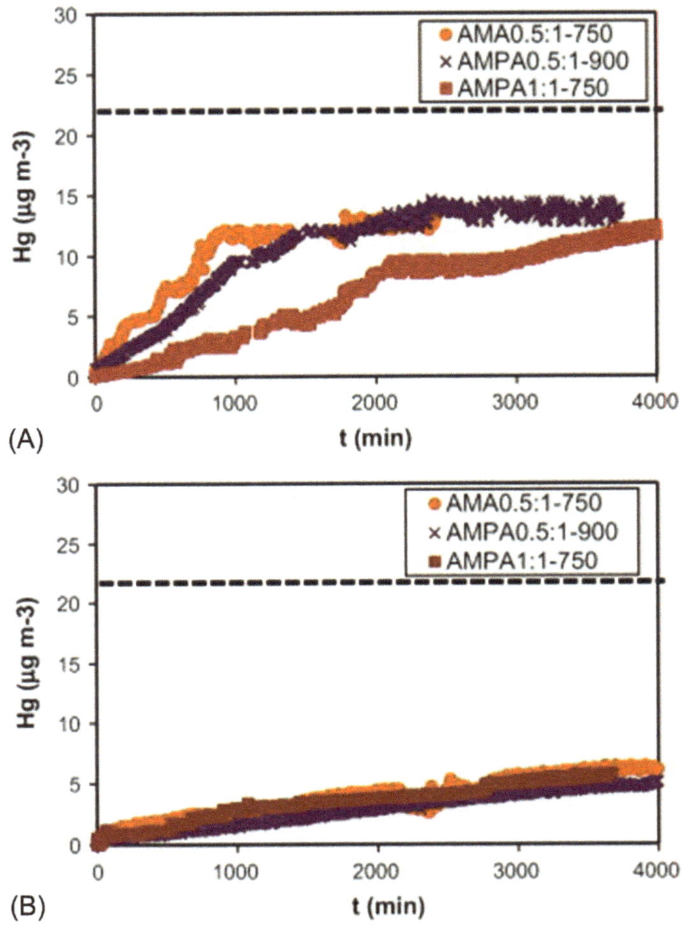

FIG. 7.12 Mercury adsorption curves corresponding to the activated carbons in (A) oxy-fuel combustion and (B) oxy-fuel combustion atmospheres free of water vapor. *(From Lopez-Anton MA, Ferrera-Lorenzo N, Fuente E, Díaz-Somoano M, Suarez-Ruíz I, Martínez-Tarazona MR, et al. Impact of oxy-fuel combustion gases on mercury retention in activated carbons from a macroalgae waste: effect of water. Chemosphere 2015;125:191–7.)*

The Hg^0 removal performance of the catalyst was preliminarily evaluated in an enriched CO_2 (70 vol%) atmosphere. After that, the Hg^0 removal performance was investigated in the simulated flue gas (SFG) atmospheres, which contained acid flue gas (SO_2, NO, and HCl) and H_2O. As shown in Fig. 7.14, relatively high Hg^0 removal efficiency was observed in the oxy-fuel combustion atmosphere (70% CO_2) compared to that in the air combustion atmosphere (12% CO_2). Within the reaction process, similar Hg^0 adsorption behaviors were observed in the air and oxy-fuel combustion atmospheres, while higher Hg^0 oxidation efficiency was attained in oxy-fuel combustion atmospheres. Furthermore, similar Hg^0 removal efficiency was attained in the atmospheres without CO_2

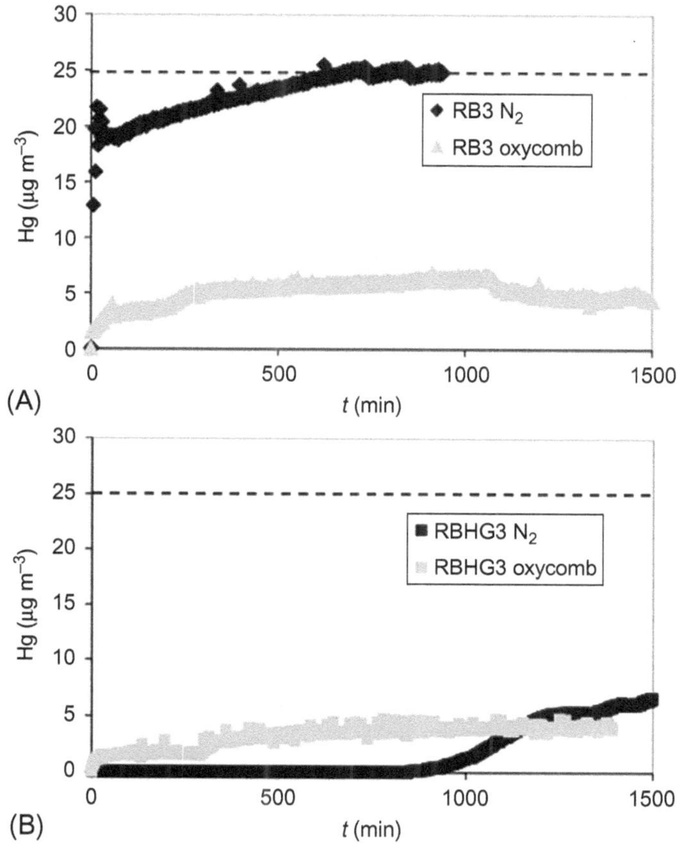

FIG. 7.13 Mercury adsorption curves obtained for (A) RB3 and (B) RBHG3-activated carbons in N_2 and oxy-fuel combustion atmospheres. *(From Lopez-Anton MA, Rumayor M, Díaz-Somoano M, Martínez-Tarazona MR. Influence of a CO2-enriched flue gas on mercury capture by activated carbons. Chem Eng J 2015;262:1237–43.)*

and with 12% CO_2, which implied that the low concentration CO_2 has little impact on Hg^0 removal. However, the enriched CO_2 in oxy-fuel combustion systems assisted the removal of mercury, which was probably due to the interactions between the enriched CO_2 and the Co-MF catalyst.

The Hg^0 removal performances of the catalyst under dry SFG atmosphere are shown in Fig. 7.15. The total mercury efficiency η_{T-i} was primarily determined by the adsorption efficiency η_{ads-i} at the first reaction stage, after which the adsorption efficiency η_{ads-i} gradually decreased, and the oxidation efficiency η_{oxi-i} increased with the reaction process. A relatively higher total Hg^0 removal efficiency was observed in oxy-fuel combustion conditions than in air combustion conditions. There was no obvious variation of adsorption efficiency in the two atmospheres. However, higher oxidation efficiency was observed in oxy-fuel combustion conditions.

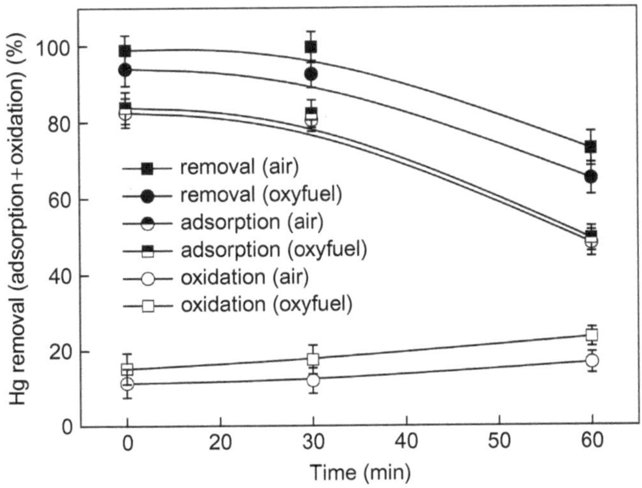

FIG. 7.14 The instantaneous Hg^0 adsorption and oxidation efficiency in air and oxy-fuel combustion atmospheres (12% and 70% CO_2, 4% O_2, balanced with N_2). *(From Yang J, Zhao Y, Chang L, Zhang J, Zheng C. Mercury adsorption and oxidation over cobalt oxide loaded magnetospheres catalyst from fly ash in oxyfuel combustion flue gas. Environ Sci Technol 2015;49:8210–8.)*

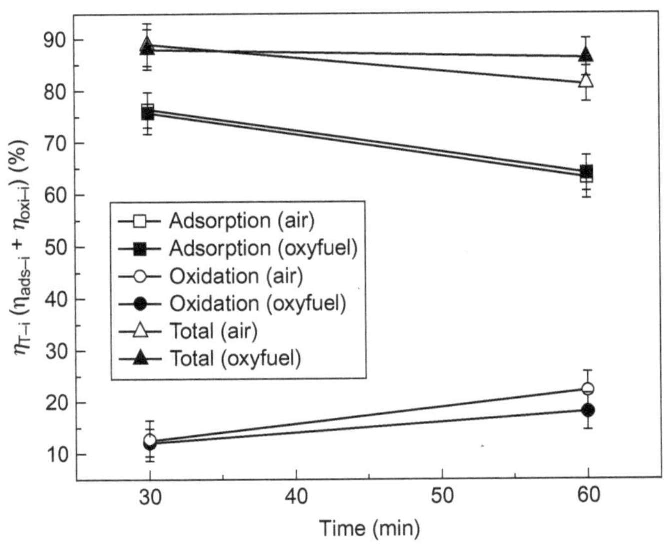

FIG. 7.15 The Hg^0 adsorption and oxidation efficiency in air and oxy-fuel SFG atmospheres. *(From Yang J, Zhao Y, Chang L, Zhang J, Zheng C. Mercury adsorption and oxidation over cobalt oxide loaded magnetospheres catalyst from fly ash in oxyfuel combustion flue gas. Environ Sci Technol 2015;49:8210–8.)*

With the addition of H_2O into the dry SFG atmosphere, serious decreases in the total mercury removal efficiency and in the adsorption efficiency were observed (Fig. 7.16), which were mainly due to the competitive adsorption with Hg^0 on the same adsorption sites. However, the Hg^0 oxidation efficiency increased from 30.9% to 35.2% when 10% H_2O was introduced into the DSFG atmosphere, while the Hg^0 oxidation efficiency further increased to 40.8% when the H_2O concentration increased to 20%. Though the presence of a high concentration of water in the flue gas seriously impeded the mercury adsorption, the greater amount of Hg^{2+} in the flue gas was favorable to the reduction of mercury emissions, since Hg^{2+} could be captured in FGD system.

Repeated mercury capture-regeneration cycles were conducted to evaluate recyclability of Co-MF catalyst under oxy-fuel combustion atmospheres. The spent Co-MF catalyst was regenerated at 400°C in air for 0.5 h, at which time the mercury adsorbed by the catalyst was almost completely released. The accumulated total Hg^0 removal efficiency η_{T-a} was selected to evaluate the regeneration performance when the catalyst attained 20% and 50% breakthrough, respectively. As shown in Fig. 7.17, the Hg^0 removal efficiency was maintained along multiple capture-regeneration cycles under oxy-fuel combustion conditions, which was similar to that in air combustion conditions [47]. This implied that after multiple capture-regeneration cycles, the catalyst did not undergo

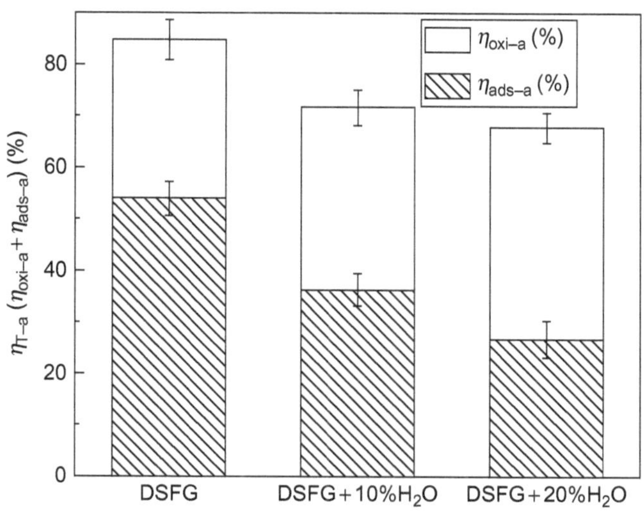

FIG. 7.16 The accumulation Hg^0 adsorption and oxidation efficiency in air and oxy-fuel combustion atmospheres (12% and 70% CO_2, 4% O_2, balanced with N_2.) (DSFG: simulated flue gas, N_2 + 4% O_2 + 70% CO_2 + 300 ppm NO + 2000 ppm SO_2 + 10 ppm HCl; the test time was 60 min.) *(From Yang J, Zhao Y, Chang L, Zhang J, Zheng C. Mercury adsorption and oxidation over cobalt oxide loaded magnetospheres catalyst from fly ash in oxyfuel combustion flue gas. Environ Sci Technol 2015;49:8210–8.)*

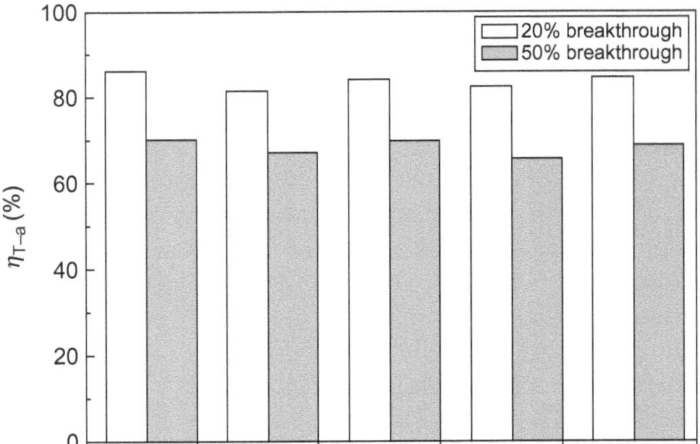

	Cycle 1	Cycle 2	Cycle 3	Cycle 4	Cycle 5
mercury capacity in 1.5 h (mg Hg/g catalyst)	0.054	0.048	0.051	0.049	0.053

FIG. 7.17 Hg^0 accumulation removal efficiency of Co-MF catalyst along repeated capture-regeneration cycles under an oxy-fuel combustion DSFG atmosphere. *(From Yang J, Zhao Y, Chang L, Zhang J, Zheng C. Mercury adsorption and oxidation over cobalt oxide loaded magnetospheres catalyst from fly ash in oxyfuel combustion flue gas. Environ Sci Technol 2015;49:8210–8.)*

significant changes. Moreover, the regeneration time was much shorter compared to the adsorption time, which is beneficial for industrial uses. Although the mercury capacity (about 0.05 mg Hg/g catalyst in 1.5 h, when the catalyst attained 50% breakthrough) of this material is lower than that of some competing technologies, such as S-AC (0.85 mg Hg/g sorbent) [52], after the removal of the mercury, the spent catalyst could be separated from fly ash and recycled after regeneration. As byproducts of coal combustion, magnetosphere catalysts have potentially huge cost-effective applications in Hg^0 removal.

7.5 EFFECT OF HG ON OXY-FUEL CO_2 COMPRESSION

Aluminum is widely used in machinery because of its heat transfer properties, high strength-to-weight ratio, and ease of fabrication. In oxy-fuel combustion power plants, aluminum alloy materials are commonly used in CO_2 purification and compression units. Because of the recycled flue gas, oxy-fuel combustion systems produce higher levels of impurity. The Hg concentration in oxy-fuel combustion systems can be up to four times higher than that in air combustion systems [53]. The higher Hg concentration is a significant concern because Hg^0 in oxy-fuel flue gas can attack aluminum heat exchangers in CO_2 purification and compression units [9].

Deposition of liquid elemental mercury in aluminum heat exchangers could compromise their structural integrity. The damage of aluminum material by mercury could be attributed to the LME and AMC [2]. LME is generally much more serious than other embrittlement processes, such as hydrogen embrittlement or stress-corrosion cracking. Once the cracks of LTE initiate, a very rapid subcritical cracking can occur, even at low stresses [54]. AMC of aluminum is similar to LME, where the oxide on the materials surface must be breached to allow physical contact between mercury and aluminum. Contact of mercury with any aluminum alloy produces amalgamation. AMC occurs spontaneously if water vapor or liquid water contacts the amalgam. The rate of AMC penetration is proportional to the humidity. The reaction scheme of AMC could be interpreted as follows:

$$Hg + Al \rightarrow Hg(Al) \text{amalgamation}$$

$$Hg(Al) + H_2O \rightarrow Al_2O_3 \cdot 3H_2O + H_2 + Hg$$

$$Hg + Al \rightarrow Hg(Al)$$

The preceding reaction scheme shows that the AMC process does not consume the mercury. As a result, the AMC process is self-propagating as long as mercury is in contact with aluminum metal and water is available. If there is sufficient moisture and mercury in the atmosphere, the aluminum structure can be penetrated rapidly. The rate of the AMC attack is limited by mass transfer. The aluminum oxide corrosion products eventually inhibit the access of moisture to the amalgam surface. The AMC reaction is not alloy-specific. All aluminum alloys can be affected.

7.6 SUMMARY

Thermodynamic modeling showed that Hg^0 is the primary species over a wide range of temperatures (800–1400°C) in oxy-fuel combustion, with a small portion of HgO (g) at low temperature. The increase of chlorine content in a combustion atmosphere, particularly at low temperature, could facilitate the formation of oxidized mercury (Hg^{2+}). A combined homogeneous-heterogeneous model showed that the mercury retention on fly ash particles depended on temperature, burnout, and particle sizes. The field measurement showed that a higher concentration of gas-phase mercury (Hg^0) is contained in oxy-fuel combustion flue gas. The conversion of Hg^0 to particulate-associated mercury (Hg^p) is enhanced in oxy-fuel combustion. Different sorbents were used to capture mercury in oxy-fuel flue gas atmospheres. The high concentration of SO_2/SO_3 and H_2O in oxy-fuel combustion atmosphere could significantly affect the Hg^0 adsorption and oxidation behaviors. The presence of Hg in flue gas will attack the aluminum materials in the CO_2 purification and compression units. Thus more attention should be paid to the removal of Hg from flue gas.

REFERENCES

[1] Nelson PF. Impacts of trace components on oxy-combustion for the callide oxy-fuel project-literature review. Sydney: Graduate School of the Environment, Macquarie University; 2013.

[2] Bessone JB. The activation of aluminium by mercury ions in non-aggressive media. Corros Sci 2006;48:4243–56.

[3] UNEP. Global mercury assessment 2013: sources, emissions, releases and environmental transport. Geneva: UNEP Chemicals Branch; 2013.

[4] Pavlish JH, Sondreal EA, Mann MD, Olson ES, Galbreath KC, Laudal DL, et al. Status review of mercury control options for coal-fired power plants. Fuel Process Technol 2003;82:89–165.

[5] Roy B, Choo WL, Bhattacharya S. Prediction of distribution of trace elements under oxy-fuel combustion condition using Victorian brown coals. Fuel 2013;114:135–42.

[6] Córdoba P, Maroto-Valer M, Delgado MA, Diego R, Font O, Querol X. Speciation, behaviour, and fate of mercury under oxy-fuel combustion conditions. Environ Res 2016;145:154–61.

[7] Smith CA, Krishnakumar B, Helble JJ. Homo- and heterogeneous mercury oxidation in a bench-scale flame-based flow reactor. Energy Fuel 2011;25:4367–76.

[8] Lopez-Anton MA, Ferrera-Lorenzo N, Fuente E, Díaz-Somoano M, Suarez-Ruíz I, Martínez-Tarazona MR, et al. Impact of oxy-fuel combustion gases on mercury retention in activated carbons from a macroalgae waste: effect of water. Chemosphere 2015;125:191–7.

[9] Ting T, Stanger R, Wall T. Oxyfuel CO_2 compression: the gas phase reaction of elemental mercury and NO_x at high pressure and absorption into nitric acid. Int J Greenhouse Gas Control 2014;29:125–34.

[10] Liu J, Qu W, Yuan J, Wang S, Qiu J, Zheng C. Theoretical studies of properties and reactions involving mercury species present in combustion fluegases. Energy Fuel 2010;24:117–22.

[11] Wilhelm SM. Risk analysis for operation of aluminum heat exchangers contaminated by mercury. Process Saf Prog 2009;28(3):259–66.

[12] Gharebaghi M, Hughes KJ, Porter RTJ, Pourkashanian M, Williams A. Mercury speciation in air-coal and oxy-coal combustion: a modelling approach. Proc Combust Inst 2011;33:1779–86.

[13] Mitsui Y, Imada N, Kikkawa H, Katagawa A. Study of Hg and SO_3 behavior in flue gas of oxy-fuel combustion system. Int J Greenhouse Gas Control 2011;5:143–50.

[14] Croiset E, Thambimuthu KV. NO_x and SO_2 emission from O_2/CO_2 recycled coal combustion. Fuel 2001;80:2117–21.

[15] Fleig D, Normann F, Andersson K, Johnsson F, Leckner B. The fate of sulphur during oxy-fuel combustion of lignite. Energy Procedia 2009;1:383–90.

[16] Stanger R, Wall T. Sulphur impacts during pulverised coal combustion in oxy-fuel technology for carbon capture and storage. Prog Energy Combust Sci 2011;37:69–88.

[17] Font O, Córdoba P, Leiva C, Romeo LM, Bolea I, Guedea I, et al. Fate and abatement of mercury and other trace elements in a coal fluidised bed oxy combustion pilot plant. Fuel 2012;95:272–81.

[18] Roy B, Chen L, Bhattacharya S. Nitrogen oxides, sulfur trioxide, and mercury emissions during oxy-fuel fluidized bed combustion of Victorian brown coal. Environ Sci Technol 2014;48:14844–50.

[19] Zhuang Y, Pavlish JH. Fate of hazardous air pollutants in oxygen-fired coal combustion with different flue gas recycling. Environ Sci Technol 2012;46:4657–65.

[20] de las Obras-Loscertales M, Izquierdo MT, Rufas A, de Diego LF, García-Labiano F, Abad A, et al. The fate of mercury in fluidized beds under oxy-fuel combustion conditions. Fuel 2016;167:75–81.

[21] Li HL, Li Y, Wu C-Y, Zhang JY. Oxidation and capture of elemental mercury over SiO_2-TiO_2-V_2O_5 catalysts in simulated lowrank coal combustion flue gas. Chem Eng J 2011;169:186–93.

[22] Zheng Y, Jensen AD, Windelin C, Jensen F. Review of technologies for mercury removal from flue gas from cement production processes. Prog Energy Combust Sci 2012;38:599–629.

[23] Wang F, Li G, Shen B, Wang Y, He C. Mercury removal over the vanadia–titania catalyst in CO_2-enriched conditions. Chem Eng J 2015;263:356–63.

[24] Fernández-Miranda N, Lopez-Anton MA, Díaz-Somoano M, Martínez-Tarazona MR. Mercury oxidation in catalysts used for selective reduction of NO_x (SCR) in oxy-fuel combustion. Chem Eng J 2016;285:77–82.

[25] O' Dowd WJ, Pennline HW, Freeman MC, Granite EJ, Hargis RA, Lacher CJ, et al. A technique to control mercury from flue gas: the thief process. Fuel Process Technol 2006;87:1071–84.

[26] Granite EJ, Freeman MC, Hargis RA, O'Dowd WJ, Pennline HW. The thief process for mercury removal from flue gas. J Environ Manag 2007;84:628–34.

[27] Lee SH, Rhim YJ, Cho SP, Baek JI. Carbon-based novel sorbent for removing gas-phase mercury. Fuel 2006;85:219–26.

[28] Hsi HC, Chen CT. Influences of acidic/oxidizing gases on elemental mercury adsorption equilibrium and kinetics of sulfur-impregnated activated carbon. Fuel 2012;98:229–35.

[29] Xie Y, Yan B, Tian C, Liu Y, Liu Q, Zeng H. Efficient removal of elemental mercury (Hg^0) by SBA-15-ag adsorbents. J Mater Chem A 2014;2:17730–4.

[30] Izquierdo MT, Ballestero D, Juan R, Garcia-Diez E, Rubio B, Ruiz C, et al. Tail-end Hg capture on au/carbon-monolith regenerable sorbents. J Hazard Mater 2011;193:304–10.

[31] Ballestero D, Gomez-Gimenez C, Garcia-Diez E, Juan R, Rubio B, Izquierdo MT. Influence of temperature and regeneration cycles on hg capture and efficiency by structured au/C regenerable sorbents. J Hazard Mater 2013;260:247–54.

[32] Wang Y, Shen B, He C, Yue S, Wang F. Simultaneous removal of NO and Hg^0 from flue gas over Mn-Ce/Ti-PILCs. Environ Sci Technol 2015;49:9355–63.

[33] Spörl R, Belo L, Shah K, Stanger R, Giniyatullin R, Maier J, et al. Mercury emissions and removal by ash in coal-fired oxy-fuel combustion. Energy Fuel 2014;28:123–35.

[34] Xu H, Qu Z, Zong C, Huang W, Quan F, Yan N. MnO_x/graphene for the catalytic oxidation and adsorption of elemental mercury. Environ Sci Technol 2015;49:6823–30.

[35] Liao Y, Xiong S, Dang H, Xiao X, Yang S, Wong PK. The centralized control of elemental mercury emission from the flue gas by a magnetic rengenerable Fe-Ti-Mn spinel. J Hazard Mater 2015;299:740–6.

[36] Li H, Wu S, Li L, Wang J, Ma W, Shih K. CuO–CeO_2/TiO_2catalyst for simultaneous NO reduction and Hg^0 oxidation at low temperatures. Catal Sci Technol 2015;5:5129–38.

[37] Xie J, Qu Z, Yan N, Yang S, Chen W, Hu L, et al. Novel regenerable sorbent based on Zr-Mn binary metal oxides for flue gas mercury retention and recovery. J Hazard Mater 2013;261:206–13.

[38] Li H, Wu S, Wu CY, Wang J, Li L, Shih K. SCR atmosphere induced reduction of oxidized mercury over CuO-CeO_2/TiO_2 catalyst. Environ Sci Technol 2015;49:7373–9.

[39] Wang Y, Duan Y. Effect of manganese ions on the structure of ca $(OH)_2$ and mercury adsorption performance of Mn^{x+}/ca$(OH)_2$ composites. Energy Fuel 2011;25:1553–8.

[40] Guo X, Zhao P, Zheng C. Theoretical study of different speciation of mercury adsorption on CaO (001) surface. Proc Combust Inst 2009;32:2693–9.

[41] Dong J, Xu Z, Kuznicki SM. Mercury removal from flue gases by novel regenerable magnetic nanocomposite sorbents. Environ Sci Technol 2009;43:3266–71.

[42] López-Antón MA, Díaz-Somoano M, Martínez-Tarazona MR. Mercury retention by fly ashes from coal combustion: influence of the unburned carbon content. Ind Eng Chem Res 2007;46:927–31.

[43] Zhao YC, Zhang JY, Liu J, Díaz-Somoano M, Abad-Valle P, Martínez-Tarazona MR, et al. Experimental study on fly ash capture mercury in flue gas [J]. Sci China Technol Sci 2010;53:976–83.
[44] Wang Y, Duan Y, Yang L, Zhao C, Shen X, Zhang M, et al. Experimental study on mercury transformation and removal in coal-fired boiler flue gases. Fuel Process Technol 2009;90:643–51.
[45] Lopez-Anton MA, Gil RR, Fuente E, Díaz-Somoano M, Martínez-Tarazona MR, Ruiz B. Activated carbons from biocollagenic wastes of the leather industry for mercury capture in oxy-combustion. Fuel 2015;142:227–34.
[46] Lopez-Anton MA, Rumayor M, Díaz-Somoano M, Martínez-Tarazona MR. Influence of a CO_2-enriched flue gas on mercury capture by activated carbons. Chem Eng J 2015;262:1237–43.
[47] Yang J, Zhao Y, Chang L, Zhang J, Zheng C. Mercury adsorption and oxidation over cobalt oxide loaded magnetospheres catalyst from fly ash in oxyfuel combustion flue gas. Environ Sci Technol 2015;49:8210–8.
[48] Dong J, Xu Z, Kuznicki SM. Magnetic multi-functional Nano composites for environmental applications. Adv Funct Mater 2009;19:1268–75.
[49] Yang S, et al. Capture of gaseous elemental mercury from flue gas using a magnetic and sulfur poisoning resistant sorbent Mn/gamma-Fe_2O_3 at lower temperatures. J Hazard Mater 2011;186:508–15.
[50] Yang S, et al. Nanosized cation-deficient Fe-Ti spinel: a novel magnetic sorbent for elemental mercury capture from flue gas. ACS Appl Mater Interfaces 2011;3:209–17.
[51] Yang S, Guo Y, Yan N, Wu D, He H, Xie J, et al. Remarkable effect of the incorporation of titanium on the catalytic activity and SO_2 poisoning resistance of magnetic Mn–Fe spinel for elemental mercury capture. Appl Catal, B 2011;101:698–708.
[52] His HC, Rood MJ, Rostam-Abadi M, Chen S, Chang R. Effects of sulfur impregnation temperature on the properties and mercury adsorption capacities of activated carbon fibers (ACFs). Environ Sci Technol 2001;35:2785–91.
[53] Wall T, Stanger R, Liu Y. Gas cleaning challenges for coal-fired oxy-fuel technology with carbon capture and storage. Fuel 2013;108:85–90.
[54] Coade R, Coldham D. The interaction of mercury and aluminium in heat exchangers in a natural gas plants. Int J Press Vessel Pip 2006;83(5):336–42.

Chapter 8

Flame Characteristics of Oxy-fuel Combustion and Burner Design

Junjun Guo, Zhaohui Liu
Huazhong University of Science and Technology, Wuhan, China

8.1 INTRODUCTION

Oxy-fuel flame characteristics in ignition, flame stability, flame shape, etc. are different from those in conventional air combustion because of the changes in the oxidants composition, which mainly consists of CO_2, H_2O, and O_2. Compared to a constant initial oxygen concentration in air combustion, the initial oxygen concentration can be controlled by the flue gas recycle ratio in oxy-fuel combustion, which makes the flame adjustment more flexible. The high CO_2 concentrations significantly effect heat capacity and gas transport properties, which strongly influences the combustion process. In addition, the oxygen injection method offers different options, such as premixed oxygen and recycled flue gas and pure oxygen injected directly into the furnace.

The burner is the key to achieving high combustion efficiency in industrial furnaces. The role of the burner is to transport the fuel and oxidants through the burners nozzles into the furnace with a certain proportion and velocity, which ensures that the fuel is fully mixed with the oxidant in the furnace and that stable combustion is maintained. At the present time, oxy-fuel burners are widely used in high-temperature industrial furnaces but are rarely used in utility coal-fired boilers. Consequently, there is no established or widely accepted design methodology for the use of oxy-fuel burners in coal-fired boilers.

In this chapter, we first review the design principles of conventional air combustion burners and then illustrate the flame features of oxy-fuel combustion. The influence of the oxy-fuel combustion conditions on coal ignition and flame stability are discussed. The design principles and methodology of oxy-coal burners intended for application in utility boilers are proposed, relative to traditional air combustion burners. For example, we analyze the new application of a 3 MW oxy-coal burner.

Oxy-fuel Combustion. http://dx.doi.org/10.1016/B978-0-12-812145-0.00008-6

8.2 DESIGN PRINCIPLES OF CONVENTIONAL PULVERIZED COAL BURNER

The combustion of pulverized coal in furnaces is closely related to the structure and arrangement of the burner. A good burner design should ensure the independent stability and efficiency of fuel combustion. A detailed discussion about the use of coal-fired burners in utility boilers is given in this section based on existing research [1,2]. In this section, we summarize the combustion modes of pulverized coal-fired boilers and the combustion characteristics of the swirl burner; then we discuss the design principles of conventional swirl burners.

8.2.1 Combustion Modes

Fig. 8.1 illustrates the combustion modes in conventional pulverized coal boilers. These modes are tangential combustion, swirling combustion (front wall-fired or opposing wall-fired), and down-fired combustion.

In tangential combustion mode, a good mixing of the oxidant and pulverized coal occurs. A highly flame-filled furnace and a uniform transfer of heat on the heating surface are achieved. The adjacent flames can ignite each other, and stable combustion and fuel adaptability are ensured. If a swing burner is used, the height of the flame in the furnace can be adjusted, but unavoidable residual swirling occurs at the furnace exit, so the flue gas temperature deviation in the superheater zone is the essential feature in tangential combustion.

The advantage of the swirling combustion mode is that the distribution of the temperature and velocity in the furnaces width direction are relatively uniform and that deviations in the flue gas temperature in the superheaters zone can be ignored. Boiler capacity can be increased by increasing the number of

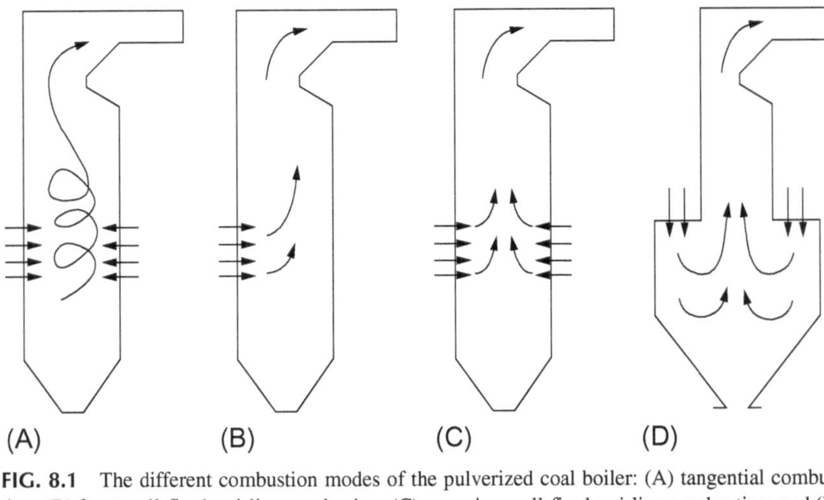

FIG. 8.1 The different combustion modes of the pulverized coal boiler: (A) tangential combustion, (B) front wall-fired swirling combusion, (C) opposing wall-fired swirling combustion, and (D) down-fired combustion.

burners along the furnaces width direction, which makes it easy to scale-up the design of the boiler. In addition, each swirl burner can be operated independently, which helps to adjust the boiler load.

A down-fired combustion boiler is suitable for burning low-volatile anthracite and lean coal. The burners are arranged in the furnaces arch. Under the furnaces arch, a carborundum is attached to the membrane wall, which helps in igniting the coal. The residence time of pulverized coal in this combustion mode greatly increases.

8.2.2 Swirl Burner Aerodynamics

Much research has been done in order to characterize the aerodynamics of swirl flow. The research shows that secondary stream is the swirling flow in a swirl burner, and the primary stream may be a swirling flow or jet flow. After passing the swirler, the stream flow swirls forward. The swirl flow is sprayed into the furnace through the burners nozzles and expands freely to form a rotating expanding airflow. An internal recirculation zone forms at the center of the burner outlet. Then the high temperature flue gas is recirculated back to heat the fuel, generally using the dimensionless swirling number N to characterize the swirl intensity. The swirl number is the ratio of the flux of angular momentum to the axial flux of axial momentum, which defined as

$$G_\phi = \rho Q \omega_q r \tag{8.1}$$

$$G_x = \rho Q \omega_x \tag{8.2}$$

$$N = G_\phi \Big/ G_x r_0 \tag{8.3}$$

where ρ is the gas density (kg/m^3), Q is the gas volume flow (m^3/s), ω_q and ω_x is the tangential velocity and axial velocity (m/s), respectively. r is gas rotation radius (m). r_0 is the characteristic scale (m), which set as the radius of the burner outlet.

The shape and structure of the expanded swirl flow depends on the swirl intensity. As shown in Fig. 8.2, there are three types of expanded swirl flow. As shown in Fig. 8.2A, a small swirl number causes a small internal recirculation zone. The swirl flow is compressed to form a closed swirl flow. The internal recirculation zone is small and wrapped by the main jet, which is characteristic of the closed swirl flow. Fig. 8.2B shows the opening swirl flow. With the increase of the swirl number, the pressure difference between the internal and external recirculation zones drops, and the volume of the internal recirculation zone increases. The swirl flow is no longer closed and forms an open structure. As shown in Fig. 8.2C, as the swirl number increases, the swirl flow expansion angle increases, and the entrainment effect is very strong in the external recirculation zone. The negative pressure in the external recirculation zone is greater than that in the internal recirculation zone. A fully expanded swirl flow is formed.

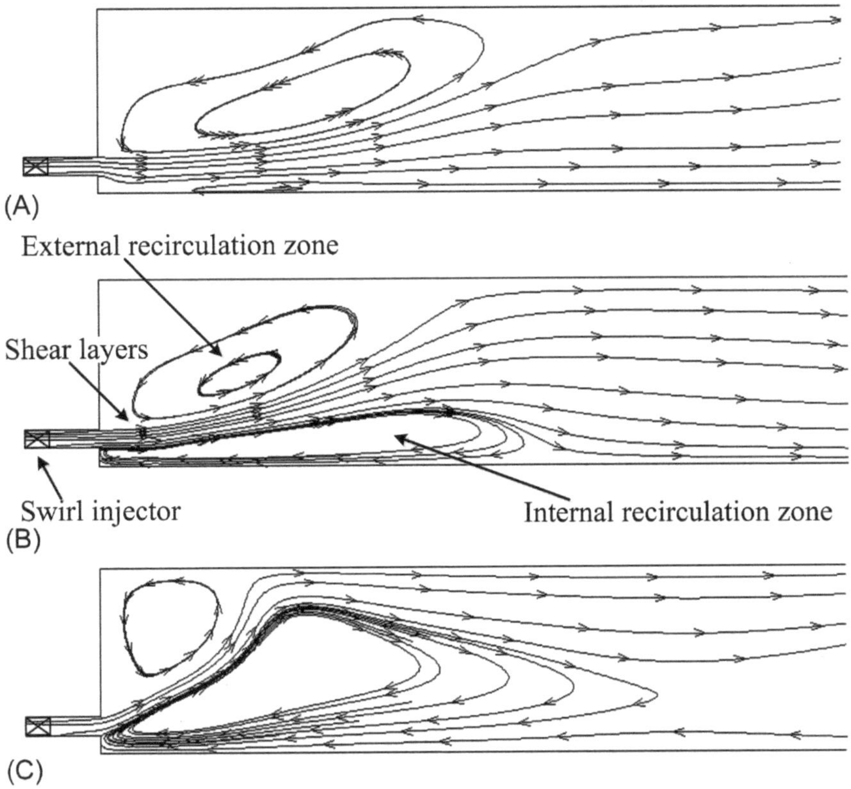

FIG. 8.2 Flow structures of a typical swirl flow: (A) closed swirl flow with low swirl number, (B) opened swirl flow with medium swirl number, and (C) expanded swirl flow with high swirl number.

The internal recirculation zone has an important effect on coal particle ignition and flame stability. An opening swirl flow is expected to ensure stable combustion. In a closed swirl flow, the internal recirculation zone is too small for coal particle ignition and stable combustion. In an expansion swirl flow, the internal recirculation zone is large enough, but the recirculated flow velocity is too small. The flame closes to the sidewall, which increases the risk of slagging.

8.2.3 Design Parameter of Swirl Burner Characteristics

For both swirl burners and jet flow burners, the thermal input of a single burner is limited by the following factors: (1) Excessive burner thermal power causes a center combustion temperature that is too high, which increases the NO_x formation. (2) The low-power burner helps to prevent slagging. (3) If the burner power is too large, a large volume of primary and secondary stream is created, which is not conducive to the mixing of oxidant and pulverized coal particle. (4) Multiple burners operate more flexibly during low loads and boiler startup and shutdown.

Table 8.1 shows the recommended values of the pulverized coal burner numbers and thermal power used in existing boilers.

We should first determine the amount of air passing through the various parts of the burner. To ensure combustion efficiency and boiler economy, the design must be strictly controlled by the equivalent ratio α. The equivalent ratio is determined according to the fuel properties and the milling system. For bituminous and lignite coal, the α is generally set at 0.83 to 0.87; for lean coal and anthracite, the α is generally set at 0.80 to 0.83.

For a swirl burner, the oxidant is injected into the furnace via primary stream nozzle, secondary stream nozzle, and tertiary stream nozzle (over-fire air). Generally, the primary stream rate r_1, secondary stream rate r_2, and tertiary stream rate r_3 are used to represent the oxidants distribution. The choice of the primary stream rate is mainly based on the coals volatility, the ignition conditions, and the milling system. In general, the velocity of primary stream has a minimum limit, due to the requirements of transporting pulverized coal, which is not < 17 m/s in industrial uses. The second stream ensures the amount of oxygen. If the amount of oxygen is insufficient, coal ignition delay is aggravated, which affects the burnout rate. If the primary stream velocity is too great, the strong flow rigidity will pass through the internal recirculation zone; then the burnout ratio will decrease due to the insufficient pulverized coal residence time. In addition, too large primary stream velocity influences the flame stability. Table 8.2 provides a reference for the selection of the primary stream ratio. The choice of secondary stream velocity ensures that the secondary stream can

TABLE 8.1 Recommended Values of the Number and Thermal Power of the Pulverized Coal Swirl Burner

		Swirl Burner (Front Wall-Fired, Side Wall-Fired, Opposing Wall-Fired)	
Boiler Thermal Input (MW)	Boiler Capacity (t/h)	Burner Number	Thermal Input Per Burner (MW)
12	66	2–3	18.6–37.2
25	120,130	4–8	18.6–28
50	220(230)	6	30
100–125	410(400)	8–12	23.2–41
200	670	18–24	23.2–41
300	1000	24–30	23.2–62
600	2000	–	–

TABLE 8.2 Selection Principles for the Primary Stream Ratio

	Anthracite	Lean Coal	Lignite	Bituminous Coal: Volatile <30%	Bituminous Coal: Volatile >30%
Exhaust gas[a]	–	–	0.20–0.45	0.20–0.30	0.25–0.35
Hot stream[b]	0.20–0.25	0.20–0.30	–	0.25–0.40	–

[a]*Transport the pulverized coal by exhaust gas.*
[b]*Transport the pulverized coal by hot stream.*

pass through the flame and quickly and evenly mix with pulverized coal. The principles for the selection of the primary and secondary stream velocities for a swirl burner are shown in Table 8.3.

The higher the secondary stream temperature, the better the pulverized coal combustion efficiency, but the higher the cost of secondary stream heating. So the choice of the secondary stream temperature is a trade-off between combustion efficiency and economy. For secondary stream temperature recommendations, see Table 8.4.

TABLE 8.3 Selection Principles for the Primary and Secondary Stream Velocity for Swirl Burners

Coal	Anthracite	Lean Coal	Bituminous Coal	Lignite
Primary stream velocity W_1 (m/s)	12–16	16–20	20–26	20–26
Secondary stream velocity W_2 (m/s)	15–22	20–25	30–40	25–35

TABLE 8.4 Recommended Values of the Secondary Stream Temperature

Fuel	Anthracite	Lean Coal	Lignite: Hot Air Drying	Lignite: Flue Gas Drying	Bituminous Coal
Temperature (K)	653–703	603–653	623–653	573–623	553–623

8.3 FLAME FEATURES OF OXY-FUEL BURNER

In oxy-fuel combustion, the change of gas composition significantly influences the coal ignition process and flame stability. Fundamental to the design of an oxy-fuel burner is an understanding of the flame features in oxy-fuel combustion. To ensure that the oxy-fuel burner is capable of operating at oxy-fuel combustion and air combustion conditions, the oxy-fuel burners design needs to follow certain basic principles. Essenhigh et al. [3], Wall et al. [4], and Annamalai et al. [5] reviewed coal ignition and flame stability in detail. In this section, the key considerations of pulverized coal ignition and flame stability are highlighted as key features of the oxy-fuel flames.

According to the ignition phenomenon, pulverized coal has three ignition modes: homogeneous ignition, multiphase ignition, and joint ignition [4,6]. Pulverized coal ignition is a particularly complex process because of the variables that impact the pulverized coal ignition characteristics, including coal type, system pressure, volatile content, gas composition, gas temperature, coal particle size, and so on. The most significant feature of oxy-fuel combustion is the change in gas composition, which significantly affects the heating rate and the chemical reaction of the pulverized coal. These factors significantly affect pulverized coal ignition.

In order to understand the ignition process, many studies have focused on the ignition temperature and ignition delay time of single particle and particle cloud ignition characteristics. At the Sandia National Laboratory, Shaddix and Molina [7,8] studied these characteristics in pulverized coal in O_2/N_2 and O_2/CO_2 atmospheres. Their results showed that replacing N_2 with CO_2 led to an increase in ignition delays under similar furnace temperatures and oxygen concentrations. At the same time, the ignition delay time was reduced by increasing the concentration of oxygen. They also found that the main mechanism for increasing ignition delays was caused by the higher specific heat capacity of CO_2 and that CO_2 also inhibited the formation of polar molecules. The increase in ignition delays may have been the result of a combination of these two factors. By observing the point of ignition in pulverized coal in a 20kW pilot-scale swirl burner, Liu et al. [9] found that the ignition delay of pulverized coal in a 21% O_2/CO_2 condition was significantly higher than in a 21% O_2/N_2 condition. In a 30% O_2/CO_2 condition, the ignition position was similar to that in a 21% O_2/N_2 condition. From the experiments in an IHI vertical furnace with a low-swirl burner, Khare et al. [10] found a slightly higher ignition delay when burning high-volatile bituminous coal in oxy-fuel conditions with an initial oxygen concentration of 27%, relative to air combustion. Zhang et al. [11] studied the combustion characteristics of a high particle concentration in Victorian lignite in O_2/N_2, and O_2/CO_2 conditions by using high-speed cameras and two-color pyrometers. Their results showed that replacing N_2 with CO_2 caused a quenching effect in single coal particle ignition, which affected the pyrolysis and ignition behaviors, as well as the temperatures of volatile flame and char particles. Because of the high concentration of CO_2, an accumulation of volatiles on particle surfaces appeared in O_2/CO_2 conditions. The ignition type of the volatile cloud was different from the ignition type of the single particle in O_2/N_2 conditions.

Flame stability reflects the ability of existing flames to maintain ignition, which is impacted by the fuel and oxidant flow, turbulence shear, fuel quality, and so on. Flame stability is an extremely important consideration for burner design and operation because the burner must maintain the ignition by controlling the fuel and oxidant flow characteristics and the heat release rates. In order to ensure the safety of the mill system and mixing of fuel and oxidant, a low initial oxygen concentration in primary stream should be applied in oxy-fuel combustion, which creates a particular challenge in designing and operating the burner [12,13].

Toporov et al. [14,15] studied the flame stability characteristics of lignite in a 40 kW swirl burner in O_2/N_2 and O_2/CO_2 conditions. When a conventional air burner operated at 21% O_2/CO_2 condition, the flame stability decreased, and a poor burnout appeared. By increasing the secondary stream swirl and redesigning the burner (to strengthen the internal recirculation zone), the flame stability was ensured in both air combustion and oxy-fuel combustion. Yamamoto et al. [16] investigated the effect of CO_2 on flame propagation velocity. They found a decrease in laminar flame propagation velocity in O_2/CO_2 conditions, relative to that in O_2/N_2 conditions. By measuring the flame propagation velocity of pulverized coal in a microgravity combustion chamber, Kiga et al. [17] and Suda et al. [18] investigated the ignition characteristics of pulverized coal in O_2/CO_2 conditions. Their results also showed that the flame propagation velocity was much lower than that in O_2/N_2 and O_2/Ar conditions and that the flame propagation velocity increased with an increasing oxygen concentration. In a 100 kW oxy-fuel combustion furnace, Hjärtstam et al. [19] studied the flame characteristics of lignite in O_2/CO_2 conditions. By changing the flue gas recycle ratio, three different oxy-fuel flames were investigated and compared with air combustion. After analyzing the O_2 and CO concentrations in the furnace and the flame images, they found that with the decrease in the flue gas recycle ratio, the ignition was more advanced, the flame was more intensive, the temperature was higher, and the flame stability increased.

With a high-speed camera, the experimental studies of ignition characteristics of pulverized coal were carried out by Huang et al. [20,21] in a flat-flame-assisted entrained flow reactor. The combustion processes with three different coals in O_2/N_2 and O_2/CO_2 conditions were studied in detail. The results showed that the presence of a high CO_2 concentration had a different effect on the ignition characteristics of pulverized coal with high volatiles and low volatiles. For pulverized coal with high volatiles, high CO_2 concentrations resulted in an increasing ignition delay and poor flame stability. In contrast, high CO_2 concentrations improved the ignition delay and flame stability for the anthracite with low volatiles. In the O_2/CO_2 condition, the increasing gas temperature and oxygen concentration shortened the ignition time and improved the ignition. The ignition delay time was inversely proportional to oxygen concentration ($Y_{O2}{}^{n}$), where n was approximately, 0.15–0.20.

8.4 OXY-FUEL BURNER DESIGN PRACTICES

8.4.1 Design Principle for Oxy-fuel Burner

In order to find the design principle for oxy-coal combustion burners, it is necessary to understand oxy-fuel environments and the requirements of oxy-fuel burners, thus the design of oxy-fuel burners is challenging.

Oxy-fuel combustion processes are quite different from those in conventional air combustion. For example: (1) A portion of flue gas is recycled in oxy-fuel combustion, whereas in air combustion, all the flue gases are emitted through the chimney. (2) Premixed or non-premixed combustion can be achieved during oxy-fuel combustion, and the initial oxygen concentrations range from 18 to 34 vol%. The oxygen partial pressure in the fuel transport stream is about 0 to 23 vol%, which follows the safety limits. The oxygen premixing and oxygen partial pressure in the fuel transport stream maintain 21 vol% in air combustion. (3) In a specific thermal input and equivalent ratio conditions, the adiabatic flame temperature ($T_{ad,\ oxy}$) changes with the change of the recycle ratio in oxy-fuel combustion, while the adiabatic flame temperature ($T_{ad,\ air}$) remains unchanged. Therefore, in an oxy-fuel combustion process, the methods to calculate the volume and composition of the oxidant and flue gas cannot simply follow those for a traditional air combustion process. It is necessary to reestablish the calculation model to calculate the theoretical combustion parameters.

An oxy-fuel burner system has four main requirements: (1) Stable oxy-fuel combustion must occur in the oxy-fuel burner system. (2) To match the convective and radiative heat transfer between oxy-fuel combustion and air combustion, similar in-furnace flow fields and furnace temperatures of oxy-fuel combustion are needed. (3) The oxy-fuel burner system must also be able to operate at air combustion conditions in order to conveniently switch between oxy-fuel and air combustion conditions. (4) The oxy-fuel burner system must avoid extremely high flame temperatures in order to suppress NO_x formation and simultaneously further decrease the NO_x level caused by NO reburning in external flue gas recirculation.

For the first requirement, stable operation, the flame stability of oxy-fuel combustion decreases because of the dilution of CO_2. Thus the design of the oxy-fuel burner system needs to improve the stability of oxy-fuel combustion. A swirl burner is designed for oxy-fuel combustion and can create an internal recirculation zone (IRZ), which attracts the hot flue gas and thus rapidly warms up the pulverized coal and oxidant. A suitable swirl number should be chosen to confirm the shape and recirculated mass flow rate of the IRZ, thereby guaranteeing combustion stability. For the second requirement, matching the convective and radiative heat transfer, the flow and thermal fields under air combustion and oxy-fuel combustion conditions should to be similar. The initial O_2 concentration in oxy-fuel combustion should be high enough to match the temperature in air combustion. However, the detailed value of the initial O_2 level should be determined for specific oxy-fuel combustion conditions by considering coal ignition, devolatilization, flame temperature, flame volume, and coal burnout. Similar coal ignition, devolatilization, and coal burnout can be guaranteed by the local O_2 concentration and the mixing properties of the primary and secondary streams. For the third requirement, compatible operation of air combustion and oxy-fuel combustion, the scalar and vector of the primary stream should be similar to ensure the transport of pulverized coal to the furnace, such as mass, kinetic energy, velocity, and momentum. For the fourth requirement, controlling the NO_x formation, the burner system aerodynamics are evaluated by theoretical analyses and computational fluid dynamic (CFD) calculations.

The design of oxy-fuel burners should maintain characteristics similar to those in the design of air combustion burners, in terms of flame structure, flame stability, pulverized coal ignition characteristics, and burnout ratio. A good oxy-fuel burner, therefore, is compatible with air combustion burners, while also having the advantages of oxy-fuel combustion.

In oxy-fuel combustion, the different primary stream criteria of the burner causes big differences in pulverized coal ignition and flame stability. In the last 10 years, some studies [10,22–24] used velocity, mass flow, momentum, and kinetic energy as the primary stream criteria to study coal ignition and flame stability. Khare et al. [10] conducted experiments in a vertical 1.2 MW pilot-scale furnace. They investigated the effect of a constant flux momentum in the primary stream on the ignition delay of oxy-fuel combustion in Type-0 flames. A greater flux momentum in the primary stream caused a more serious ignition delay. In addition, in low-load operation conditions, the ignition delay was likewise further enhanced.

A series of numerical investigations of oxy-fuel burner design were performed by Liu et al. [25]. The effects of the blockage ratio, swirl number, flue gas recycle ratio, and initial oxygen concentration in the primary stream on the flame stability, type, shape, and structure were evaluated. They found that the internal recirculation zone improved the flame stability but that the central dark primary core destroyed the oxy-coal flame. The swirl number and recycle ratio had a greater effect on the volume of the dark primary core than on the other two factors. However, the internal recirculation zone was most affected by the blockage ratio. Their study contributed to the understanding of flame stability, shape structure, and type of oxy-fuel combustion.

Looking at 40 kW coaxial turbulent diffusion oxy-coal flames, Zhang et al. [24] studied the influence of burner inlet conditions on oxy-flame stability, such as the initial oxygen concentration in the primary stream, the secondary stream temperature, the initial oxygen concentration in the secondary stream, and the transport medium. They found that these factors influenced oxy-flame stability, while the oxygen concentration in the primary stream was the first quantifiable effect on oxy-flame stability. The flame stability has been compared in experiments with different transporting media of primary streams (O_2/CO_2 mixture and O_2/N_2 mixtures). The results showed that coal ignition mechanisms are dependent on the molecular diffusion of oxygen and heat transfer from the secondary stream to particle surfaces.

In a 1.5 MW pilot-scale furnace, Fry et al. [22,23] conducted an experimental study of the different flame behaviors between air and oxy-fuel conditions in a pulverized coal oxy-fuel burner. Three operational strategies related to air and oxy-fuel burners were investigated: the matching oxy-fuel burner primary mass, the momentum, and the velocity in air-fired conditions. Fig. 8.3 shows the flame pictures from different operation strategies. In an oxy-fuel condition, the flame stability was ensured by using the matching primary mass or momentum with that in an air-fired condition. However, a detached flame with an ignition delay resulted from matching the primary velocity with air combustion conditions. In order to obtain a stable flame in the quarl, which was similar to air combustion condition, it was necessary to decrease the primary velocity of 13% in oxy-fuel combustion.

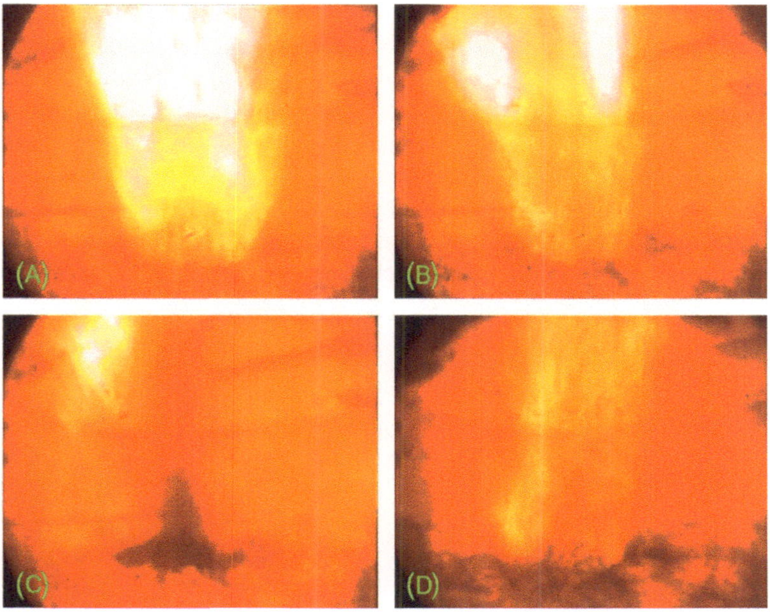

FIG. 8.3 Retrofit flame pictures for (A) air combustion baseline, (B) oxy combustion with matched burner primary momentum, (C) oxy combustion with matched burner primary velocity, and (D) oxy combustion with no primary O_2 enrichment.*(From Fry A, Adams B, Paschedag A, Kazalski P, Carney C, Oryshchyn D, et al. Principles for retrofitting coal burners for oxy-combustion. Int J Greenhouse Gas Control 2011;5(12):S151–8.)*

8.4.2 The Realistic Oxy-Coal Burners

Several realistic oxy-coal burners have been evaluated in terms of their oxy-fuel combustion operations. In the Vattenfall Schwarze Pumpe Project (Germany), which used lignite, several swirl burners with different structures were tested—those manufactured by Alstom Power [26–28], Hitachi Power Europe [28,29], Doosan Babcock Energy Ltd. [28], and BBS Bilfinger [28]. Multiple front-wall-fired burners with six swirl burners were achieved in the Callide Oxyfuel Project (Australia) [30,31]. In addition, Babcock and Wilcox (B&W) has made many contributions to the research of oxy-fuel burners [32]. The previously mentioned oxy-fuel burners were all swirl burners. It is worth noting that low NO_x tangentially fired burners were used in a 15 MW oxy-fuel project by Alstom [33].

A research group at Huazhong University of Science and Technology (China) did systematic studies on oxy-fuel combustion. A design method for oxy-fuel swirl burners has been developed and applied. A series of swirl burners from 0.3 MW [34], to 3 MW [35], to 12 MW [36] have been used in realistic oxy-fuel combustion boilers. In this section, a detailed description of 3 MW oxy-coal burner is used to illustrate the oxy-fuel burner design method. A 3 MW full-flow oxy-fuel combustion test system will be discussed in Chapter 10. In addition, numerical results of a 200 MW tangentially fired oxy-fuel burner were reported in our previous study [37].

Fig. 8.4 shows a cross-sectional view of the swirl burners outlet. This oxy-fuel burner consists of four nozzles: secondary stream nozzle, dam ring, primary stream nozzle, and oil gun. The design of the oil gun structure ensures that oil is injected from the burners center. In addition, the structure of the oil gun helps in the formation of the internal recirculation zone. The oil gun size is determined by the oil gun type. For this design, the oil gun size was set at $\Phi 45 \times 3$. The dotted line in Fig. 8.4 depicts the projection of the bluff body, which was connected to the oil gun. The bluff body was used mainly to reduce the rigidity of the primary stream, which can help form the internal recircula-

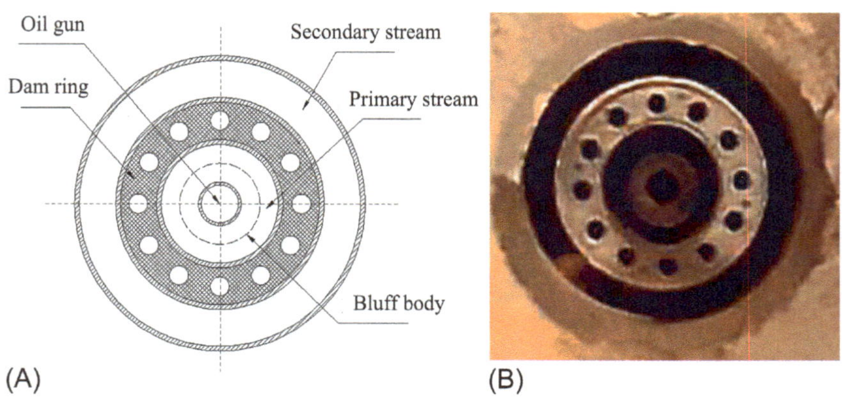

FIG. 8.4 Cross-sectional structure of the outlet of the 3 MW swirl burner: (A) geometric structure of the burner outlet and (B) photo of the burner outlet.

tion zone and ensure flame stability. In addition, the bluff body helped the primary stream and the secondary stream mix faster and better. The numerical results showed that when the coverage of the bluff body was <50%, the internal recirculation zone increased as the bluff body coverage increased. When the coverage of the bluff body was >50%, the internal recirculation zone did not change as the bluff body coverage increased, but the external recirculation zone increased significantly, which was not conducive to combustion. So, in this design, the bluff body taper was 45°, and the bluff body coverage in the primary stream nozzle was 50%.

Twelve oxygen jet pipes were evenly arranged in the dam ring. The premixed and non-premixed combustion modes were investigated separately. The role of the 12 oxygen jet pipes was to provide high-speed, relatively uniform oxygen jets. The high oxygen jet velocity benefitted oxygen diffusion in the flue gas. For this design, the oxygen jet velocity was 50 m/s, and the diameter of the oxygen jet pipe was 20 mm. The thickness of the dam ring provided enough space to arrange the oxygen jet pipes. The dam ring delayed the mixing time of primary air and secondary air, which maked the flame surface backward and protects the burner nozzles. In addition, the NO_x emissions were expected to reduce because coal combusts at lower oxygen conditions. The detailed structure parameters of this oxy-coal burner are shown in Table 8.5.

The choice of a swirl number for the secondary stream is mainly based on the flame shape and the fullness of the furnace. As discussed in Section 8.2, a swirl number that is too high or too low is not conducive to pulverized coal combustion. From the CFD results, we found that when the swirling number was >0.7, the internal recirculation zone formed effectively. With an increase in the swirling number, the internal recirculation flow changed only slightly. But a significant increase was obtained in the external recirculation zone. So the swirling number of this burner was set at 1.0.

TABLE 8.5 Design of the Swirl Burner Size

Content	Design (mm)	Actual Value (mm)
Inner diameter of oil gun	39	39
Outer diameter of oil gun	45	45
Inner diameter of primary stream	129	123
Outer diameter of primary stream	139	133
Inner diameter of dam ring	219	209
Outer diameter of dam ring	229	219
Inner diameter of secondary stream	302	295
Outer diameter of secondary stream	312	305

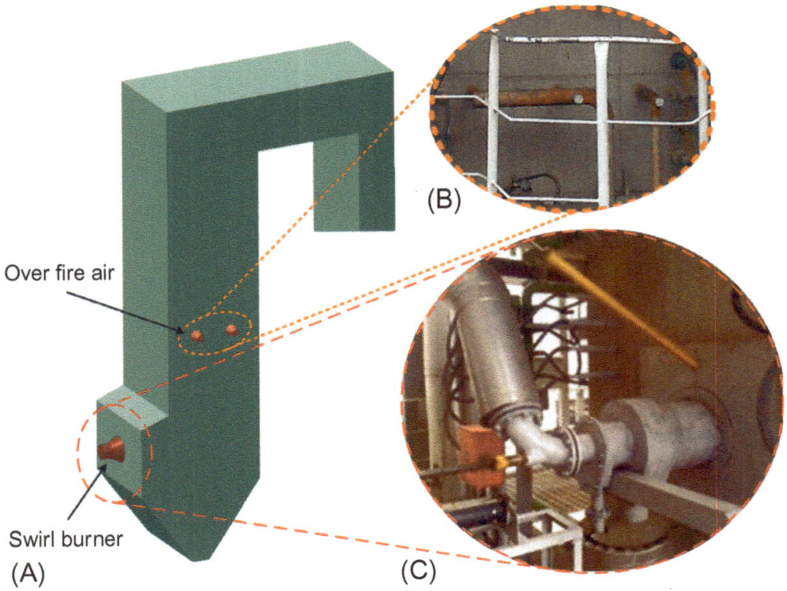

FIG. 8.5 Description of the furnace and oxy-fuel burner: (A) geometric structure of the furnace, (B) view of the over-fire air nozzles, and (C) installation diagram of the burner.

The geometry of the furnace is shown in Fig. 8.5A. This 3 MW swirl burner was installed on the front wall of the furnace, as shown in Fig. 8.5B. Five tertiary stream nozzles were arranged about two meters higher than the burner on the two side walls so that three nozzles were on the left and two were on the right to form an opposing jet flow, which favored gas mixing. In oxy-fuel combustion, the total oxidant volume decreased because of the increase in the initial oxygen concentration. During the air combustion experiment, in order to obtain a similar oxidant volume in burner nozzles, a portion of the oxidant was injected into the furnace via the tertiary stream nozzles, which accounted for about 20% of total oxidant volume.

In air combustion, the primary stream velocity is set at approximately 23 m/s (refer to Table 8.3 in Section 8.2). In order to ensure flame stability and a similar flow field and flame shape between air combustion and oxy-fuel combustion, the primary stream momentum in oxy-fuel combustion must be similar to that in air combustion. So the primary stream velocity decreases to approximately 20 m/s in oxy-fuel combustion.

During the experiment, air combustion and oxy-fuel combustion were performed. In oxy-fuel combustion, the different flue gas recycle ratios and the different thermal load conditions were investigated. The detailed flame images captured by CCD camera at different flue gas recycle ratios were reported in Ref. [35]. Stable flames were ensured by this specific oxy-coal burner. In oxy-fuel combustion, a similar burnout ratio of pulverized coal (approximately 96%–98%) was obtained, relative to air combustion.

8.5 SUMMARY

The different physical and chemical environments in oxy-coal combustion significantly influenced pulverized coal ignition and flame stability. Relative to air combustion, the high CO_2 concentration in oxy-fuel combustion increased the ignition delay because of the higher molar heat capacity. The adiabatic flame temperature decreased in oxy-fuel combustion. In addition, the laminar flame speed of coal combustion was lower, sometimes much lower in oxy-fuel combustion. These two effects decreased flame stability.

Flame stability and combustion efficiency are the basic targets of the burner design. Mature design principles for coal-fired air burners are used in industrial boilers. However, to our knowledge, there are no generally accepted design principles for oxy-coal burners. In order to match the convective and radiative heat transfer between air combustion and oxy-fuel combustion conditions, the initial O_2 concentration of oxy-fuel combustion must be higher in order to match the similar temperature in air combustion. A high O_2 concentration helped to counteract the ignition delay, but the oxidant volume decreased, which impacted the burners aerodynamics. Different oxygen mixing modes and adjustable flue gas recycle ratios make oxy-coal burners highly flexible, and also present significant design challenges.

A 3 MW oxy-coal burner was designed and tested in an oxy-fuel combustion plant. For this burner design, matching the burners primary momentum in oxy-fuel conditions with that in air combustion was seen as the design criteria. Simultaneously, in order to match the flow rate in the burners inlet, the OFA nozzles were turned off under oxy-fuel combustion. From the test results, a compatible, stable operation of air combustion and oxy-fuel combustion was achieved. The system could be easily switched between oxy-fuel and air combustion conditions.

REFERENCES

[1] Wall TF. The combustion of coal as pulverized fuel through swirl burner. In: Lawn CJ, editor. Principles of combustion engineering for boilers. London: Academic Press; 1987. p. 197–335.

[2] Stultz SC, Kitto JB. Steam: its generation and use. Barberton, OH, USA: The Babcock & Wilcox Company; 1992.

[3] Essenhigh RH, Misra MK, Shaw DW. Ignition of coal particles: a review. Combust Flame 1989;77(1):3–30.

[4] Wall TF, Gupta RP, Gururajan VS, Zhang DK. The ignition of coal particles. Fuel 1991;70(9):1011–6.

[5] Annamalai K, Ryan W. Interactive processes in gasification and combustion. Part I: liquid drop arrays and clouds. Prog Energy Combust Sci 1992;18(3):221–95.

[6] Zhang DK, Wall TF. Ignition of coal particles: the influence of experimental technique. Fuel 1994;73(7):1114–9.

[7] Shaddix CR, Molina A. Particle imaging of ignition and devolatilization of pulverized coal during oxy-fuel combustion. Proc Combust Inst 2009;32(2):2091–8.

[8] Molina A, Shaddix CR. Ignition and devolatilization of pulverized bituminous coal particles during oxygen/carbon dioxide coal combustion. Proc Combust Inst 2007;31(2):1905–12.

[9] Liu H, Zailani R, Gibbs BM. Comparisons of pulverized coal combustion in air and in mixtures of O_2/CO_2. Fuel 2005;84(7–8):833–40.

[10] Khare SP, Wall TF, Farida AZ, Liu Y, Moghtaderi B, Gupta RP. Factors influencing the ignition of flames from air-fired swirl pf burners retrofitted to oxy-fuel. Fuel 2008;87(7):1042–9.

[11] Zhang L, Binner E, Qiao Y, Li CZ. In situ diagnostics of Victorian brown coal combustion in O_2/N_2 and O_2/CO_2 mixtures in drop-tube furnace. Fuel 2010;89(10):2703–12.

[12] Kimura N, Omata K, Kiga T, Takano S, Shikisima S. The characteristics of pulverized coal combustion in O_2/CO_2 mixtures for CO_2 recovery. Energy Convers Manag 1995;36(6):805–8.

[13] Chui EH, Douglas MA, Tan Y. Modeling of oxy-fuel combustion for a western Canadian subbituminous coal. Fuel 2003;82(10):1201–10.

[14] Toporov D, Bocian P, Heil P, Kellermann A, Stadler H, Tschunko S, et al. Detailed investigation of a pulverized fuel swirl flame in CO_2/O_2 atmosphere. Combust Flame 2008;155(4):605–18.

[15] Heil P, Toporov D, Stadler H, Tschunko S, Förster M, Kneer R. Development of an oxycoal swirl burner operating at low O_2 concentrations. Fuel 2009;88(7):1269–74.

[16] Yamamoto A, Suda T, Okazaki K. Mechanism of ignition delay in O_2/CO_2 pulverized coal combustion. In: Proceedings of the annual international Pittsburgh coal conference. Pittsburgh, PA; 2004.

[17] Kiga T, Takano S, Kimura N, Omata K, Okawa M, Mori T, et al. Characteristics of pulverized-coal combustion in the system of oxygen/recycled flue gas combustion. Energy Convers Manag 1997;38(96):S129–34.

[18] Suda T, Masuko K, Sato JI, Yamamoto A, Okazaki K. Effect of carbon dioxide on flame propagation of pulverized coal clouds in CO_2/O_2 combustion. Fuel 2007;86(12–13):2008–15.

[19] Hjärtstam S, Andersson K, Johnsson F, Bo L. Combustion characteristics of lignite-fired oxy-fuel flames. Fuel 2009;88(11):2216–24.

[20] Huang X, Li J, Liu Z, Yang M, Wang D, Zheng C. Ignition and devolatilization of pulverized coals in lower oxygen content O_2/CO_2 atmosphere. In: Qi H, Zhao B, editors. Cleaner combustion and sustainable world. Berlin Heidelberg: Springer; 2013. p. 99–104.

[21] Huang X. Oxyfuel combustion characteristics of pulverized coal based on flat-flame assisted entrained flow reactor (Dissertation for the Doctoral Degree). Wuhan: Huazhong University of Science and Technology; 2013.

[22] Fry A, Adams B, Davis K, Cremer M, Swensen D, Munson S, et al. Topics in oxy-coal retrofit of utility boilers–burner principles and fire-side corrosion. In: NETL CO_2 capture technology meeting. Pittsburgh, PA, USA; 2010.

[23] Fry A, Adams B, Paschedag A, Kazalski P, Carney C, Oryshchyn D, et al. Principles for retrofitting coal burners for oxy-combustion. Int J Greenhouse Gas Control 2011;5(12):S151–8.

[24] Zhang J, Kelly KE, Eddings EG, Wendt JOL. Ignition in 40kW co-axial turbulent diffusion oxy-coal jet flames. Proc Combust Inst 2011;33(2):3375–82.

[25] Liu J, Liu Z, Chen S, Santos SO, Zheng C. A numerical investigation on flame stability of oxy-coal combustion: effects of blockage ratio, swirl number, recycle ratio and partial pressure ratio of oxygen. Int J Greenhouse Gas Control 2017;57:63–72.

[26] Marion J, Brautsch A, Kluger F, Larsson M, Pourchot T. Overview of a manufacturer's efforts to commercialize oxy-combustion for steam power plants. In: 2nd Oxyfuel Combustion Conference. Queensland, Australia; 2011.

[27] Kluger F, Mönckert P, Bäck A, Wang W, Grubbström J, Levasseur A, et al. Oxy-combustion testing in 30MWth pilot plant Schwarze Pumpe. In: 2nd oxyfuel combustion conference. Queensland, Australia; 2011.

[28] Burchhardt U, Giering R, Weiß G. Overview of burner tests in Vattenfall's oxyfuel pilot plant. In: 3rd oxyfuel combustion conference. Ponferrada, Spain; 2013.
[29] Rehfeldt S, Schiffer FP, Weckes P, Bergins C, Tigges KD. Oxyfuel combustion with Hitachi's DST-burner at Vattenfall's 30 MWth Pilot Plant at Schwarze Pumpe. In: 2nd oxyfuel combustion conference. Queensland, Australia; 2011.
[30] Spero C, Yamada T, Nelson P, Morrison T, Bourhy-Weber C. Callide Oxyfuel Project: combustion and environmental performance. In: 3rd oxyfuel combustion conference. September, Ponferrada, Spain; 2013.
[31] Spero C. Callide Oxyfuel Project: from concept to commissioning. In: 2nd oxyfuel combustion conference. Queensland, Australia; 2011.
[32] McCauley K, Farzan H, Alexander K, McDonald D, Varagani R, Prabhakar R, et al. Commercialization of oxy-coal combustion: applying results of a large 30MWth pilot project. Energy Procedia 2009;1(1):439–46.
[33] Levasseur A, Turek D, Kenney J, Chapman P, Edberg C, Kang S, et al., editors. Oxy-fired tangential boiler development and large-scale (15 MWth) validation. In: 2nd oxyfuel combustion conference. Queensland, Australia; 2011.
[34] Li P, Wang F, Tu Y, Mei Z, Zhang J, Zheng Y, et al. Moderate or intense low-oxygen dilution oxy-combustion characteristics of light oil and pulverized coal in a pilot-scale furnace. Energy Fuel 2014;28(2):15–20.
[35] Luo Z, Xiang Q, Cheng Q, Lou C. Changing characteristics of flame images under different oxy-fuel atmospheres in a 3 MW pilot-scale furnace. IEEE Trans Instrum Meas 2016;65:1–7.
[36] Guo J, Liu Z, Huang X, Zhang T, Luo W, Hu F, et al. Experimental and numerical investigations on oxy-coal combustion in a 35 MW large pilot boiler. Fuel 2017;187(1):315–27.
[37] Guo J, Liu Z, Wang P, Huang X, Li J, Xu P, et al. Numerical investigation on oxy-combustion characteristics of a 200 MWe tangentially fired boiler. Fuel 2015;140(2):660–8.

Chapter 9

Heat Transfer During Oxy-fuel Combustion and Boiler Design

Junjun Guo, Zhaohui Liu
Huazhong University of Science and Technology, Wuhan, China

9.1 INTRODUCTION

The transfer of heat is a key process in an industrial combustion boiler. The thermal efficiency of a boiler is determined by the heat transfer processes between hot combustion products and a boilers membrane wall. In order to accurately predict heat transfer processes, it is necessary to have an in-depth understanding of the principles behind the heat transfer process. There are three primary modes of heat transfer: conduction, convection, and radiation. For combustion products with high temperatures and turbulence in industrial boilers, radiation is the dominant mechanism, though convection also plays an important role.

In oxy-fuel combustion, by using a mixture of recycled flue gas and oxygen to replace air, a high CO_2 concentration is achieved in flue gas. Compared to conventional air combustion, the heat transfer characteristics under oxy-fuel combustion are significantly altered by increased portions of CO_2, and H_2O and by relatively higher concentrations of particulate media such as char, fly ash, and so on. Experimental and numerical investigations of heat transfer in oxy-fuel combustion have been performed in laboratory-scale [1–3], pilot-scale [4–7], and demonstration-scale facilities [8–10]. In numerical studies of oxy-fuel combustion, some sub-models developed for conventional air combustion, especially emissivity models, are not valid in CO_2-rich environments.

In this chapter, we first introduce the calculation method of convective heat transfer coefficient for an oxy-fuel combustion facility. Then we provide an in-depth review of the development of the gaseous and particle radiative property model.

Seeking to accurately predict total emissivity, we compared a number of new parameters of the weighted-sum-of gray-gases (WSGG) model as presented by different authors, during which we utilized line-by-line (LBL) model benchmark calculations, the HITEMP-2010 database, 0–50 m path lengths, and temperatures of 500, 1500, and 2000 K. Moreover, the radiative source terms

Oxy-fuel Combustion. http://dx.doi.org/10.1016/B978-0-12-812145-0.00009-8

and heat flux were compared in the nonisothermal and nonhomogeneous gas mixture. In order to investigate the heat transfer properties in a practical oxy-fuel combustion facility, the experimental and numerical investigations of heat transfer in a 35 MW oxy-coal combustion boiler were performed. This oxy-fuel boiler was a typical industrial front wall-fired boiler. Three 12 MW pulverized coal burners and two over-fire air (OFA) nozzles were arranged in the front wall.

9.2 CONVECTIVE HEAT TRANSFER

Convection heat transfer is the thermal energy transfer between the fluid and the contacted solid wall, which describes a comprehensive heat transfer process of macroscopic heat convection and microscopic heat conduction. According to the form of fluid flow, the convective heat transfer can be divided into two types: natural convection and forced convection. Forced convection plays a more important role for convection heat transfer in industrial combustion boilers.

The basic interaction in convection heat transfer is

$$Q = hA\left(T_{w} - T_{\infty}\right) \tag{9.1}$$

where Q is the heat transfer (W), A is the area of the object (m^2), h is the convective heat transfer coefficient (W/m^2K), T_w is the solid wall temperature (K), and T_∞ is the fluid temperature (K). When calculating the convective heat transfer, the key is to determine the convective heat transfer coefficient, which is dependent on the physical properties of the fluid and the physical situation.

In a utility boiler, the heat exchangers, such as superheaters, reheaters, economizers, and air preheaters, are located in the convective section. The principal mode of thermal transfer in this section is forced convection. For a conventional air combustion boiler, the O_2 and CO_2 concentrations in the flue gas are only slightly changed in different combustion conditions, so the calculation of the convective heat transfer coefficient is based on the average flue gas composition. The viscosity, thermal conductivity, and other physical parameters of the flue gas depend on the volume fraction of water vapor. In oxy-fuel combustion conditions, the volume fraction of carbon dioxide and water vapor in flue gas sharply increases. The convection heat transfer coefficient changes significantly, relative to that in conventional air combustion. The general computational method for the convective heat transfer coefficient is

$$h = X\frac{\lambda}{d}\left(\frac{\omega d}{\upsilon}\right)^{m} Pr^{n} \tag{9.2}$$

where X is the structure constant, λ is the thermal conductivity coefficient, d is the pipe diameter, ω is the flue gas velocity, υ is the kinematic viscosity, and Pr is the Prandtl number. These parameters of the mixture gas are calculated as follows:

$$\lambda = \sum y_i \lambda_i M_i^{1/3} \Big/ \sum y_i M_i^{1/3} \tag{9.3}$$

$$\mu = \sum y_i \mu_i M_i^{1/2} \Big/ \sum y_i M_i^{1/2} \tag{9.4}$$

$$\upsilon = \mu \frac{R_g T}{P} \tag{9.5}$$

$$Pr = \frac{\mu c_p}{\lambda} \tag{9.6}$$

The thermal conductivity coefficient λ, dynamic viscosity μ, and specific heat capacity C_p were selected according to the NIST database. To calculate the convective heat transfer coefficient of flue gas, after calculating the mixture gas parameters, the convective heat transfer coefficient was obtained by Eq. (9.2) for oxy-fuel combustion conditions.

9.3 RADIATIVE PROPERTIES OF GASES AND PARTICLES

Radiative heat transfer is caused by electromagnetic waves, so it does not require a medium for its transfer. As is well known, radiative heat transfer rates are generally proportional to the fourth power of the temperature, i.e.,

$$q \propto T^4 - T_\infty^4 \tag{9.7}$$

where q is the radiative heat flux, T is the radiative source temperature, and T_∞ is a reference temperature. Therefore radiative heat transfer is more important and may be dominant in thermal transfers at high temperatures.

A great amount of energy is released when fuel is burned in boilers, resulting in high flue gas temperatures. In pulverized coal combustion furnaces, because of the high flame temperature and the intensive absorption and emission characteristics of the flame, the heat transferred to the membrane wall is done primarily via radiative heat transfer, with convective heat transfer delivering only a small fraction (~5%). The larger the furnace, the greater the radiative heat transfer. In addition, the proportion of radiant heat in the coal particles ignition heat could cause an increase from 10% of small particles to 50% of large particles in the furnace. Therefore radiant heat significantly influences the temperature rise rate and ignition distance. It can be seen that radiative heat transfer has a direct or indirect influence on ignition, temperature distribution, and the generation of pollutants.

Generally, the interaction of thermal radiation accounts for absorbing, emitting, and scattering media. For a coal combustion boiler, the triatomic gas of CO_2 and H_2O and particulate media, such as char, fly ash, and so on, are generally considered for the radiative heat transfer. For a radiative participating medium, the radiative transfer equation (RTE) is

$$\frac{dI_\eta\left(\vec{r},\hat{s}\right)}{ds} = \alpha_g \frac{\sigma T_g^4}{\pi} + \sum_i \varepsilon_{p,i} n_i A_{p,i} \frac{\sigma T_{p,i}^4}{\pi} - \left(\alpha_g + \alpha_p + \sigma_p\right) I_\eta\left(\vec{r},\hat{s}\right) + \frac{\sigma_p}{4\pi} \int_{4\pi} I_\eta\left(\vec{r},\hat{s}'\right) \phi\left(\hat{s}',\hat{s}\right) d\Omega \tag{9.8}$$

The left term is the change of the radiative intensity. The first and second terms on right are gas and particle emission, respectively. The third term is the absorption and scattering away from the direction $\hat{s}$. The last term is the gain from scattering to the direction of $\hat{s}$.

9.3.1 The Gas Radiative Models

The gas radiative properties strongly and rapidly change throughout the spectrum. For the gas radiation calculation, the common spectral emissivity models can be divided into detailed models and simplified models. The LBL model [11] is the most accurate model. Generally, the LBL model is used only to verify the accuracy and validity of other models because of its high computational burden. The statistical narrow band (SNB) model [12,13] is also highly accurate, but the number of calculations required decrease its acceptability. It yields the spectral transmissivity rather than the spectral absorption coefficient [14]. The exponential wide band (EWB) model [15] simplifies the spectral models, but the computational accuracy decreases.

The full-spectrum *k*-distribution (FSK) model [16], the WSGG model [17], and the spectral line-based WSGG (SLW) model [18] are simplified models. They are useful for computationally demanding applications, such as the simulation of coal combustion because they can be coupled with any RTE [19–21]. The SLW and FSK models are based on the absorption line blackbody distribution function (ALBDF) and blackbody intensity weighted full-spectrum *k*-distribution, respectively. The cumulative *k*-distribution in the FSK model is similar to the ALBDF associated with the SLW model [22]. The relationship between the FSK and SLW model was discussed in detail by Modest [23], Solovjov and Webb [24], and Chu et al. [22]. The WSGG model is very straightforward and assumes several gray gases with different absorption coefficients to replace the non-gray gases. The computing efficiency is greatly improved, though the accuracy is likely to be lower than in the SLW and FSK models. The WSGG model is easy to incorporate into a CFD code, and considering computational costs, it may be the best selection in 3D full-scale simulations [25].

The widely used WSGG model used by Smith et al. [17] is developed for conventional air combustion and is suited only to H_2O/CO_2 ratios of 1: 2. In recent years, improving its accuracy and making it applicable to oxy-fuel combustion became important research topics. The published WSGG model parameters for oxy-fuel combustion are summarized in Table 9.1. Yin et al. [26] proposed new parameters for the WSGG model with four gray gases and emissivity data calculated by the EWB model at atmospheric pressure. The H_2O/CO_2 ratios encompassed both air and oxy-fuel combustion. When compared to the standard WSGG model, the new parameters provided similar simulation results in a small 0.8 MW furnace and remarkably different results in a 609 MW furnace in oxy-fuel combustion. Johansson et al. [27] proposed new parameters for the WSGG model using four gray gases with emissivity data that were computed

TABLE 9.1 Summary of Reported WSGG Model Parameters for Oxy-Fuel Combustion

Authors	Temperature (K)	Mole Fraction Ratio	Pressure Path Length (bar m)	Reference Model
Yin et al. [26]	500–3000	0.125,0.25, 0.5,0.75,1,2, 4, three set for air combustion	0.001–60	EWBM
Johansson et al. [27]	500–2500	0.125–2	0.01–60	SNBM
Kangwanpongpan et al. [28]	400–2500	0.125–4	0.001–60	LBL
Bordbar et al. [29]	300–2400	0.01–4	0.01–60	LBL
Guo et al. [30]	600–2500	0.05–2	0.01–50	FSK

via the SNB model [15,16] and the EM2C database [31]. They used a more accurate emissivity database than Yin et al. used. [26] Both Kangwanpongpan et al. [28] and Bordbar et al. [29] used the most accurate emissivity database, which generated data from LBL calculations in the HITEMP-2010 database [32]. Then new parameters for the WSGG model were obtained according to the emissivity database. Generally, the published parameters of the WSGG model were determined by the mathematical fitting for the total emissivity, which calculated by a reference model.

Based on the FSK model, Guo et al. [30] recently developed new WSGG parameters that are directly derived from the spectral RTE. By using the Gaussian integral, the absorption coefficient and weight factor were obtained from the integral points in the *k*-distribution that represent the actual properties. The absorption coefficient is considered as a function of temperature and mole fraction, but only as a function of mole fraction in other WSGG models. This consideration is more applicable to non-isothermal gas mixtures.

The prediction accuracy of the total emissivity from different WSGG model parameters was evaluated in a series of cases. The total pressure was 1 bar, and the path length ranged from 0 m to 50 m. Two molar ratios were included: MR = 0.125 and MR = 1.0. The total mole fraction of CO_2 and H_2O was 0.9 in the case with MR = 0.125, which referred to the dry recycle oxy-fuel conditions. For MR = 1.0, the total mole fraction of CO_2 and H_2O was set at 0.96 to represent the wet recycle oxy-fuel conditions [33]. In this section, the results using the improved WSGG model by Johansson et al. [27] are denoted as Johansson 2011, by Yin et al. [26] as Yin 2010, and by Kangwanpongpan et al. [28] as

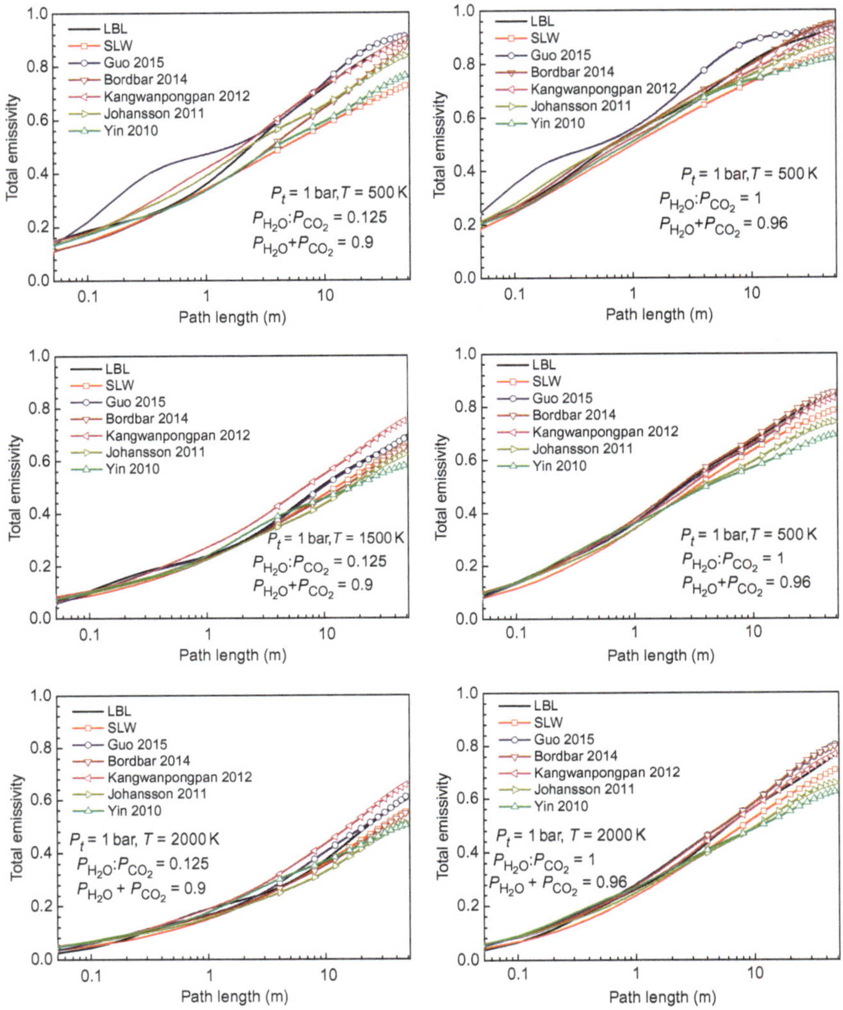

FIG. 9.1 The total emissivity calculated by different models for dry recycle (left column) and wet recycle (right column) oxy-fuel combustion. *(From Guo J, Li X, Huang X, Liu Z, Zheng C. A full spectrum k-distribution based weighted-sum-of-gray-gases model for oxy-fuel combustion. Int J Heat Mass Transf 2015;90:218–26.)*

Kangwanpongpan 2012, by Bordbar et al. [29] as Bordbar 2014, and by Guo et al. [30] as Guo 2015.

Fig. 9.1 shows the total emissivity calculated from different models for MR = 0.125 (dry recycle) and MR = 1.0 (wet recycle) at three temperatures: 500, 1500, and 2000 K, respectively. As shown in Fig. 9.1, the Guo 2015 provided the most accurate prediction of emissivity at a temperature close to flame temperature (1500 K) in both dry recycle and wet recycle oxy-fuel combustion.

At a high temperature (2000 K), the Guo 2015 was also most accurate among all the models in dry recycle, and was slightly lower than Kangwanpongpan 2012 in wet recycle. The maximum relative error of Guo 2015 was about 10% at a temperature of 2000 K. At a lower temperature (500 K), which rarely appears in-furnace, the Guo 2015 could also provide a good prediction of total emissivity at path lengths >10 m. This improved model introduced additional errors at path lengths <10 m, because this temperature is outside the appropriate temperature range (600–2500 K).

Considering all the cases with path lengths ranging from 0 m to 50 m, the WSGG models with the parameters from Yin et al. [26] (Yin 2010) and Johansson et al. [27] (Johansson 2011) led to a considerable error when the total emissivity was calculated, and the WSGG models with the parameters from Guo et al. [30] (Guo 2015), Bordbar et al. [29] (Bordbar 2014), and Kangwanpongpan et al. [28] (Kangwanpongpan 2012) provided better accuracy. The SLW model resulted in the prediction with the lowest accuracy at a temperature of 500 K, implying that its parameters cannot well represent the radiation characteristics of H_2O and CO_2 in a low temperature.

In order to evaluate the effectiveness of solving the RTE, a nonisothermal, nonhomogeneous case was investigated. In this case, the total pressure was 1.0 atm, the path length ranged from 0 to 10 m, and the wall was assumed to be black ($\varepsilon_w = 1$). The temperature was similar to a cosine profile, which is given by

$$T = 1250 - 750\cos\left(2\pi x / L\right) \tag{9. 9}$$

where x is the length coordinate along the space between the infinite slabs, and L is the distance between the plates. The minimum temperature is equal to 500 K at the wall, and the maximum temperature is equal to 2000 K. The mole fraction of H_2O is denoted as the following profile:

$$Y_{H_2O} = 0.12 + 0.04\cos\left(2\pi x / L\right) \tag{9. 10}$$

The MR ranges from 0.1 to 0.2, and $Y_{CO2} = 0.8$.

Fig. 9.2 shows that the maximum relative errors of the radiation source terms and heat fluxes were 27% and 11% for SLW, 28% and 11% for Guo 2015, 20% and 20% for Bordbar 2014, 70% and 30% for Kangwanpongpan 2012, 35% and 23% for Johansson 2011, and 54% and 34% for Yin 2010 in nonisothermal, nonhomogeneous gas mixtures, respectively. For the prediction of radiative source terms, the accuracy of SLW, Guo 2015, and Bordbar 2014 were obviously higher than the other WSGG models. For the prediction of heat flux, the accuracy of Guo 2015 was similar to SLW. The WSGG model parameters from Guo et al. [30] showed a great advantage in calculating radiation in nonisothermal, nonhomogeneous gas mixtures, without introducing additional errors. So the WSGG model from Guo et al. [30] is recommended for use in large-scale CFD simulations for oxy-fuel and air combustion.

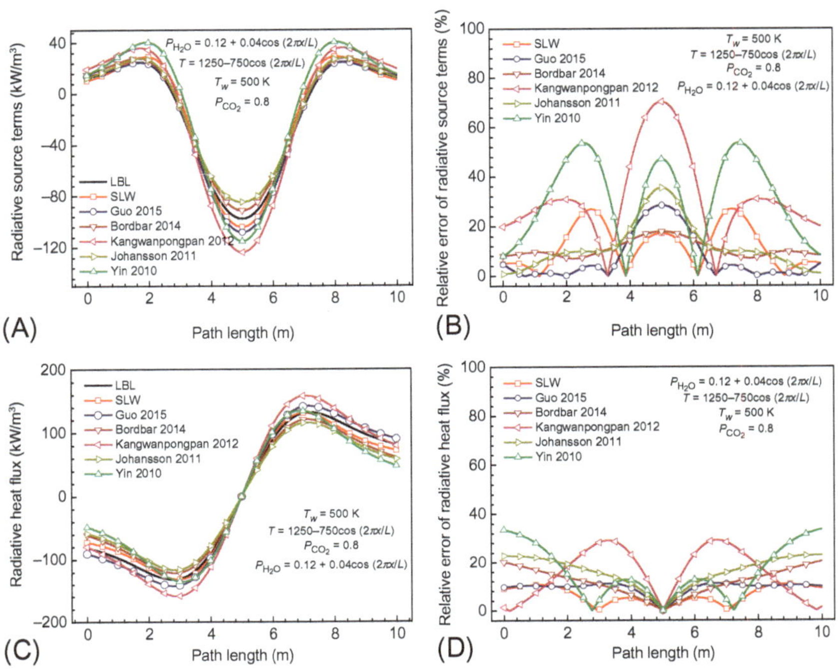

FIG. 9.2 The values and relative errors of radiative source terms and radiative heat fluxes calculated by different models in nonisothermal nonhomogeneous conditions. (A) Radiative source terms calculated by different models, (B) relative errors of radiative source terms, (C) radiative heat fluxes calculated by different models, (D) relative errors of radiative heat fluxes. *(From Guo J, Li X, Huang X, Liu Z, Zheng C. A full spectrum k-distribution based weighted-sum-of-gray-gases model for oxy-fuel combustion. Int J Heat Mass Transf 2015;90:218–26.)*

9.3.2 The Particle Radiation

Radiative intensity may be attenuated by the absorption and/or scattering of particles. The radiation characteristics of particles vary with the wavelength. During pulverized coal combustion, the relatively large coal and fly ash particles are very important, as well as the small soot particles.

The Mie scattering theory is one of the most commonly used and widely accepted methods in the calculation of particle radiative properties. It has a long history of development and application. Based on the Mie theory, for spherical particles, the radiation characteristics of the particles are related only to the particle size parameter α and the optical constants m of the particles. The size parameter and optical constants are defined as

$$\alpha = 2\pi D / \lambda \tag{9.11}$$

$$m = n + ik \tag{9.12}$$

where size parameter α represents the relative particle size to the wavelength; n is the refractive index, which is determined by the speed of the wave propagation in the material; and k is the absorption factor of the material, which determines by the decay rate of the wave propagation in the material.

The detailed analytical process of the Mie theory can be found in a number of books on optical scattering and absorption of particles, including those by Bohren and Huffman [34], Kerker [35], van de Hulst [36], and Gouesbet and Gréhan [34,37]; therefore, we provide only the main formulas for calculation of coal particle radiation.

Particle scattering, extinction, and absorption factor are calculated as follows:

$$Q_{sca} = \frac{2}{x^2}\sum_{n=1}^{\infty}(2n+1)\left(|a_n|^2 + |b_n|^2\right) \tag{9.13}$$

$$Q_{ext} = \frac{2}{x^2}\sum_{n=1}^{\infty}(2n+1)Re\{a_n + b_n\} \tag{9.14}$$

$$Q_{abs} = Q_{ext} - Q_{sca} \tag{9.15}$$

The phase function is as follows:

$$\phi(\theta) = \frac{\left(|S_1|^2 + |S_2|^2\right)}{Q_{sca}x^2} \tag{9.16}$$

In the preceding formulas, Re represents the real part of the complex a_n and b_n are defined as

$$a_n = \frac{m\psi_n(mx)\psi_n'(x) - \psi_n(x)\psi_n'(mx)}{m\psi_n(mx)\xi_n'(x) - \xi_n(x)\psi_n'(mx)} \tag{9.17}$$

$$b_n = \frac{\psi_n(mx)\psi_n'(x) - m\psi_n(x)\psi_n'(mx)}{\psi_n(mx)\xi_n'(x) - m\xi_n(x)\psi_n'(mx)} \tag{9.18}$$

where $\xi_n(x)=\psi_n(x)-\chi_n(x)$, $\psi_n(x)$, and $\chi_n(x)$ are the Riccati-Bessel equation, which satisfies the following recursive equation [38]:

$$\psi_0(x) = \sin x, \quad \psi_1(x) = \frac{\sin x}{x} - \cos x, \quad \chi_0 = \cos x, \quad \chi_1 = \frac{\cos x}{x} + \sin x$$

The expression for the scattering amplitude functions S_1 and S_2 is as follows:

$$S_1 = \sum_{n=1}^{\infty}\frac{2n+1}{n(n+1)}(a_n\pi_n + b_n\tau_n) \tag{9.19}$$

$$S_2 = \sum_{n=1}^{\infty}\frac{2n+1}{n(n+1)}(a_n\tau_n + b_n\pi_n) \tag{9.20}$$

π_n and τ_n are the scattering angle function, which is defined as

$$\pi_n(\cos\theta) = P_n'(\cos\theta) \tag{9.21}$$

$$\tau_n(\cos\theta) = \cos\theta\pi_n(\cos\theta) - \sin^2\theta\pi_n'(\cos\theta) \tag{9.22}$$

where $\pi_0 = 0$, $\pi_1 = 1$, and P_n are the Legendre polynomial; "′" means the derivation of the variable.

After obtaining the radiation characteristic of the single particle, we get the radiation characteristic of the particle clouds. For the particle clouds with discontinuous particle size distribution, its spectral radiative properties are

$$\beta = 1.5\sum_{i=1}^{n} Q_{e,i}\frac{f_{v,i}}{D_i} \tag{9.23}$$

$$\sigma_s = 1.5\sum_{i=1}^{n} Q_{s,i}\frac{f_{v,i}}{D_i} \tag{9.24}$$

$$k = \beta - \sigma_s \tag{9.25}$$

$$\Phi(\theta) = \frac{1}{\sigma_s}\sum_{i=1}^{n}\frac{\pi}{4}D_i^2 N_i Q_{s,i}\Phi_{p,i}(\theta) \tag{9.26}$$

The subscripts i, e, and s denote the ith particle, the attenuation factor, and the scattering factor, respectively. f is the volume fraction, N is the particle number density (m^{-3}), and ϕ is the scattering phase function.

The particle radiation characteristics vary with the particle size. When the particle size is <0.8 μm, the absorption and scattering efficiency of char and fly ash is very small. With an increase in particle size, the absorption efficiency of char quickly reaches maximum, but it begins to decrease slowly when the particle size is larger than 4 μm. However, the absorption efficiency of fly ash continuously increases with an increase in particle size. The monochromatic radiation characteristic of a single particle varies greatly with the wavelength. Because of the harmonic characteristics between large and small particles, the variation between the monochromatic radiation characteristics of a particle cloud and the wavelength is much smaller than that of a single particle; and it is also much smaller than the variations of the monochromatic radiation characteristics under different particle sizes.

The total radiation of char and a fly ash particle cloud will change with an increase in temperature. For most coal types, the absorption efficiency is high at low temperatures, and the scattering efficiency and forward factor are small, which is determined by the nongray properties of the particle radiation characteristics. With an increase in temperature, the proportion of radiation energy increases in a short-wave band, and the influence of the monochromatic radiation characteristics increase in the corresponding bands. Fly ash comprises mainly SiO_2, CaO, Fe_2O_3, and Al_2O_3, and its absorption factor is the weighted sum of absorption factor of these components, so the optical constant of fly ash will affected by its chemical composition.

The Mie scattering theory is not widely used for the calculation of large-scale industrial boilers because its process is complex and computationally expensive. Generally, the coal particle is treated as a gray particle. The particle absorption coefficient α_p and particle scattering coefficient σ_p are respectively evaluated using Eqs. (9.27), (9.28), [39]

$$\alpha_p = \lim_{V \to 0} \sum_{n=1}^{N} \left(\varepsilon_{pn} A_{pn} / V \right) \tag{9.27}$$

$$\sigma_p = \lim_{V \to 0} \sum_{n=1}^{N} \left[\left(1 - f_{pn}\right)\left(1 - \varepsilon_{pn}\right) A_{pn} / V \right] \tag{9.28}$$

where ε_{pn}, A_{pn}, and f_{pn} are the emissivity, projected area, and scattering factor of particle n, respectively. Char and fly ash have different radiative properties. So, in pulverized coal combustion, the particle emissivity and scattering factor generally are seen as relating to the emissivity and scattering factor of char and fly ash [39–42]. However, many studies have used a further simplified method with average constants for particle emissivity (0.7–0.9) [43–46].

The radiative properties of unburned char and their effect on radiative heat transfer were investigated by Liu et al. [47] in pulverized coal furnaces. According to the effective optical constant, the calculation of radiative properties was carried out based on three assumptions about the structure of unburned char: the constant diameter model, the shrinking core model, and a transient model. The change of char particle absorption during burnout is shown in Fig. 9.3. For small char particles ($D = 1\,\mu m$), the absorption property varied linearly with the burnout ratio. But for large char particles ($D > 10\,\mu m$), the absorption in the initial stage did not decrease, while a sharp decrease occurred near the burnout ($B > 0.9$). A 300 MW tangentially fired furnace was studied with the preceding particle radiative model. The results showed that the furnace exit temperature was higher, because the residual char shielded radiative heat transfer along the radial direction in the upper part of the furnace. By ignoring the additional absorption of residual char, the radiative heat flux was underpredicted in the superheater area.

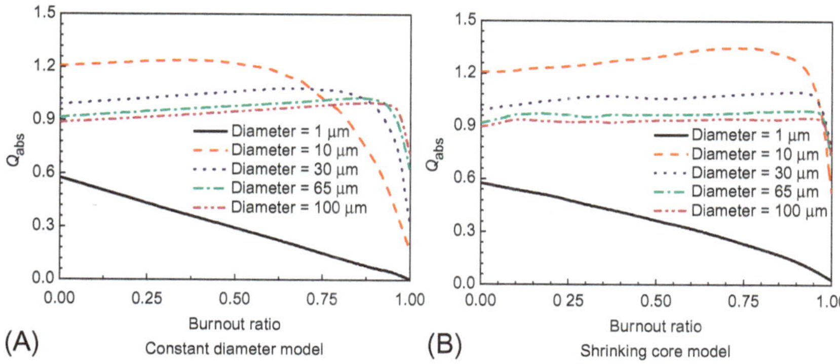

FIG. 9.3 Development of absorption of char particle during burnout. (A) Absorption factor calculated by Constant diameter model, (B) absorption factor calculated by Shrinking core model. *(From Liu Z, Zheng C, Xing H. Radiative properties of residual char and its effects on radiative heat transfer in pulverized coal fired furnaces. Asia Pac J Chem Eng 2000;8(3–4):269–79.)*

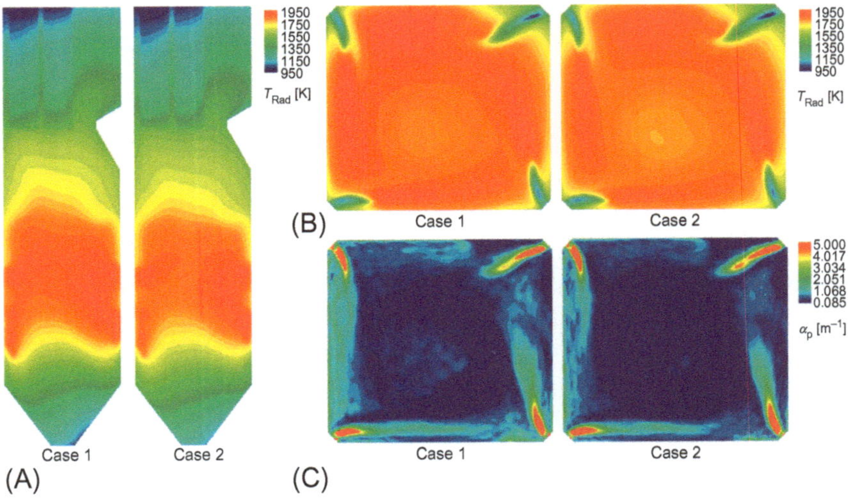

FIG. 9.4 (A) Radiation temperature on the middle cross-section between the two side walls, (B) radiation temperature on a horizontal cross-section, and (C) particle absorption coefficient on a horizontal cross-section. *(From Yin C. On gas and particle radiation in pulverized fuel combustion furnaces. Appl Energy 2015;157(15):59–70.)*

Yin et al. [48] evaluated the effects of emissivity and the scattering factor by comparing the constant and conversion-dependent emissivity and the scattering factor in a 609 MW pulverized coal utility boiler. The constant particle emissivity and scattering factor were 0.9 and 0.6 in case 1. The mutative emissivity and scattering factor were set as

$$\varepsilon_p = 0.4U_c + 0.6 \tag{9.29}$$

$$f_p = 0.9U_{VM,C} + 0.6\left(1 - U_{VM,C}\right) \tag{9.30}$$

where U_c is the fraction of unburnt char, and $U_{VM,C}$ is the fraction of unburnt combustibles. Fig. 9.4 shows the predicted radiation temperature on the vertical cross-section and horizontal cross-section of the burner. A slight difference can be found between the results of conversion-dependent particle radiative properties (case 2) and constant particle radiative properties (case 1).

9.4 0-DIMENSIONAL CALCULATION METHOD

The 0-dimensional method of calculating heat transfer is one of the most important calculations in boiler design, as it deals with the results of the furnace exit temperature and the heat transfer in the furnace. This method is based on the following assumptions: (1) The heat transfer and combustion processes are assumed to be two separate processes; (2) only the radiation heat transfer is considered in the furnace, and the convective heat transfer, which accounts for ~5%, is ignored; (3) the furnace is assumed to be a 0-dimensional system, so the temperature, gas composition, and particle concentration are uniform in the furnace; and the parameters used for 0-dimensional calculation come from the results in furnace exit; and (4) for a thermal calculation of a boiler design, the heat transfer calculations of the superheater and membrane wall are carried out separately, and the results of the furnace exit temperature and flue gas composition from the 0-dimensional calculation method is the basis for the superheaters design.

Based on the preceding assumptions, the radiation heat flux at the membrane wall can be simplified as

$$q = \varepsilon_f \sigma T_f^{\,4} - \left[\varepsilon_w \sigma T_w^{\,4} + (1-\varepsilon_w)\varepsilon_f \sigma T_f^{\,4}\right] \tag{9.31}$$

where ε_f is the total emissivity of the flame, σ is the Stefane-Boltzmann constant, T_f is the entire flame temperature, ε_w is the wall emissivity, and T_w is the wall temperature. It can be seen that the flame temperature, emissivity, and wall boundary conditions control the radiative heat transfer to the wall.

The flame emissivity represents the radiant capacity of the high temperature flame in the furnace, which will vary throughout the furnace. However, an average flame emissivity is used in the 0-dimensional calculation method, which is based on the flue gas composition and temperature in the furnace exit. According to Section 9.3, the radiation medium in the furnace includes triatomic molecules (H_2O and CO_2), char, fly ash, and soot particle. For the calculation of flame emissivity, the flame is considered as being gray medium. The flame emissivity is calculated by Eq. (9.32),

$$\varepsilon_f = 1 - e^{-kpS} \tag{9.32}$$

$$k = k_{\text{gas}}\gamma_{\text{gas}} + k_{\text{ash}}\gamma_{\text{ash}} + 10C_1C_2 \tag{9.33}$$

where k represents the total absorption coefficient; γ_{gas}, γ_{ash} represent the triatomic molecules' fraction and the fly ash concentration; C_1, C_2 are the dimensionless number, which takes into account the effect of char particle concentration—C_1 depends on the fuel type, and C_2 depends on the combustion mode. The gas absorption coefficient k_{gas} can calculate by the weight sum of the gray gas model. The fly ash absorption coefficient k_{ash} is mainly determined by the particles temperature, concentration, and size.

The wall emissivity is affected by the membrane walls degree of fouling and the furnace size. For a new boiler, little ash deposition is observed on the membrane wall, so the wall emissivity is small, for example, 0.45 [7]. For boilers used continuously over time, the wall emissivity will be greater, ~0.7–0.9 [8,10,49–51]. The membrane wall temperature is determined according to the saturation temperature of steam, and a temperature difference of 50–200 K between the steam and wall surface should be considered.

The radiation heat transfer obtained by the 0-dimensional calculation method is a complete and uniform heat transfer method. But the flame temperature and emissivity are inhomogeneous, so the heat transfer distribution varies along the furnace height. In order to determine the actual heating load of the heating surface of a certain area of the furnace, a heat load non-uniformity coefficient is introduced.

$$q_{fi} = \eta_i q_f \quad (9.34)$$

Here η_i is the heat load nonuniformity coefficient, which is selected in the monograph according to the specific condition.

9.5 HEAT TRANSFER IN OXY-FUEL COMBUSTION FACILITIES

In order to investigate the heat transfer properties in a practical oxy-fuel combustion facility, experimental and numerical investigations were carried out in a 35 MW oxy-coal combustion boiler. A subbituminous coal with medium ash of 23.94% and high volatile content of 23.77% was used as fuel. The oxy-fuel combustion furnace was a typical industrial front wall-fired furnace and was originally used under the air combustion conditions for power generation. The furnace was 3533, 4733, and 15,150 mm in depth, width, and height, respectively. Three 12 MW pulverized coal burners and two OFA nozzles were arranged in the front wall. Several groups of superheaters were installed in the upper part of the furnace.

During the experiment, one air combustion (Case Air) and two oxy combustions, that is, wet recycle (Case WetO28) and dry recycle (Case DryO28), were considered. The thermal loads in these three cases were ~30 MW. The initial O_2 level in the experiments under wet and dry recycle oxy combustions was maintained at ~28% by volume. The results of heat transfer across each heating surface were obtained from the statistics based on the monitoring data of distributed control systems according to the real-time monitoring of the temperature of water pipe and vapor pressure.

For the numerical simulation, the modified WSGG model by Guo et al. [30] was used for air combustion and oxy-fuel combustion. The flue gas and particles were treated as a single gray medium. The gas absorption coefficient k was calculated as

$$k = -\text{In}\left(1 - \varepsilon_{\text{total}}\right) / L \quad (9.35)$$

$$L = 3.6V / A \quad (9.36)$$

where L was the mean beam length of the furnace, V represented furnace volume, and A was the wall surface area. The experimental and numerical conditions were described in detail in a previous paper [7].

Fig. 9.5A shows the net heat flux (NHF) on the membrane wall for all cases from the CFD results. The trends in the NHF distribution were similar in all cases. The membrane wall was divided into 40 horizontal sections, and the area-averaged NHF area was calculated to make a more accurate comparison. Fig. 9.5B shows the profiles of the average NHF along the furnace height. The NHF in WetO28 was higher than in Air, while the thermal input in the burner

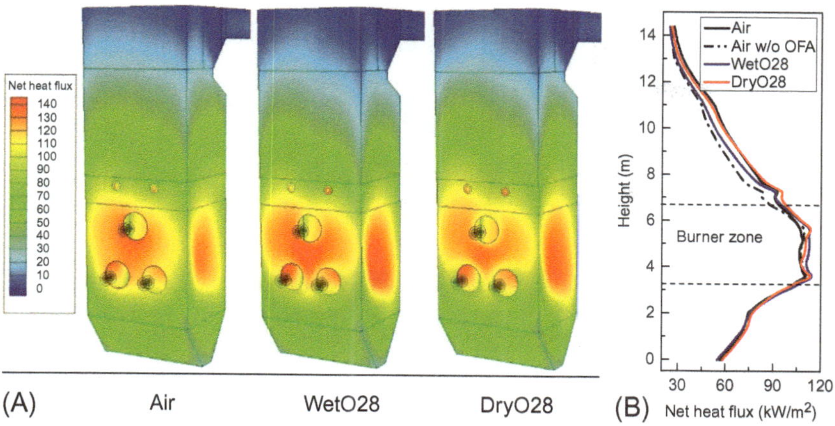

FIG. 9.5 Predicted net heat flux (kW/m^2) on the membrane wall for the different cases: (A) distributions on the walls and (B) average net heat flux profiles along the furnace height. *(From Guo J, Liu Z, Huang X, Zhang T, Luo W, Hu F, et al. Experimental and numerical investigations on oxy-coal combustion in a 35MW large pilot boiler. Fuel 2017;187:315–27.)*

zone was 28 MW in Air and WetO28. This difference was caused by the difference in the radiative property and temperature uniformity. The heat-transfer capability was enhanced by the increase in the absorption coefficient of gas and particles in oxy combustion. The difference in the NHF distribution between dry and wet recycles depended on the average temperature profile. One hypothetical air combustion case (Air w/o OFA) was computed without pulverized coal in exhaust gas to study the effect of the thermal input (~ 3 MW) in the exhaust gas from the mill system. This case indicated that the NHF in the upper part of the furnace under oxy conditions was between that in Air and Air w/o OFA. The NHF distribution along the furnace height was more uniform in oxy combustion when the impact of OFA was disregarded. This characteristic resulted from the enhanced radiative capacity of gas and particles.

The thermal inputs in the three cases slightly differed in the experiment. The transferred heat was normalized by the thermal input for comparison. Experimental and numerical results are summarized in Fig. 9.6. The numerical results agreed well with the experiment. Importantly, although the peak flame temperature of the air combustion was higher than that of the oxy combustion, the total heat transfer of the oxy combustion increased by ~9% and 6% for wet and dry recycle oxy-fuel conditions, respectively, relative to that of air combustion. This phenomenon was due to the extremely high CO_2 level and the high particle concentration in the radiative heat transfer. Compared with air combustion, the increased transferred heat to superheaters of oxy combustion resulted from the increased convective and radiative heat transfers. The transferred heat in the membrane wall increased by ~4% under the wet recycle condition compared with the dry recycle condition. This behavior resulted from the combined effect of the increase in average temperature in the burner zone and the increase in the gas radiative intensity.

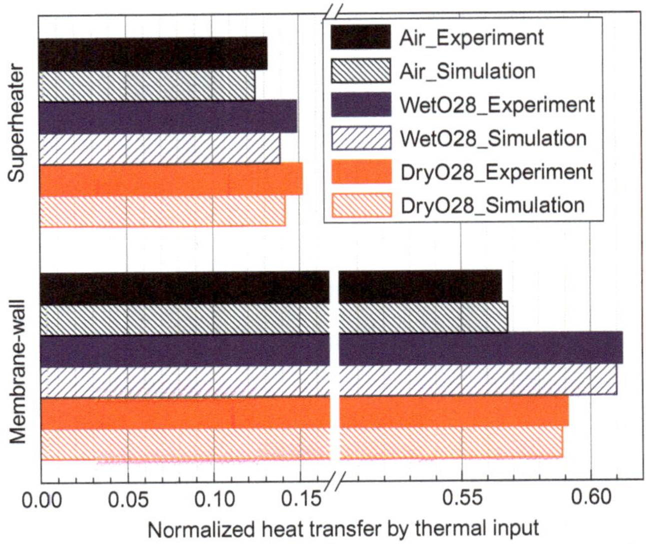

FIG. 9.6 Normalized heat transfer by thermal input of different cases. *(From Guo J, Liu Z, Huang X, Zhang T, Luo W, Hu F, et al. Experimental and numerical investigations on oxy-coal combustion in a 35MW large pilot boiler. Fuel 2017;187:315–27.)*

The O_2 concentration in the oxidant could be adjusted by controlling the flue gas recycle ratio, which was the ratio of the recycled gas volume flow rate to the total flue gas volume flow rate. The flow rate of recycled flue gas in the experiment increased from 31 to 35 t/h by adjustment of the blowers damper. Fig. 9.7A shows the normalized transferred heat of oxy combustion with O_2 levels of 28% and 27% by volume for the flue gas dry recycle ratios of 0.71 and 0.73, respectively. This characteristic indicated that the transferred heat in the membrane wall decreased, while the transferred heat in superheaters slightly increased. The flow rate of the main steam dropped (from 30.96 to 29.75 t/h) because of the decrease in the transferred heat in the membrane wall. Simultaneously, the degree of superheater overheating increased given the reduced flow rate of the superheated steam. Thus the superheater needed more desuperheating water (from 2.76 to 3.14 t/h). Fig. 9.7B shows that the decrease in the NHF appeared in the burner zone with an increase in the recycle ratio. This decrease was mainly due to the decrease in the flame temperature. The increased CO_2 concentration in the flue gas and the decreased O_2 concentration in oxidant led to the decrease in the flame temperature. The radiative heat transfer decreased with the increase in the recycle ratio. There was an optimal recycle ratio, as well as O_2 concentration, to match the transferred heat between oxy combustion and air combustion. In this study, with the initial O_2 concentration of 27% by volume, the combined effect of a slightly lower flame temperature and a higher flame emissivity achieved a transferred heat in the membrane wall similar to that in air combustion. Although the transferred heat ratio in the membrane wall

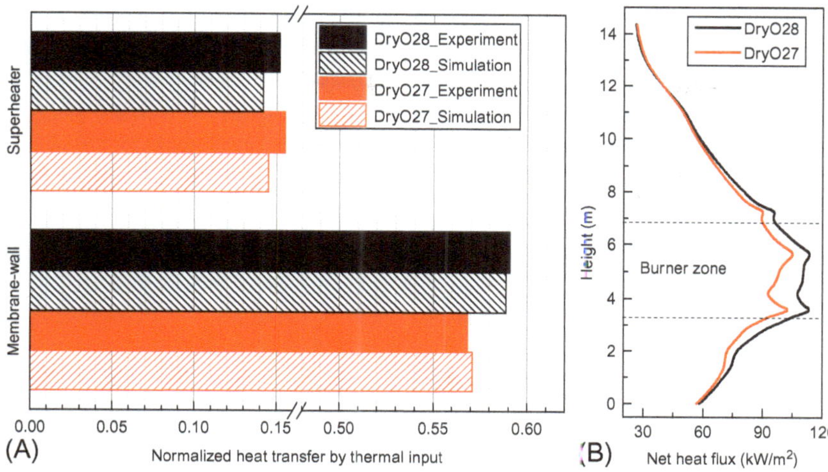

FIG. 9.7 (A) Normalized heat transfer by thermal input and (B) average net heat flux profile of oxy-combustion with different flue gas recycle ratios. *(From Guo J, Liu Z, Huang X, Zhang T, Luo W, Hu F, et al. Experimental and numerical investigations on oxy-coal combustion in a 35MW large pilot boiler. Fuel 2017;187:315–27.)*

in DryO27 was closer to that of Air, the observed desuperheating water flow rate increased, which led to the boilers reduced performance. Thus we recommend the use of a slightly higher O_2 concentration in the oxidant for practical applications—for example, DryO28, which was conducive to the increase in the main steam flow rate—and a reduction in the desuperheating water flow rate, thereby enhancing boiler performance. Flame stability will be affected by the decrease in the oxidant volume if the O_2 concentration is increased further. Therefore the choice of the optimized O_2 concentration is a trade-off between boiler performance and the flame stability of burners.

9.6 SUMMARY

Heat transfer plays an important role in industrial combustion processes. The selected models, input parameters, and boundary conditions have an important effect on the accuracy of heat transfer calculations. In oxy-fuel combustion facilities, the convective heat transfer coefficient calculation method should be modified according to the flue gas composition. For the consideration of heat transfer in oxy-fuel combustion, the radiative properties of gas, fly ash, char, and soot are very important. In oxy-fuel combustion, the gas radiation intensity increases due to the high concentration of the CO_2 and H_2O in the mixture gas. Compared to air combustion, the particle radiation intensity also increases in the oxy-fuel combustion cases, resulting from the increase in particle concentration, which is caused by the decrease in the flue gas volume.

In a 35 MW oxy-fuel combustion facility, at an initial O_2 level of ~28% by volume, the heat transfer of oxy-fuel combustion is greater than that of air combustion because of the large amount of CO_2 that participates in radiative heat transfer. The total heat transfer increases by ~9% and 6% for wet and dry recycle oxy-fuel conditions, respectively, relative to air combustion. The heat transfer under oxy-fuel combustion can be controlled by the flue gas recycle ratio. There is a specific flue gas recycle ratio to match the convective and radiative heat transfer with that of air combustion. The choice of the optimized flue gas recycle ratio, as well as the O_2 concentration, is a trade-off between boiler performance and the flame stability of burners.

REFERENCES

[1] Tan Y, Croiset E, Douglas MA, Thambimuthu KV. Combustion characteristics of coal in a mixture of oxygen and recycled flue gas. Fuel 2006;85(4):507–12.
[2] Andersson K, Johansson R, Hjärtstam S, Johnsson F, Bo L. Radiation intensity of lignite-fired oxy-fuel flames. Exp Thermal Fluid Sci 2008;33(1):67–76.
[3] Smart JP, O'Nions P, Riley GS. Radiation and convective heat transfer, and burnout in oxy-coal combustion. Fuel 2010;89(9):2468–76.
[4] Anheden M, Burchhardt U, Ecke H, Faber R, Jidinger O, Giering R, et al. Overview of operational experience and results from test activities in vattenfall's 30 MWth oxyfuel pilot plant in Schwarze Pumpe. Energy Procedia 2011;4:941–50.
[5] Fujimori T, Yamada T. Realization of oxyfuel combustion for near zero emission power generation. Proc Combust Inst 2013;34(2):2111–30.
[6] Lupion M, Alvarez I, Otero P, Kuivalainen R, Lantto J, Hotta A, et al. 30 MWth CIUDEN oxy-CFB boiler-first experiences. Energy Procedia 2013;37:6179–88.
[7] Guo J, Liu Z, Huang X, Zhang T, Luo W, Hu F, et al. Experimental and numerical investigations on oxy-coal combustion in a 35 MW large pilot boiler. Fuel 2017;187(1):315–27.
[8] Al-Abbas AH, Naser J, Dodds D. CFD modelling of air-fired and oxy-fuel combustion in a large-scale furnace at Loy Yang a brown coal power station. Fuel 2012;102:646–65.
[9] Black S, Szuhánszki J, Pranzitelli A, Ma L, Stanger PJ, Ingham DB, et al. Effects of firing coal and biomass under oxy-fuel conditions in a power plant boiler using CFD modelling. Fuel 2013;113(2):780–6.
[10] Guo J, Liu Z, Wang P, Huang X, Li J, Xu P, et al. Numerical investigation on oxy-combustion characteristics of a 200 MWe tangentially fired boiler. Fuel 2014;140(2):660–8.
[11] Arnold JO, Whiting EE, Lyle GC. Line by line calculation of spectra from diatomic molecules and atoms assuming a voigt line profile. J Quant Spectrosc Radiat Transf 1969;9(6):775–98.
[12] Goody RM. A statistical model for water vapor absorption. Q J R Meteorol Soc 1952;78(336):165–9.
[13] Malkmus W. Random lorentz band model with exponential-tailed S-1 line-intensity distribution function. J Opt Soc Am 1967;57(3):323–9.
[14] Becher V, Clausen S, Fateev A, Spliethoff H. Validation of spectral gas radiation models under oxyfuel conditions. Part a: gas cell experiments. Int J Greenhouse Gas Control 2011;5:S76–99.
[15] Edwards DK, Balakrishnan A. Thermal radiation by combustion gases. Int J Heat Mass Transf 1973;16(1):25–40.
[16] Modest MF, Zhang H. The full-spectrum correlated-k distribution for thermal radiation from molecular gas-particulate mixtures. J Heat Transf 2002;124(1):30–8.

[17] Smith TF, Shen ZF, Friedman JN. Evaluation of coefficients for the weighted sum of gray gases model. ASME J Heat Transf 1982;104:602–8.

[18] Denison MK, Webb BW, Denison MK. A spectral line-based weighted sum of gray gases model for arbitrary RTE solvers. J Heat Transf 1993;115(4):1004–12.

[19] Pierrot L, Soufiani A, Taine J. Accuracy of narrow-band and global models for radiative transfer in H_2O, CO_2, and H_2O/CO_2 mixtures at high temperature. J Quant Spectrosc Radiat Transf 1999;62(5):523–48.

[20] Goutiere V, Liu F, Charette A. An assessment of real-gas modelling in 2D enclosures. J Quant Spectrosc Radiat Transf 2000;64(3):299–326.

[21] Coelho PJ. Numerical simulation of radiative heat transfer from non-gray gases in three-dimensional enclosures. J Quant Spectrosc Radiat Transf 2002;74(3):307–28.

[22] Chu H, Liu F, Consalvi JL. Relationship between the spectral line based weighted-sum-of-gray-gases model and the full spectrum k -distribution model. J Quant Spectrosc Radiat Transf 2014;143(8):111–20.

[23] Modest MF. Radiative heat transfer. 2nd ed. New York: Academic Press; 2003.

[24] Solovjov VP, Webb BW. Global spectral methods in gas radiation: the exact limit of the SLW model and its relationship to the ADF and FSK methods. ASME J Heat Transf 2011;133(4):1–9.

[25] Becher V, Goanta A, Spliethoff H. Validation of spectral gas radiation models under oxy-fuel conditions—part C: validation of simplified models. Int J Greenhouse Gas Control 2012;11:34–51.

[26] Yin C, Johansen LCR, Rosendahl LA, Kær SK. New weighted sum of gray gases model applicable to computational fluid dynamics (CFD) modeling of oxy-fuel combustion: derivation, validation, and implementation. Energy Fuel 2010;24(12):6275–82.

[27] Johansson R, Leckner B, Andersson K, Johnsson F. Account for variations in the H_2O to CO_2 molar ratio when modelling gaseous radiative heat transfer with the weighted-sum-of-grey-gases model. Combust Flame 2011;158(5):893–901.

[28] Kangwanpongpan T, França FHR, Silva RCD, Schneider PS, Krautz HJ. New correlations for the weighted-sum-of-gray-gases model in oxy-fuel conditions based on HITEMP 2010 database. Int J Heat Mass Transf 2012;55(25–26):7419–33.

[29] Bordbar MH, Węcel G, Hyppänen T. A line by line based weighted sum of gray gases model for inhomogeneous CO_2–H_2O mixture in oxy-fired combustion. Combust Flame 2014;161(9):2435–45.

[30] Guo J, Li X, Huang X, Liu Z, Zheng C. A full spectrum k -distribution based weighted-sum-of-gray-gases model for oxy-fuel combustion. Int J Heat Mass Transf 2015;90:218–26.

[31] Scutaru D, Rosenmann L, Taine J. Approximate intensities of CO_2 hot bands at 2.7, 4.3, and 12 microns for high temperature and medium resolution applications. J Quant Spectrosc Radiat Transf 1994;52(6):765–81.

[32] Rothman LS, Gordon IE, Barber RJ, Dothe H, Gamache RR, Goldman A, et al. HITEMP, the high-temperature molecular spectroscopic database. J Quant Spectrosc Radiat Transf 2010;111(15):2139–50.

[33] Kangwanpongpan T, Silva RCD, Krautz HJ. Prediction of oxy-coal combustion through an optimized weighted sum of gray gases model. Energy Fuels 2012;41(1):244–51.

[34] Bohren CF, Huffman DR. Absorption and scattering of light by small particles. New York: Wiley; 1998.

[35] Kerker M. The scattering of light and other electromagnetic radiation. New York: Academic Press; 1969.

[36] van de Hulst HC. Light scattering by small particles. New York: Dover Publications; 1981.

[37] Gouesbet G, Gréhan G. Generalized Lorenz-Mie Theories. Berlin, Heidelberg: Springer; 2011.

[38] Fischer J, Gamache RR, Goldman A, Rothman LS, Perrin A. Total internal partition sums for molecular species in the 2000 edition of the HITRAN database. J Quant Spectrosc Radiat Transf 2003;82(1):401–12.

[39] Chui EH, Hughes PMJ, Raithby GD. Implementation of the finite volume method for calculating radiative transfer in a pulverized fuel flame. Combust Sci Technol 1993;92(4–6):225–42.

[40] Lockwood FC, Rizvi SMA, Shah NG, Lockwood FC, Rizvi SMA, Shah NG. Comparative predictive experience of coal firing. Proc Inst Mech Eng C J Mech Eng Sci 1986;200(2):79–87.

[41] Asotani T, Yamashita T, Tominaga H, Itaya Y, Mori S. Prediction of ignition behavior in a tangentially fired pulverized coal boiler using CFD. Fuel 2008;87(4):482–90.

[42] Tian ZF, Witt PJ, Schwarz MP, Yang W. Numerical modeling of Victorian brown coal combustion in a tangentially fired furnace. Energy Fuel 2010;24(9):4971–9.

[43] Yu MJ, Baek SW, Park JH. An extension of the weighted sum of gray gases non-gray gas radiation model to a two phase mixture of non-gray gas with particles. Int J Heat Mass Transf 2000;43(10):1699–713.

[44] Ma L, Gharebaghi M, Porter R, Pourkashanian M, Jones JM, Williams A. Modelling methods for co-fired pulverised fuel furnaces. Fuel 2009;88(12):2448–54.

[45] Al-Abbas AH, Naser J, Dodds D. CFD modelling of air-fired and oxy-fuel combustion of lignite in a 100 KW furnace. Fuel 2011;90(5):1778–95.

[46] Zhang J, Prationo W, Zhang L, Zhang Z. Computational fluid dynamics modeling on the air-firing and oxy-fuel combustion of dried victorian brown coal. Energy Fuel 2013;27(8): 4258–69.

[47] Liu Z, Zheng C, Xing H. Radiative properties of residual char and its effects on radiative heat transfer in pulverized coal fired furnaces. Asia Pac J Chem Eng 2000;8(3–4):269–79.

[48] Yin C. On gas and particle radiation in pulverized fuel combustion furnaces. Appl Energy 2015;157:554–61.

[49] Lockwood FC, Rizvi SMA, Lee GK, Whaley H. Coal combustion model validation using cylindrical furnace data. Symp Combust 1985;20(1):513–22.

[50] Zheng C, Liu Z, Duan X, Mi J. Numerical and experimental investigations on the performance of a 300 MW pulverized coal furnace. Symp Combust 2002;29(1):811–8.

[51] Henriksen M, Aaboe J, Gravennielsen T, Bliddal H, Langberg H. Experimental and numerical investigations on a swirl oxycoal flame. Appl Therm Eng 2012;49:161–9.

Chapter 10

Pilot and Industrial Demonstration of Oxy-fuel Combustion

Zhaohui Liu, Tai Zhang
Huazhong University of Science and Technology, Wuhan, China

10.1 INTRODUCTION

Oxy-fuel combustion technology was first proposed by Abraham in 1982 with the aim of generating CO_2 for enhanced oil recovery (EOR). With the intensification of global warming and climate change, CO_2 has gradually attracted worldwide attention as a major factor in greenhouse gas emissions. Therefore oxy-fuel combustion technology as the most potential combustion technology to reduce carbon dioxide emissions has become a hot topic for global researchers.

Fig. 10.1 shows the historical progress, from the 1980s onward, of oxy-fuel combustion technology, moving from the lab-scale level to that of commercial application [1]. At present, oxy-fuel combustion technology is mainly developed and studied in the United States, China, Japan, Canada, Australia, Britain, Spain, France, The Netherlands, and so on. The major research institutions and companies include Energy & Environmental Research Center (EERC), Argonne National Laboratory (ANL), B&W and Air Liquide, Alstom, IHI, CANMET, IFRF, BHP, Newcastle, CS Energy, CIUDEN, Doosan Babcock, and Vattenfall.

10.2 SMALL PILOT FACILITIES

The term "small pilot facilities" refers to laboratory-scale facilities. Developed countries such as Japan, Canada, the United Kingdom, the Netherlands, France, Germany, and Sweden have invested heavily in the research and development of oxy-fuel combustion methods and technologies since the early 1980s. The research work on the O_2/CO_2 recycle combustion experiment has focused on the fundamental issues of technical application, especially on combustion characteristics, coal reactions, heat transfer, and pollutant emissions. Buhre et al. [2] and Toftegaard et al. [3] described the experimental studies on

Oxy-fuel Combustion. http://dx.doi.org/10.1016/B978-0-12-812145-0.00010-4

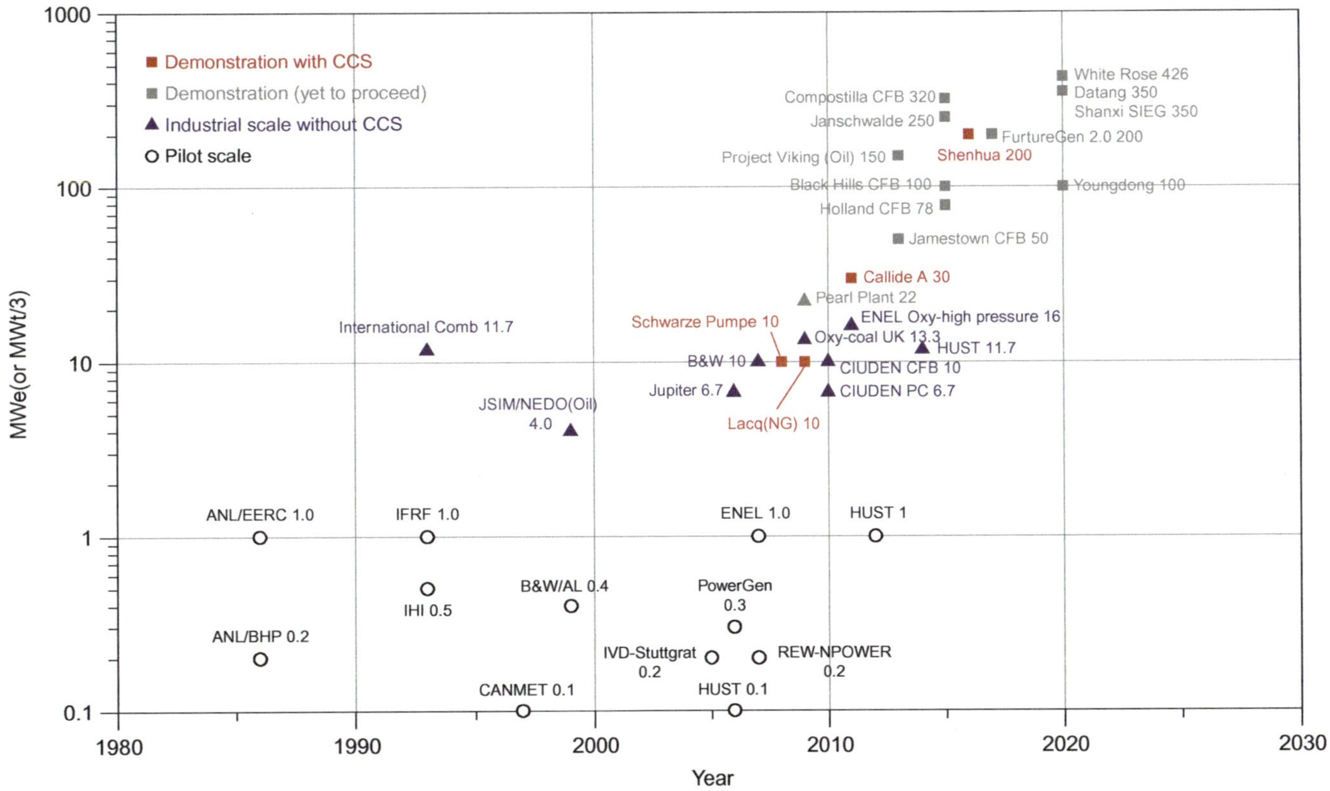

FIG. 10.1 Historical progression of oxy-fuel combustion technology. *(Modified from Liu, 2016; Stanger R, Wall T, Spörl R, Paneru M, Grathwohl S, Weidmann M, et al. Oxyfuel combustion for CO2 capture in power plants. Int J Greenhouse Gas Control 2015;40:55–125. http://dx.doi.org/10.1016/j.ijggc.2015.06.010.)*

O_2/CO_2 recycle combustion in detail, as shown in Table 10.1, which also includes information on the most important parameters and study results of small pilot-scale facilities.

Next some important progress that has been made in recent years is presented.

In 1988 ANL and EERC established the world's first O_2/CO_2 recycle combustion test facility, which proved the feasibility of O_2/CO_2 recycle combustion technology, and conducted quantitative observation of operational changes such as flame stability, emission of pollutants, and burnout [4–6].

In the 1990s CO_2 capture and storage technology began to receive extensive attention, and most of the research focused on the reform of existing small-scale power stations. From 1993 to 1995, the International Flame Research Foundation (IFRF) organized a research consortium that included Mitsui Babcock, Air Products, International Combustion, Ulster University, Imperial College of Technology, and the University of Naples. The researchers studied the combustion characteristics of oxy-fuel combustion, including ignition, flame stability, heat transfer, combustion efficiency, and pollutant emission.

From 1992 to 2000 IHI in Japan used NEDO funds to organize a research society on O_2/CO_2 recycle combustion that covered a wide range of topics, from technical and economic studies to flame characteristics and pollutant emissions [7–11].

In 1999 Canadas CANMET Energy Technology Center built a 300kW vertical reactor to carry out O_2/CO_2 recycle combustion research. The research investigating emissions from oxy-fuel combustion, validating modeling tools, and examining burner and boiler design [12–14].

In 2002 B&W and Air Liquide established a 1.5MW oxy-fuel combustion facility with an air-staged combustion system and studied the impact of the oxy-fuel combustion process on pollutant emissions and boiler efficiency, demonstrating the technical feasibility of conversion from air firing to oxy-fuel combustion in large-scale boilers.

In 2006 Huazhong University of Science and Technology (HUST) carried out numerous, productive research and development projects in a small pilot facility and built a $300\,kW_{th}$ oxy-fuel combustion facility, which was the first small pilot-scale oxy-fuel combustion facility in China.

In 2012 HUST developed the first $3\,MW_{th}$ facility for oxy-fuel coal combustion in China, which is shown in Fig. 10.2. The distinct feature of the $3\,MW_{th}$ oxy-coal test facility is that it is a full-scope integration from air separation, to oxy firing boiler, to GPU and production of compressed CO_2. The facilitys design concept took root in March 2010, construction began in February 2011, and the facility was completed in December 2011. It was the worlds largest and most advanced test performance system for complete processing of oxygen-enriched combustion. Therefore it acted as an easy-to-operate and lower-cost alternative to larger scale 30 and 50 MWe pilot test oxy-fuel plants (such pilot plants have been built at Schwarz Pumpe, Lacq, and Callide.) The 3MWt facility is designed

TABLE 10.1 Oxy-fuel Groups Working in Pilot Scale

Group	Test Unit (MW_{th})	Unit Description	Recycle: Wet/Dry	Coal Type	Aim of Work (Interpretation of the Authors)
ANL and EERC	3	Tower furnace, single swirl burner	Wet/dry	Subbituminous	Determine the feasibility of oxy-fuel combustion technology
IFRF	2.5	Horizontal furnace, air-staged swirl burner	Wet	Bituminous	Implementation of oxy firing
IHI	1.2	Vertical, down-fired furnace, swirl burner	Wet/dry	Bituminous	Implementation of oxy firing; direct injection of O_2; ignition characteristics; effect of load changes; ash impacts
CANMET	0.3	Vertical combustor, down-fired swirl burner	Wet/dry	Bituminous, subbituminous, lignite, NG	Emissions from oxy-fuel combustion; modeling tool validation; burner and boiler design
B&W and AL	1.5	Pilot-scale boiler, air-staged combustion system	Dry	Subbituminous	Demonstrating technical feasibility of conversion from air firing to oxy-fuel combustion for large-scale boiler
IVD	0.5	Vertical furnace, swirl burner	Wet	Bituminous, lignite	Identification of variations in combustion behavior; impact on facility components other than furnace; continuous operation for 100 h
HUST	0.3	Vertical combustor, down-fired swirl burner	Wet/dry	Subbituminous	Validation of oxy-fuel concept achieved; completed design and optimization of low NO_x oxy-fuel burner
HUST	3	Front wall furnace, swirl burner	Wet/dry	Subbituminous	

From Toftegaard MB, Brix J, Jensen PA, Glarborg P, Jensen AD. Oxy-fuel combustion of solid fuels. Prog Energy Combust Sci 2010;36(5):581–625. http://dx.doi.org/10.1016/j.pecs.2010.02.001.

FIG. 10.2 3 MW_{th} oxy-fuel coal combustion facility in Wuhan, China.

to provide flexibility to test and validate several critical components of oxy-fuel pants, including

- ASU and CO_2 compression
- oxy-coal burner
- boiler material
- gas processing unit
- humidity condenser
- system integration and process optimization

The 3 MWt facility verified the oxy-fuel combustion technology. The results had important implications for amplifying the design of large pilot-scale and demonstration oxy-fuel combustion facilities. The prominent results included the following: (1) the whole system operated stably for a long period of time; (2) switches between air to oxy-fuel and oxy-fuel to air were safe and stable; (3) enrichment of high concentration CO_2 was achieved in the flue gas, and the CO_2 concentration of product gas reached 95% from the outlet of the compression purification unit; and (4) the characteristics and control methods of system heat transfer and pollutant emission under oxy-fuel combustion were obtained.

The studies in the small pilot facilities indicated that

(1) Oxy-fuel combustion technology of pulverized coal is feasible.
(2) There were no large technical obstacles in the pilot-scale experiments.
(3) Oxy-fuel combustion technology is suitable for the transformation of existing boilers.

10.3 LARGE PILOT FACILITIES

The results of laboratory-scale studies hold theoretical implications in terms of their effects on combustion characteristics. But there are also some limitations for the commercial applications of oxy-fuel combustion technology. And these need to be studied and validated on large pilot facilities, such as heat transfer characteristics and the formation of pollutants. Air Liquide, CANMET, IFRF, and IHI conducted pilot studies at or above the pilot-scale level and conducted feasibility studies and economic evaluation of oxy-fuel combustion technology on these facilities. A feasibility study compared oxy-fuel combustion technology to flue gas separation technology and IGCC technology, and the results showed that oxy-fuel combustion had obvious advantages. With the increasing interest in the research of oxy-fuel combustion technology, many large pilot facilities are developing around the world, as shown in Table 10.2.

10.3.1 Vattenfall's Schwarze Pumpe Pilot Plant (Germany)

Vattenfall's 30 MW_{th} oxy-fuel pilot plant has operated since 2008. It is located next to the Schwarze Pumpe Power Plant in Brandenburg, Germany. It has operated for about 18,000 h, with >13,000 h in oxy-fuel combustion mode [15]. The facility was designed for technical verification of the commercial oxy-fuel plant and the integrated operation of the whole system, including boiler, flue gas cleaning system (FGC), air separation unit (ASU) and compression and purification unit (CPU).

The main focus of the plant is to explore and optimize the flue gas recirculation in oxy-fuel combustion, as well as the feasibility of the main components of the oxy-fuel process, including the air separation system, the boiler system with flue gas recirculation—including ESP, flue gas desulfurization (FGD), and flow gas condenser (FGC)—and the CO_2 compression and liquefaction systems. At the same time, the direct injection of sorbent for in-furnace SO_x removal and the injection of ammonia in the SCR for NO_x removal can be tested on the boiler.

In oxy-fuel conditions, the primary recycled flue gas is taken after the FGC, and the secondary recycled flue gas is taken after the ESP. It can be seen that using wet recycle conditions should improve overall efficiency and that high humidity and SO_x content in wet recycled flue gas is not significantly impacted due to corrosion after >10,000 h of operation [16].

10.3.2 Total's Lacq Pilot Plant (France)

Total's 30 MW_{th} gas-fired oxy-fuel pilot plant was the first large-scale facility in the world dedicated to storing CO_2 taken from an oxy-fuel combustion plant in a depleted gas field [17]. The plant was retrofitted with oxy-fuel combustion from a 35 MW_{th} power plant using 4 × 8 MW_{th} low NO_x burners provided by Air Liquide [17–19]. The CO_2 capture facility also comprised a steam generation plant, a small, standard ASU, a flue gas condenser, and a CO_2 processing unit. The Lacq plant operated >11,000 h and successfully stored ~51,000 tonnes of CO_2. To improve the safety of operation, the gas distribution system had a special circuit for pure oxygen.

TABLE 10.2 Large Worldwide Pilot Facilities

	Name of Demonstration/Plant	MW_{th}	New/ Retrofit	Start-Up Time	Fuel	Power	CPU	CO_2 Purity (%)	Flue Gas Purification
1	Vattenfall (Germany)	30	New	2008	Coal	No	Yes	99.90	SCR ESP
2	Total (France)	30	Retrofit	2009	Natural gas	Yes	Yes	99.90	FGD
3	CIUDEN (Spain)	30	New	2010	Coal	Yes	Yes		SCR FF
4	Callide (Australia)	30 MWe	Retrofit	2010	Coal	Yes	Yes		FF
5	HUST (China)	35	Retrofit	2011	Coal	No	No	85	EPS

The flue gas condenser included a venturi scrubber and a water spray tower that could cool the flue gas from 200°C to 30°C before being fed to the CO_2 compressor. The gaseous CO_2, compressed and separated by a three-stage reciprocating compressor, was delivered to the pipeline nearly bone dry. A CO_2 booster compressor (compressing the gas to 50 bar) and a single injection well were at the storage site. The CO_2 storage reservoir was a depleted gas field with a depth of 4500 m with 3% porosity. Post-monitoring of the storage site is currently ongoing.

10.3.3 CIUDEN's Technology Development Facilities (Spain)

The aim of CIUDEN's CCS Technology Development Plant (TDP) is to drive CCS technologies toward commercialization, as part of a Spanish government initiative. The test facility for CO2 capture is located beside Endesa's Compostialla Thermal Power Station near Ponferrada, Spain.

There are three large-scale test rigs for oxy-PC (20 MW_{th}) and oxy-CFB (15–30 MW_{th}) and a small-scale biomass gasifier (1 MW_{th}) in the plant. Also, a number of flue gas cleaning units (SCR, bag filter, FGD), a CO_2 processing unit, and a closed loop CO_2 pipeline transport test rig are associated with these facilities.

CIUDEN's TDP CO_2 capture project developed and demonstrated the use of the large-scale oxy-CFB pilot plant. The 30 MW_{th} CFB facility is a first-of-its-kind, world-class facility; the CFB boiler with natural circulation comprises a combustion chamber, a cyclone, loop seals, a fluidized bed heat exchanger, superheaters, evaporators, and an economizer. Limestone and made-up sand is used to control SO_x emissions and maintain the proper bed characteristics. To reduce the NO_x, there is NH3 injection at the top of the cyclone. The impurities of the flue gas are removed by the bag filter and flue gas condenser. Then the flue gas is compressed and further purified by the CO_2 compression and CPU. The CPU is capable of handling 160 TPD of CO_2 in the warm part and 10 TPD of CO_2 in the cold part of the CPU.

10.3.4 COSPL's Callide Oxyfuel Project (Australia)

The Callide Oxyfuel Power Plant is the world's first and largest demonstration coal-fired power plant to operate with oxy-fuel combustion technology.

The Callide-A Power Plant was built and commissioned in 1965 and is located near Biloela, Queensland (Australia). This plant includes four units, each with a nominal capacity of 30 MWe. Unit 4 was retrofitted with oxy-fuel combustion and successfully commissioned in 2012. It has achieved >10,000 h of operation and around 5500 h in oxy-fuel combustion mode, delivering electricity and capturing CO_2 in a real power plant scenario.

The boiler is equipped with three bitter mills to feed the coal to the burners. The boiler has six burners in 3×2 front wall arrangements (with four burners in service). Two new IHI burners were installed to replace the old burners in the middle row. The boiler is a two-drum type boiler without a reheater and produces steam at 460°C and 4.3 MPa (~43 bar), and the steam turbine delivers electricity at 132 kV.

In order to achieve oxy-fuel combustion, the flue gas recirculation system was retrofitted, including additional air heaters (in existing air heaters), piping systems, new FD & ID fans, coolers/feed water heaters (at the inlet of the baghouse), new control systems, and additional power systems for flue gas treatment and storage of CO_2. The CPU at 75 TPD CO2 is designed to process up to 17% of the flue gas. The size was determined by the smallest centrifugal compressor available on the market.

10.3.5 HUST Industrial Project for 35 MW_{th} Oxyfuel Pilot Plant (China)

Yingcheng 35 MW_{th} oxy-fuel combustion industrial power plant [20], located in Central China, is the largest oxy-fuel combustion facility in the world and is the only industrial operating project in China. It is a newly built plant and is designed to enable both air combustion and oxy-fuel combustion. This plant was successfully commissioned in 2015. A simplified process flow diagram of the Yingcheng 35 MW_{th} oxy-fuel pulverized coal boiler is illustrated in Fig. 10.3. The boiler island contains a medium-pressure drum boiler, an intermediate storage ball mill pulverizing system, and a gas cleaning system. The gas cleaning system contains a bag filter, a FGD unit, and a flue gas condenser.

Three flue gas recirculation pipes are built after and before the flue gas condenser and the FGD for dry recycle oxy-fuel combustion, for wet recycle oxy-fuel combustion, and for wet recycle oxy-fuel combustion with an SO_2 recycle, respectively. The recirculation flue gas is divided into primary air and secondary air, both of which are preheated to a special temperature by the air preheater. The ASU and flue gas recycle system are installed to achieve oxy-fuel combustion. The ASU is a cryogenic ASU with a maximum capacity of 7200 Nm^3/h that supplies high purity oxygen (>97%) to the boiler. Oxygen from the ASU is mixed with recirculated flue gas and is directly injected into the primary and secondary air in the downstream and upstream of the air preheater via O_2 injectors, respectively. It's noteworthy that an intermediate storage ball mill pulverizing system was chosen to reduce air leakage instead of positive-pressure direct-firing pulverizing system.

Stable air combustion and oxy-fuel combustion were successfully achieved, and a reliable switching strategy was adopted for a smooth and stable shift process. Both four-step and simultaneous one-step transition strategies proved to be reliable. Under oxy-fuel combustion with dry recycle, a CO_2 purity of 82.7% by volume in dry flue gas was reached, and the boiler efficiency was higher than

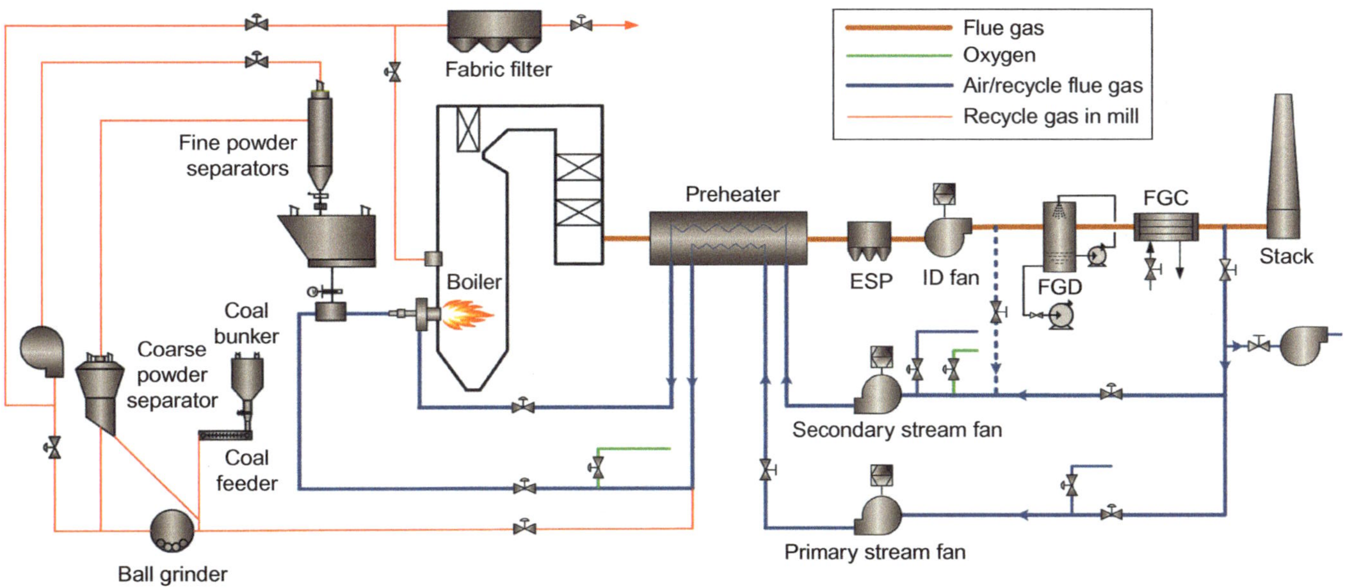

FIG. 10.3 Simplified process flow diagram of Yingcheng 35 MW_{th} oxy-fuel pulverized coal boiler. *(From Guo J, Liu Z, Huang X, Zhang T, Luo W, Hu F, et al. Experimental and numerical investigations on oxy-coal combustion in a 35 MW large pilot boiler. Fuel 2017;187(1):315–27. http://dx.doi.org/10.1016/j.fuel.2016.09.070.)*

that under air combustion conditions. The matched heat transfer distribution under air combustion and oxy-fuel combustion conditions showed that this design can successfully retrofit exiting boilers.

10.4 LEARNING FROM SMALL AND LARGE PILOT PLANTS OXY-FUEL COMBUSTION

With the successful operation of a series of small and large pilot power plants, validation of oxygen-enriched, combustion carbon capture technology in the entire pilot plant process was basically completed. It was verified that the technology had the capacity to build a 200–600 MW commercial-scale technically sound power plant.

10.4.1 Emission of NO_x and SO_2

To understand the effects of oxy-fuel combustion on the emission of NO_x and SO_2, measurements of NO_x and SO_2 under different conditions were carried out in all small and large pilot facilities with oxy-fuel combustion. The emissions of energy based on the mg/MJ of NO_x and SO_2 under different conditions from different oxy-fuel facilities are shown in Figs. 10.4 and 10.5.

As can be seen from Fig. 10.4, the NO_x emission under oxy-fuel mode was 0.27 to 0.35 times that under air mode, which showed low NO_x characteristics

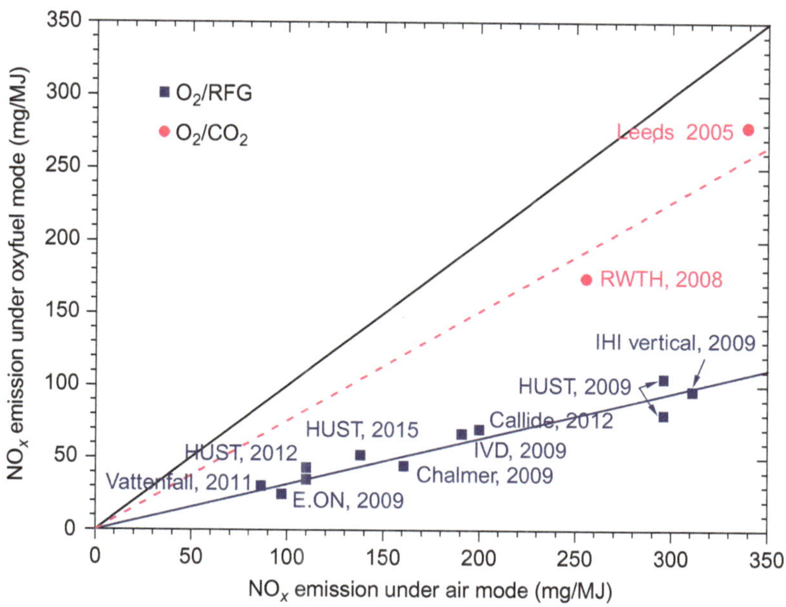

FIG. 10.4 NO_x emissions for air combustion and oxy-fuel combustion.

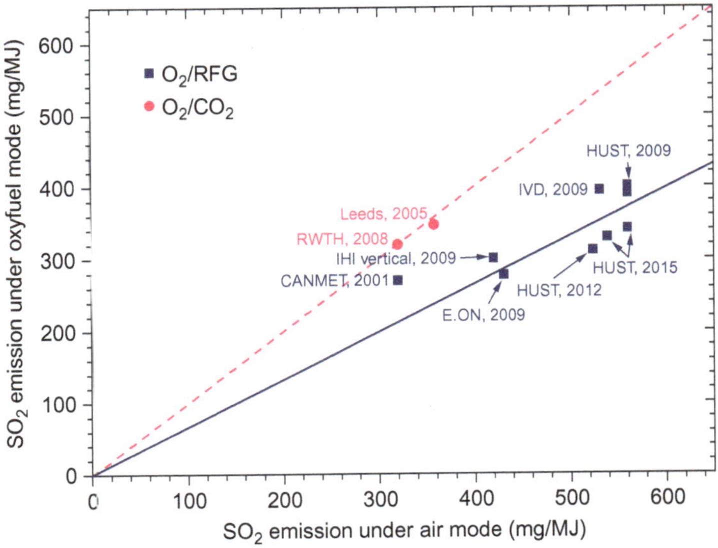

FIG. 10.5 SO_2 emissions for air combustion and oxy-fuel combustion.

in oxy-fuel combustion. From Fig. 10.5, it can be seen that typical SO_2 emissions from the boiler were generally the same in both air and oxy-fuel combustion. All results indicated that a high CO_2 concentration in the flue gas under oxy-fuel combustion had no effect on the desulfurization efficiency of WFGD [10–20].

10.4.2 Recommendations for Amplification

Currently, worldwide feasibility studies and preliminary work on commercial-scale demonstration projects with unit capacities of 160–600 MW_{th} have been carried out. The unit capacity of planning and construction was 168–426 MW_{th}, and the unit parameters were from subcritical to ultra-supercritical. The following factors were garnered from small and large pilot projects on oxy-fuel combustion:

(1) A single burner is generally used in 30–40 MW_{th}-scale pilot plants, which apply a 1:1 ratio at the 600 MW level, and wall combustion is the mode of most pilot plants. The combustion mode of Alstom's 15 MW pilot plant is tangential combustion with four DC burners, and the experimental studies show that no difference exists between the two combustion modes in terms of combustion stability and performance, indicating that both combustion modes adapt to oxy-fuel combustion.

(2) The reducing of net efficiency (absolute value) of an entire plant can be controlled in 6%–10% with the optimization of a fully integrated system for commercial-scale oxy-fuel combustion carbon capture coal-fired power plants. Optimizing system integration is the main way to improve efficiency. For example, the efficiency of Vattenfalls 250 MW supercritical demonstration plant is 36%, which is equivalent to 342 g/kWh coal consumption.

(3) Large-scale air separation plants progression and technologies have matured to the point of fully meeting commercial-scale requirements. Because of the high consumption of water in air separation systems, air-cooling systems can be adopted, but the investment for doing so is costly, and the overall efficiency of the plant is reduced. The seawater DC circulation system can also improve the overall efficiency of a plant, but again the cost of investment increases.

(4) In the current research on commercial-scale demonstration power plant technology, the system design is seen as being compatible with both air combustion and oxy-fuel combustion. This design can alleviate high operational costs, but the system configuration is complex, and the investment is expensive.

At present no commercial-scale projects are under construction. Rather than technological barriers, the main constraints on the development of large-scale demonstration facilities include the lack of funds and government policy support, local carbon sequestration regulations, unestablished transmission networks, licensing, and so on.

10.5 SUMMARY

This chapter summarized the progress of small- and large-scale pilot facilities across the globe. These small- and large-scale projects are critical steps toward developing large-scale demonstration facilities, and as a result, significant progress has been achieved in demonstration plants worldwide. Over the past 5 years, the Schwarze Pumpe, CUIDEN, Callide, and Yingcheng projects confirmed the oxy-fuel combustion concept. Studies of small- and large-scale pilot oxy-fuel combustion facilities show the maturity and readiness of oxy-fuel combustion technology and that it can help reduce CO_2 emissions. However, the construction of demonstration power plants is subject to a variety of factors that cause uncertainty about their practical, real-world feasibility. Nevertheless, it is clear that with further planning and work, oxy-fuel combustion technology will be implemented eventually in large-scale demonstration plants.

REFERENCES

[1] Stanger R, Wall T, Spörl R, Paneru M, Grathwohl S, Weidmann M, et al. Oxyfuel combustion for CO_2 capture in power plants. Int J Greenhouse Gas Control 2015;40:55–125.

[2] Buhre BJP, Elliott LK, Sheng CD, Gupta RP, Wall TF. Oxy-fuel combustion technology for coal-fired power generation. Prog Energy Combust Sci 2005;31(4):283–307.

[3] Toftegaard MB, Brix J, Jensen PA, Glarborg P, Jensen AD. Oxy-fuel combustion of solid fuels. Prog Energy Combust Sci 2010;36(5):581–625.

[4] Wang CS, Berry GF, Chang KC, Wolsky AM. Combustion of pulverized coal using waste carbon dioxide and oxygen. Combust Flame 1988;72(3):301–10.

[5] Yamada T, editor. Development of CO2 recovery type pulverized coal fired plant applied oxygen and recycled flue gas combustion. 10th Japan/Australia joint technical meeting on coal proceedings, Fukuoka, Japan; 2000.

[6] Wolsky AM. In: A new method of CO2 recovery. 79th annual meeting of the air pollution control association, Minneapolis, MN; 1986.

[7] Kimura N, Omata K, Kiga T, Takano S, Shikisima S. The characteristics of pulverized coal combustion in O2/CO2 mixtures for CO2 recovery. Energy Convers Manag 1995;36(6–9): 805–8.

[8] Wall T, Liu Y, Spero C, Elliott L, Khare S, Rathnam R, et al. An overview on oxyfuel coal combustion—state of the art research and technology development. Chem Eng Res Des 2009;87(8):1003–16.

[9] Kiga T, Takano S, Kimura N, Omata K, Okawa M, Mori T, et al. Characteristics of pulverized-coal combustion in the system of oxygen/recycled flue gas combustion. Energy Convers Manag 1997;38(96):S129–34.

[10] Wall TF. In: Performance of PF boilers retrofitted with oxy-coal combustion: understanding burnout, coal reactivity, burner operation, and furnace heat transfer. 3rd workshop of the IEA GHG International Oxy-Combustion Network, Yokohama, Japan; 2008.

[11] Wall TF, Elliott L, Khare S, Liu Y, Yamada T, Tamura M, et al, editors. Ash impacts in oxy-fuel combustion. Proceedings: impacts of fuel quality on power production, UT, USA; 2006.

[12] Eric C, Kelly T, Allan P. Coal combustion in O2/CO2 mixtures compared with air. Can J Chem Eng 2000;78(2):402–7.

[13] Croiset E, Thambimuthu KV. NOx and SO2 emissions from O 2/CO 2 recycle coal combustion. Fuel 2001;80(14):2117–21.

[14] Tan Y, Douglas MA, Thambimuthu KV. CO2 capture using oxygen enhanced combustion strategies for natural gas power plants. Fuel 2002;81(8):1007–16.

[15] Strömberg L, Lindgren G, Jacoby J, Giering R, Anheden M, Burchhardt U, et al. Update on Vattenfall's 30 MWth oxyfuel pilot plant in Schwarze Pumpe. Energy Procedia 2009;1(1): 581–9.

[16] Anheden M, Burchhardt U, Ecke H, Faber R, Jidinger O, Giering R, et al. Overview of operational experience and results from test activities in Vattenfall's 30 MWth Oxyfuel pilot Plant in Schwarze Pumpe. Energy Procedia 2011;4:941–50.

[17] Copin D. In: The storage dimension of the oxy-combustion based integrated CCS project of Lacq and Rousse. 3rd oxyfuel combustion conference, Ponferrada, Spain; 2013.

[18] Marcano N, Recourt P, Tsiava R, Laurent J, Lethier S, Deveaux M, et al. In: Oxy-combustion at Lacq CCS pilot plant: preliminary analysis of burner-boiler performance. 2nd oxyfuel combustion conference (OCC2); Yeppoon, Australia; 2011.

[19] Pham J, Lethier S, Marcano N, Recourt P. In: Operating experience and performance characteristics of gas-oxy combustion technology at total's carbon capture and storage demonstration plant. Power-Gen International, Fl, USA; 2012.

[20] Guo J, Liu Z, Huang X, Zhang T, Luo W, Hu F, et al. Experimental and numerical investigations on oxy-coal combustion in a 35 MW large pilot boiler. Fuel 2017;187(1):315–27.

Chapter 11

System Integration and Optimization for Large Scale Oxy-fuel Combustion Systems

Bo Jin*, Haibo Zhao†, Zhaohui Liu†
*Hunan University, Hunan, China
†Huazhong University of Science and Technology, Wuhan, China

11.1 INTRODUCTION

Oxy-fuel combustion, as a promising pathway to capturing enormous amounts of CO_2 emissions at power plants, takes place in a CO_2/O_2 atmosphere rather than in a N_2/O_2 atmosphere and produces flue gas that is enriched from about 80 mol.% to about 90 mol.% CO_2. Compared to conventional air combustion, an oxy-fuel combustion plant's configuration is more complex because an oxygen-supplying system, a flue gas recirculation system, and a CO_2 production system must be considered. However, additional systems, especially the air separation unit (ASU) and CO_2 compression and purification unit (CPU), lead to high energy penalties and economic costs. To reduce such adverse effects on its commercial use, great efforts have been made to improve its thermodynamic behavior and to verify its economic feasibility. This chapter gives a brief overview on these issues, mainly focusing on steady-state simulation, exergy analysis, system integration, process optimization, and techno-economic evaluation of large-scale oxy-fuel combustion power plants.

11.2 PROCESS CHARACTERISTICS

Process simulation is a powerful tool for understanding process characteristics because it can be used to identify thermodynamic behavior and manifest the effects of operating parameters on a system's operating performance. Based on the process flow diagram of the studied system, first a steady-state process model was established to obtain the stream and energy flows, which laid the foundation for thermodynamic analysis. Then steady-state simulation was conducted to investigate the influence of the operating parameters on the system's

Oxy-fuel Combustion. http://dx.doi.org/10.1016/B978-0-12-812145-0.00011-6

operation. Moreover, the simulation results provided important data for implementing sophisticated analyses such as exergy analysis, thermoeconomic evaluation, and techno-economic assessment. In this section, steady-state simulation for a 600 MWe oxy-fuel combustion system [1,2] is used to illustrate the process characteristics.

11.2.1 Steady-State Model

As shown in Fig. 11.1, the analyzed oxy-fuel combustion power plant presented consisted of four subsystems: ASU, boiler island, steam cycle, and CPU. Before building the steady-state model in Aspen Plus (V7.1), some necessary assumptions were established: (1) Because the cryogenic ASU was the only available commercial oxygen production technology for the application of oxy-fuel combustion [3,4], it would be adopted to supply the required oxygen for the oxy-fuel boiler island. (2) In the boiler island, a simplified furnace model to capture the main macro-thermodynamic features of this complicated combustion process would be considered [5]. The generated flue gas with a high temperature would flow through the heat transfer section in the furnace to convert feed water into steam; then oxidant streams would be heated by an air heater. A water wall (WW), a superheater (SH), a reheater (RH), and an economizer (ECO) would be used to characterize the preceding heat transfer process. Before part of the flue gas entered the CPU, it would be treated by an air quality control system (AQCS), where the electrostatic precipitator (ESP), flue gas desulphurization (FGD), and flue gas cooler (FGC) would be considered. Another part (about 70% in volume) of the dried flue gas would be recycled and mixed with oxygen from the ASU to be sent to the oxy-fuel burner, a process that could be defined as dry recirculation. (3) For the CPU, a distillation type [1] would be employed to obtain CO_2 products with 99 mol.% purity. (4) To minimize retrofit costs, a 30 mol.% oxygen concentration [3,6,7] in the mixture of oxygen and recycled flue gas (RFG) would be maintained in order to match the temperature profile similar to that of a conventional air combustion. (5) The air in-leakage and excess oxygen coefficient would be estimated at 2% of the total gas entering the burner [1] and 1.05 [8], respectively. (6) The required coal flow rate and theoretic oxygen for combustion would be calculated from Eqs. (11.1) and (11.2),

$$M_C = \left(M_{68}\left(h_{40} - h_{68}\right) + M_{43}\left(h_{44} - h_{43}\right)\right) / \left(\eta_b LHV\right) \tag{11.1}$$

$$\upsilon_O = \left(C_{ar}/12 + H_{ar}/4 + S_{ar}/32 - O_{ar}/32\right) \times 22.4 \tag{11.2}$$

where, M meant mass flow rate (kg/s), h stood for enthalpy (kJ/ kg), η_b was thermal efficiency of the boiler, subscript C represented coal, v_O was unit theoretic volumes of oxygen (m^3/kg), and numerical subscripts illustrated stream numbers, as shown in Fig. 11.1. The proximate and ultimate analyses of coal are listed in Table 11.1.

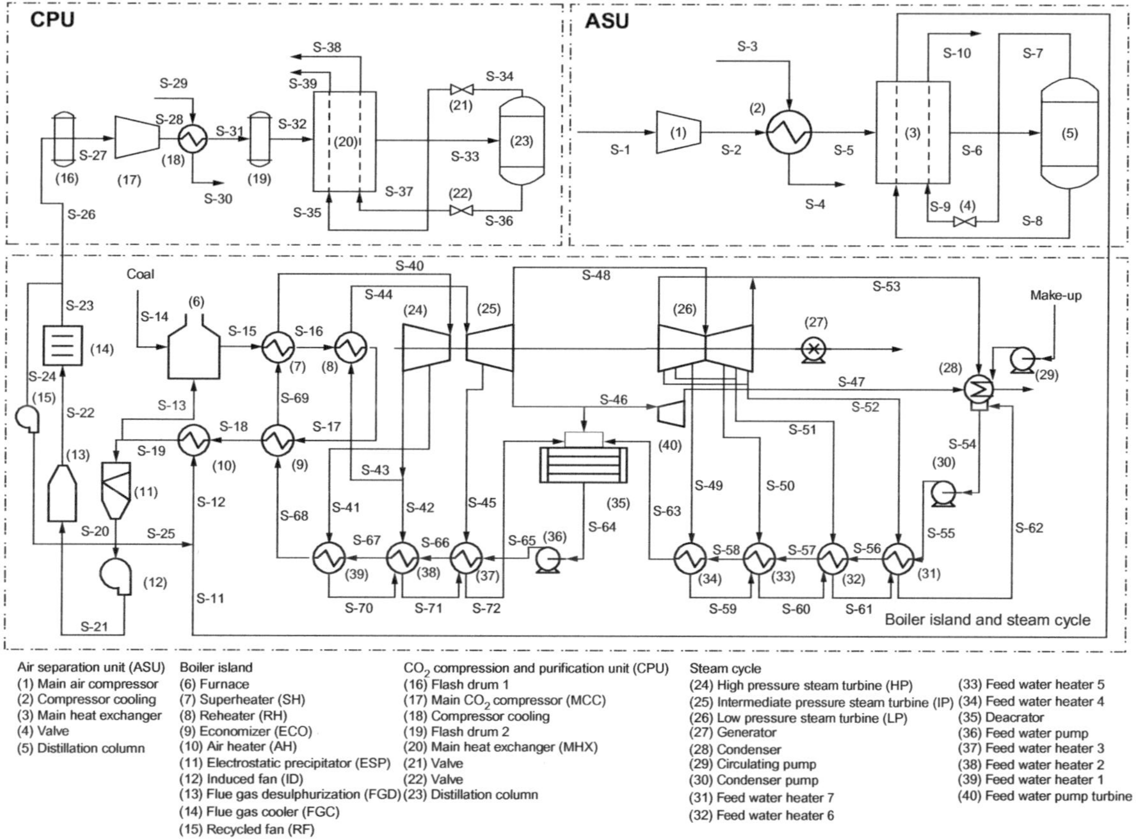

FIG. 11.1 Schematic diagram of the studied oxy-fuel combustion system. *(Reproduced with permission from Xiong J, Zhao H, Zheng C. Exergy analysis of a 600 MWe oxy-combustion pulverized-coal-fired power plant. Energy Fuels 2011;25(8):3854–3864. Copyright (2011) American Chemical Society.)*

TABLE 11.1 The Composition of Shenhua Coal (As-Received Basis)

Proximate Analysis (wt.%)		Ultimate Analysis (wt.%)	
Moisture	13.80	Carbon	60.51
Volatile matter	26.20	Hydrogen	3.62
Ash	11.00	Nitrogen	0.70
Fixed carbon	49.00	Sulfur	0.43
LHV (MJ/kg)	22.77	Oxygen	9.94

11.2.2 Simulation Results

Based on the steady-state simulation results, thermodynamic analysis and the effects of operating parameters on system operation were further conducted. The energy performance of an oxy-fuel combustion power plant was compared to that of the reference case (identical scale air combustion power plant), as shown in Table 11.2. The net efficiency of oxy-fuel combustion was about 10.1 percentage points lower than that of conventional air combustion, because the total power consumption of oxy-fuel combustion increased about 149.32 MW in the ASU and CPU [2]. For subsystems, the specific O_2 energy consumptions were 0.25 MWh/t of O_2, while the specific CO_2 energy consumption and CO_2 recovery rate were 0.32 MWh/t of CO_2 and 96.12%, respectively. When compared to the oxy-steam combustion (oxy-steam, taking steam as dilute agent) case, it was found that the net efficiency of traditional oxy-fuel combustion was 0.90 percentage points lower because less energy from the ASU, the CPU and auxiliary devices were required in an oxy-steam combustion power plant [9]. If solar parabolic trough collectors (SPTCs) were to be integrated into oxy-steam combustion, the net efficiency would be further increased [9].

Based on the steady-state model, the influence of the operating parameters (recycle ratio, oxygen supply purity, excess oxygen coefficient, air in-leakage, SOx and NOx, and recycle position) [1] on system performance can also be identified by sensitivity analysis, which would provide useful information for engineers and operators. The oxygen concentration in the gas mixture entering the oxy-fuel burner decreased when the recycle ratio increased and the purity of the oxygen supply decreased. Therefore selecting the correct recycle ratio and oxygen supply purity would be important to match the temperature profile in furnace between oxy-fuel combustion and air combustion. The decrement of the excess oxygen coefficient benefitted the system's thermodynamic behavior and reduced the ASU's demand for oxygen [9]. Because air in-leakage could not be avoided and would have a harmful impact, a potential choice was to operate the boiler under a positive pressure [1]. Because the amount of NOx decreased and

TABLE 11.2 Energy Performance for Oxy-fuel Combustion and Reference Case

Power Consumption (MW)	Oxy-fuel Combustion	Reference Case
ASU	102.24	–
CPU	47.13	–
Circulating pump	4.26	4.26
Condenser pump	0.75	0.75
Total power consumption	154.33	5.01
Total power production	599.43	599.43
Coal feed rate (kg/s)	60.52	61.58
Oxygen stoichiometric amount (kmol/s)	3.42	3.48
Fuel heat input (LHV)	1377.92	1402.05
Gross efficiency (LHV), %	43.50	42.75
Net efficiency (LHV), %	32.30	42.40
Specific O_2 energy consumption (MWh/t of O_2)	0.25	–
Specific CO_2 energy consumption (MWh/t of CO_2)	0.32	–
CO_2 recovery rate (%)	96.12	–

Reproduced with permission from Xiong J, Zhao H, Zheng C. Exergy analysis of a 600 MWe oxy-combustion pulverized-coal-fired power plant. Energy Fuels 2011;25(8):3854–3864. Copyright (2011) American Chemical Society.

the SOx concentration increased in the flue gas, it was not considered necessary to further treat the NOx and injecting limestone into the furnace and activating unreacted calcium (LIFAC) might be suitable for the removal of SOx [1]. There were various positions for recirculation [7,10], and the choice might depend on the plant type and customer demand.

11.3 EXERGY ANALYSIS

Exergy analysis is commonly used to determine the amount of energy used and to find the source of thermodynamic irreversibility. More specifically, it is used to measure the exergies of stream and energy flows, to locate the distributions of exergy destructions for unit devices, and to determine the exergy efficiencies for equipment, subsystems, and an entire plant. To render the thermodynamic performance efficient, exergy analysis was employed at the oxy-fuel combustion power plant under study. In the upcoming subsections, we first introduce the exergy analysis method and then use it to discover where the exergy destructed.

11.3.1 Methodology

As a concept of the Second Law of Thermodynamics, exergy measures a system's departure from its initial state to its environmental state [2,11,12] and characterizes the quantity and quality of energy. Since kinetic exergy (E^{KN}) and potential exergy (E^{PT}) are equal to 0, the sum of physical exergy (E^{PH}) and chemical exergy (E^{CH}) is the stream exergy (E). E^{PH}, defined as the acquired maximum theoretical useful work when a system changes from the initial state (T and P) to the environment state (T_0 and P_0), can be calculated by Eq. (11.4). E^{CH} arises from the departure of the chemical composition of a system from the environment. Because of the presence of different stream forms, the calculation of E^{CH} corresponds to different equations.

Eqs. (11.5)–(11.7) were used to determine the E^{CH} for a pure gaseous component, a gaseous mixture, and nonexistent substances in the referenced environment, respectively, whereas the sum of the E^{CH} for all the components was used for nongaseous mixtures. To implement the exergy analysis, the substance model proposed by Gaggioli and Petit [13] was chosen as the reference state.

$$E = E^{PH} + E^{CH} \tag{11.3}$$

$$E^{PH} = m\left[(h - h_0) - T_0(s - s_0)\right] \tag{11.4}$$

$$E^{CH} = m\left[-RT_0 \ln x^0\right] \tag{11.5}$$

$$E^{CH} = m\left[\sum x_k e_k^{CH} + RT_0 \sum x_k \ln x_k\right] \tag{11.6}$$

$$E^{CH} = m\left[-\left\{\sum_{p} v_p g_p - \sum_{r} v_r g_r\right\} + \left\{\sum_{p} v_p e_p^{CH} - \sum_{r} v_r e_r^{CH}\right\}\right] \tag{11.7}$$

After determining the stream exergy, some well-defined exergy indicators were introduced to implement the exergy assessment. First, to account for the thermodynamic irreversibility, exergy destruction (E_D) and exergy loss (E_L) were considered. The former aspect was the internal exergy loss resulting from thermodynamic or economic constraints, while the latter one was the external exergy loss transmitted to the environment owing to heat loss, cooling, by-products, ash, waste water, and so on. Then fuel exergy (E_F) and product exergy (E_P) were used to quantify the system's input and output in terms of exergy, respectively. Finally, the exergy destruction ratio (y_D), exergy destruction distribution (ε), and exergy efficiency (η) were further determined in order to describe the system's exergy performance. The first aspect signifies how much one process or device contributes to the total fuel exergy ($E_{F,tot}$) of the system, the second one reflects the distribution of exergy destruction in the system, and the last one provides a true measure of the system's energy performance.

$$y_D = \frac{E_D}{E_{F,tot}} \tag{11.8}$$

$$\varepsilon = \frac{E_D}{E_{D,tot}} \tag{11.9}$$

$$\eta = \frac{E_P}{E_F} = 1 - \frac{E_D + E_L}{E_F} \tag{11.10}$$

11.3.2 Exergy Assessment

Adopting the preceding exergy method, we were able to locate and identify the thermodynamic inefficiency for unit devices, subsystems, and entire oxy-fuel combustion systems. Fig. 11.2 shows that the exergy performance of unit devices in the oxy-fuel boiler was more favorable than that of the air-fired boiler. The exergy destruction of heat transfer in the furnace was 0.60% greater, while the exergy losses from flue gas and radiation were 6.38% and 1.23% greater than those in the air-fired boiler. The total exergy destruction for the oxy-fuel boiler was 1.01% lower than that in an air-fired boiler, even though the additional

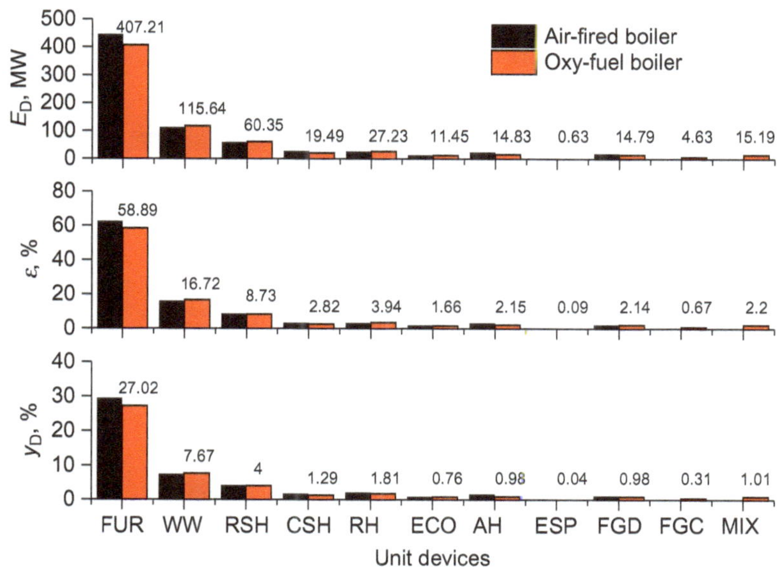

FIG. 11.2 The exergy analysis results of oxy-fuel and air-fired boilers. *(Reproduced with permission from Xiong J, Zhao H, Zheng C. Exergy analysis of a 600 MWe oxy-combustion pulverized-coal-fired power plant. Energy Fuels 2011;25(8):3854–3864. Copyright (2011) American Chemical Society.)*

exergy destructions were required by the FGC and mixture of oxygen and RFG. Calculated via Eq. (11.10), the exergy efficiency of oxy-fuel boiler was 48.39% and was 0.46 percentage points higher than that of the air-fired boiler. Fig. 11.3 shows the exergy performance of the ASU, the CPU, and steam cycle. For the ASU, the main exergy destruction came from the compression process and distillation process, which accounted for about 40.72% and 40.08% of the ASU's total exergy destruction, respectively. A possible solution for improving the ASU's thermodynamic behavior may be focused on the equipment used in these two processes. On the other hand, a combination of flue gas cleaning, compression, and cooling processes, rather than the distillation process, led to the primary exergy destruction in the CPU, and the turbine contributed to the maximum irreversibility in the steam cycle. Using Eq. (11.10), we determined that the exergy efficiency of these three subsystems was 15.84%, 73.45%, and 81.51%, respectively [2]. With respect to the overall oxy-fuel combustion system, the distribution of exergy destruction was different from that of conventional the air combustion system. The addition of the ASU and CPU lead to about 7.95% of the total exergy destructions/losses which accounted for 7.71% of the total fuel exergy in oxy-fuel combustion [2]. Therefore the exergy efficiency of the oxy-fuel combustion power plant was calculated to be 37.13%,

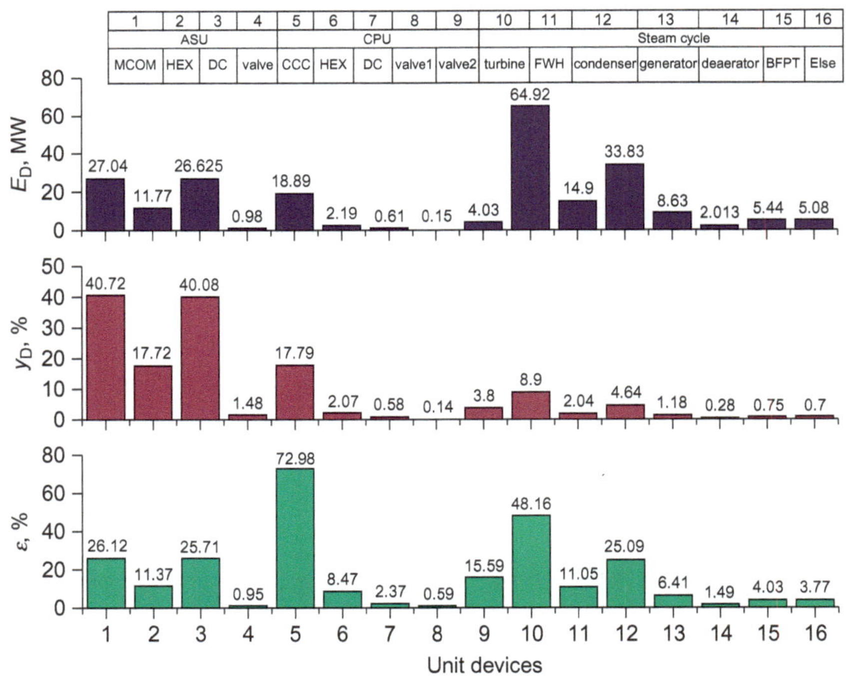

FIG. 11.3 The exergy analysis results of ASU, CPU, and steam cycle. *(Reproduced with permission from Xiong J, Zhao H, Zheng C. Exergy analysis of a 600 MWe oxy-combustion pulverized-coal-fired power plant. Energy Fuels 2011;25(8):3854–3864. Copyright (2011) American Chemical Society.)*

which was 4.08% lower than that of the air combustion power plant. When oxy-steam combustion and oxy-steam combustion were integrated with SPTCs, their exergy efficiencies further increased about 1.10 and 1.64 percentage points [9].

11.4 PROCESS INTEGRATION AND OPTIMIZATION

11.4.1 Process Integration

Process integration is an effective way to improve the thermodynamic efficiency of oxy-fuel combustion systems. Usually, a heat integration approach can make the best use of the available heat or energy in a complex energy conversion system. Through adiabatic compression, oxygen preheating to approximately 15°C, and condensates preheating to approximately 150°C, the energy penalty in oxy-fuel combustion plants decreased about 1.3% [14]. Taking into account the heat integration between the compression duties in the ASU and the steam cycle for preheating the boiler feed water, the thermal efficiencies for a three-stage compressor case and an adiabatic-stage compressor case increased about 0.38% and 0.47%, respectively [15]. An investigation of the compression duties in a double-column ASU and a two-stage flash drum CPU integrated with a steam cycle [16] showed that thermal efficiency increased 0.38%, 0.27%, and 0.72% for the ASU only, the CPU only, and both, respectively.

Taking an oxy-fuel combustion power plant for example [17], the following approaches were considered in process integration: (1) flue gas exited from the boiler was used to heat the generation gas of the molecular sieve in the ASU; (2) the MCOM was cooled down by the cooling water from the air cooling tower; (3) waste heat from the CPU was adopted to heat the main condensate water or feed water in the steam cycle; and (4) the compressor was directly driven by steam. By using the fisrt method, the power consumption for regeneration of the molecular sieve saved about 2.5 MW. The second method contributed to the result that compression power requirement decreased about 21.17 MW when the discharge temperature in the compressor was decreased 10°C and the isentropic efficiency was selected as 0.85. For the last approach, the energy consumption from the steam driving compressor reduced about 4.97 MW. More importantly, the net efficiency would increase about 3.11 percentage points if process integration was coupled with the studied oxy-fuel combustion system.

11.4.2 Process Optimization

Process optimization is often used to improve system performance by obtaining the desirable operating parameters and arrangements. Typically, sensitivity analysis is first carried out to identify the key operating parameters, and single-variable or multivariable optimization based on an objective function follows. In the application of oxy-fuel combustion, several optimizing strategies, which means the optimizations on the operating performance of subsystems, the plant sturcture and arrangement, and the process characteristics of overall system, were considered to attain energy and economic efficiencies. To minimize the impact on the product purity, distillation column size, investment cost, and regeneration heat duties of the ASU, the total number of stages, the reflux ratio, and the feed stage for the distillation column were optimized to be 28, 1.044, and 15, respectively [1]. The trade-off between the ASU and the CPU enhanced the oxygen supply purity and reduced the total power demanded by the ASU and the CPU. The desired operating conditions for the CPU were determined by minimizing the specific power consumption as an objective function, varying the four objective variables (compressor discharge pressure, compressor exit temperature, and operating temperature for the first and second flash separators), and selecting three operating constraints (CO_2 product purity, CO_2 recovery rate, and stream temperature at the downstream of the second flash separator) [12,18]. After optimization, it was revealed that the CO_2 recovery rate increased 0.9 percentage points, the specific power consumption decreased 2.58%, the exergy efficiency increased 1.9 percentage points, the investment cost reduced $0.04 million, and the maximum increment of the unit thermoeconomic cost was 5.18%.

Regarding to the overall oxy-fuel combustion system, some modifications of the system structure and arrangement were made to optimize the retrofitting of a conventional air combustion system [19]. These modifications involved replacing the air heater with economizers, recovering low-grade heat to heat up condensate, driving the CO_2 compressor by an auxiliary steam turbine, and firing more to produce more power. The purpose was to minimize the power losses and the CO_2 removal penalty. It was found that the CO_2 mitigation cost and the power penalty reduced to approximately $20/t$CO_2$ and 11%, respectively.

Vent gas power recovery and recirculation for CO_2 recovery can be incorporated in a greenfield project to minimize the penalty for CO_2 removal. From the viewpoint of system design and operation, a multivariable, gradient-based optimization approach [20] can be applied to optimize the overall pressurized oxy-fuel combustion system. The essential idea is to obtain maximum thermal efficiency subjected to 16 optimization variables (operating combustor pressure, combustor's thermal leaks, deaerator pressure, etc.) and 10 operating constraints (CO_2 recovery rate, CO_2 product purity, minimum internal temperature approach for heat exchanger, etc.). The optimization gained a maximum thermal efficiency of 34.48% (LHV) which was 0.82% higher than that of the reference case. Additionally, the addition of air is proposed as an alternative way to optimize system performance. Air can be added to primary, secondary, and ternary flows for operational safety [21]. The results showed that the ASU's parasitic power demand, with 65% additional air and the base case (0% air addition), varied between approximately 35 MW and 65 MW. Similarly, the capital cost varied between $84 million and $154 million. With a 37% air addition in the overall plant, the energy savings was 13 MW [21].

11.5 TECHNO-ECONOMIC EVALUATION

Before deciding whether to adopt a novel technology, a study of its technical and economic prospects should be carried out. Usually, techno-economic evaluations are investigated once a technology is confirmed as being viable. Much research has been conducted in an attempt to justify the economic feasibility of oxy-fuel combustion. When retrofitted from an identical air combustion power plant, oxy-fuel combustion seems to be more attractive than post-combustion with MEA scrubbing [22–25]. This section highlights a method that calculates the investment cost, cost of electricity, CO_2 avoidance cost, and CO_2 capture cost and applies that information to oxy-fuel combustion power plants.

11.5.1 Calculation Method and Assumptions

The total investment cost for oxy-fuel combustion power plant can be calculated from the cost types listed in Table 11.3. For power generation cost, it is the sum of fuel cost (C_1), loan interest (C_2), operation and maintenance cost (C_3), depreciation cost (C_4), amortization cost (C_5), pollutants' emission tax (C_6),

TABLE 11.3 Cost Calculation Equations for Oxy-fuel Combustion Power Plant [9,25,26]

Cost Type	Equation
Fuel cost (C_1)	$C_1 = m_{F,0} \times c_F \times W \times H \times \eta_e$
Load interest cost (C_2)	$C_2 = C_{IT,1} \times p_{loan} \times \xi = \begin{pmatrix} C_{IT,\ base,\ 0} + C_{I,\ boiler,\ 0} \times 7\% \\ +C_{IT,\ S,\ 0}/3 + C_{ASU} + C_{CPU} \end{pmatrix} \times p_{loan} \times \xi$
Operation and maintenance cost (C_3)	$C_3 = \left(C_{IT,\ base,\ 0} + C_{I,\ boiler,\ 0} \times 7\%\right) \times p_{OM,base,0} + C_{OM,S,0}/3 + C_{ASU} \times p_{OM,ASU} + C_{CPU} \times p_{OM,CPU}$
Depreciation cost (C_4)	$C_4 = C_{IT,1} \times p_{fa} \times (1 - p_{lv})/Y_d$
Amortization cost (C_5)	$C_5 = C_{IT,1} \times p_{ia}/Y_a$
Pollutants emission cost (C_6)	$C_6 = E_{S,0} \times T_S + E_{N,0} \times T_N$
Personnel wages (C_7)	$C_7 = \left(N_{base,0} + N_{S,0} + N_{N,0} + N_{ASU} + N_{CPU}\right) \times c_{pay} \times (1 + r_w)$
Material cost (C_8)	$C_8 = p_{m,0} \times W \times H$
Other costs (C_9)	$C_9 = p_{o,0} \times W \times H$
By-products revenue (C_{10})	$C_{10} = M_{CO_2} \times c_{CO_2}$
Annualized cost (CT)	$C_T = \sum_{i=1}^{9} C_i - C_{10}$
Cost of electricity (c_{COE})	$c_{COE} = C_T/(W_{net} \times H)$
CO_2 avoidance cost (c_{CAC})	$c_{CAC} = \dfrac{c_{COE,1} - c_{COE,0}}{e_{CO_2,0} - e_{CO_2,1}} = \dfrac{c_{COE,1} - c_{COE,0}}{\dfrac{E_{CO_2,C}}{W_{net,0}H} - \dfrac{E_{CO_2,1}}{W_{net,1}H}}$
CO_2 capture cost (c_{CCC})	$c_{CCC} = \dfrac{c_{COE,1} - c_{COE,0}}{m_{CO_2,1} - m_{CO_2,0}} = \dfrac{c_{COE,1} - c_{COE,0}}{m_{CO_2,1}} = \dfrac{c_{COE,1} - c_{COE,0}}{\dfrac{M_{CO_2,1} r_{CO_2}}{W_{net,1}H}}$

personnel wages (C_7), material cost (C_8), and other costs (C_9) [25]. The annualized cost (C_T) is considered as "hard" costs because the financial rules are complicated, and it is the result of power generation cost being subtracted from the other costs (C_{10}). Then combined with net power output and annual operating hours, the cost of electricity (c_{COE}) for oxy-fuel combustion power plants can be determined. Finally, based on unit CO_2 emission, unit CO_2 capture, and the cost index of the corresponding air combustion power plant, the CO_2 avoidance cost (c_{CAC}) and CO_2 capture cost (c_{CCC}) can be calculated to reflect the environmental impact of oxy-fuel combustion. Their difference lies in which denominator is used: unit CO_2 emission for the first parameter while unit CO_2 capture for the second parameter.

Because oxy-fuel combustion is different from air combustion, certain assumptions were addressed to calculate every cost term: (1) Because of the absence of commercial oxy-fuel combustion systems, the corresponding air combustion system was chosen as the reference case for performing the techno-economic assessment. (2) As presented in the previous sections, it was assumed that the boiler's thermal efficiency would increase [25]. (3) The stream and energy flows from simulation results were used to determine the net power output, unit CO_2 emission, and unit CO_2 capture. (4) The investment cost of the ASU was estimated from the industrial data, while the investment cost of the CPU was expected to be 2.5% [27] of the total investment cost of the air combustion power plant. (5) Different from that in air combustion, the captured CO_2, rather than gypsum, would be the by-product and that revenues from CO_2 sales should be calculated. (6) The calculations for other costs were similar to those in air combustion and were based on the data, mainly data in the book, "Reference Costing Indexes in Quota Design for Coal-Fired Power Plant Projects" [28].

11.5.2 Results Analysis

Based on the cost calculation equations and assumptions, the investment cost and the cost distributions for oxy-fuel combustion power plants were determined. The results depended on the plant scale, configuration, type, and data source. For an 865 MW oxy-fuel combustion power plant studied by the Chalmers University of Technology [29], it was found that the costs for electricity and CO_2 avoidance were \$64.3/kW and \$26/t, respectively. Alstom [30] declared that the unit investment cost and the CO_2 avoidance cost for the analyzed 450 MW oxy-fuel combustion system were \$823/kW and \$42/t, while they were \$35/t and \$791/kW (400 MW oxy-fuel combustion power plant) as shown by CANMET [22], respectively. When compared to that of post-combustion with MEA scrubbing, the results of EDF [31] indicated that the CO_2 avoidance cost was 29% lower for oxy-fuel combustion power plants (1200 MW).

A 600 MW conventional oxy-fuel combustion power plant is explained as follows. As the basis for identifying the economic property of oxy-fuel combustion, the costs of electricity and unit CO_2 emission of air combustion were identified as \$47.72/(MWh) and 911.47 kg/(MWh) [25]. When compared to that of air combustion, the cost of electricity for oxy-fuel combustion increased about 37.94% because the ASU and CPU were added to supply oxygen and produce CO_2 [25]. The cost distribution of conventional oxy-fuel combustion (called Oxy-CO_2) was similar and is shown in Fig. 11.4. The primary contributions forming the total investment costs were fuel cost, depreciation cost, load interest cost, and operation and maintenance cost. Compared to that of oxy-CO_2, the total investment cost and cost of electricity of oxy-steam combustion decreased about 3.77% and 6.49% [9], respectively. To further the use of the SPTCs to heat recycled condensate water would require the total investment cost and the cost of electricity of approximately 3.00% and $1.98/(MWh) more than that of oxy-steam combustion [9].

From the simulation results, net power output, unit CO_2 emission, and CO_2 recovery rate were used to calculate the CO_2 avoidance cost and CO_2 capture cost. As summarized in Table 11.4, the two oxy-steam combustion cases appeared to perform better than the oxy-CO_2 combustion case. Because of the smaller amount of CO_2 emissions from the two oxy-steam combustion cases, the CO_2 avoidance cost was decreased by $5.81/t and $3.58/t [9]. Although the specific CO_2 capture was almost the same, two oxy-steam combustion cases also required about 23.94% and 12.62% less CO_2 capture cost than oxy-CO_2 combustion did [9].

The economic property of the oxy-fuel combustion power plant was affected by key operating parameters, such as fuel price, excess oxygen coefficient, power consumption from ASU, and CO_2 recovery rate [24,25]. Because the fuel cost accounted for more than 60% of the total investment cost, the increment of fuel price would lead to the increases of the cost of electricity, CO_2 avoidance cost and CO_2 capture cost. The effects of the ASU's excess oxygen coefficient

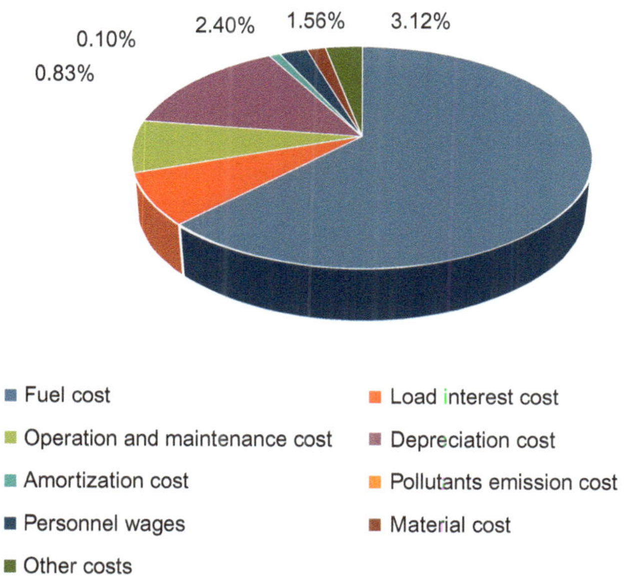

FIG. 11.4 The cost distribution for oxy-fuel combustion power plant.

TABLE 11.4 Techno-Economic Analysis Results for Three Oxy-fuel Combustion Systems

Cost	Oxy-CO_2	Oxy-Steam	Oxy-Steam With SPTCs
Unit CO_2 emission, kg/(MWh)	67.69	27.43	27.50
CO_2 avoidance cost, $/t	21.46	15.65	17.88
Unit CO_2 capture, kg/(MWh)	1128.26	1134.78	1128.12
CO_2 capture cost, $/t	16.05	12.19	14.01

and power consumption on the cost of electricity were greater while their effects on CO_2 avoidance cost and CO_2 capture cost were less than that of the fuel price. As shown in Table 11.3, the calculation equation for the CO_2 recovery rate had no impact on the cost of electricity, while a decrease in it would reduce the CO_2 avoidance cost and the CO_2 capture cost.

11.6 SUMMARY

This chapter included comprehensive investigations on identifying the thermodynamic behavior and economic performance of larger oxy-fuel combustion power plants. Based on the process simulation, exergy assessment located the position of inefficiency, system integration and optimization sought possible solutions to achieve energy savings, and techno-economic evaluation provided an important recommendation for adoption of the technology. The results provide a theoretical basis that engineers can use to deeply understand the process characteristics. However, larger oxy-fuel combustion power plants still face great challenges, even though significant efforts have been made to reduce the energy penalty and economic cost. In order to promote the use of oxy-fuel combustion in power plants, more effective approaches are required to improve system performance, to properly integrate the subsystems, and to optimize its economic feasibility.

REFERENCES

[1] Xiong J, Zhao H, Chen M, Zheng C. Simulation study of an 800 MWe oxy-combustion pulverized-coal-fired power plant. Energy Fuels 2011;25(5):2405–15.

[2] Xiong J, Zhao H, Zheng C. Exergy analysis of a 600 MWe oxy-combustion pulverized-coal-fired power plant. Energy Fuels 2011;25(8):3854–64.

[3] Buhre BJP, Elliott LK, Sheng CD, Gupta RP, Wall TF. Oxy-fuel combustion technology for coal-fired power generation. Prog Energ Combust Sci 2005;31(4):283–307.

[4] Darde A, Prabhakar R, Tranier J-P, Perrin N. Air separation and flue gas compression and purification units for oxy-coal combustion systems. Energy Proc 2009;1(1):527–34.
[5] Jin B, Zhao H, Zheng C. Dynamic simulation for mode switching strategy in a conceptual 600 MWe oxy-combustion pulverized-coal-fired boiler. Fuel 2014;137(2014):135–44.
[6] Davidson RM, Santos SO. Oxyfuel combustion of pulverised coal. London: IEA Clean Coal Centre; 2010.
[7] Toftegaard MB, Brix J, Jensen PA, Glarborg P, Jensen AD. Oxy-fuel combustion of solid fuels. Prog Energ Combust Sci 2010;36(5):581–625.
[8] Wall TF. Combustion processes for carbon capture. Proc Combust Inst 2007;31(1):31–47.
[9] Jin B, Zhao H, Zou C, Zheng C. Comprehensive investigation of process characteristics for oxy-steam combustion power plants. Energ Convers Manage 2015;99:92–101.
[10] Nakayama S, Noguchi Y, Kiga T, Miyamae S, Maeda U, Kawai M, et al. Pulverized coal combustion in O_2/CO_2 mixtures on a power plant for CO_2 recovery. Energ Convers Manage 1992;33(5–8):379–86.
[11] Bejan A, Tsatsaronis G, Moran M. Thermal design & optimization. New York: John Wiley & Sons; 1996.
[12] Jin B, Zhao H, Zheng C. Optimization and control for CO_2 compression and purification unit in oxy-combustion power plants. Energy 2015;83(2015):416–30.
[13] Gaggioli R, Petit PJ. Use the second law, first. CHEMTECH (US) 1977;7(8).
[14] Tranier JP, Dubettier R, Perrin N. Air separatiom unit for oxy-coal combustion systems. In: 1st oxy-fuel combustion conference; 2009.
[15] Fu C, Gundersen T. Integrating the compression heat in oxy-combustion power plants with CO_2 capture. Chem Eng 2012;29(1):781–6.
[16] Fu C, Gundersen T. Exergy analysis and heat integration of a coal based oxy-combustion power plant. Energy Fuels 2013;27(11):7138–49.
[17] Kong H, Liu Z, Chen S, Zhang Q, Zheng C. Process simulation and optimization of a 600 MW O_2/CO_2 power plant. Proc CSEE 2012;32(2):53–60.
[18] Jin B, Zhao H, Zheng C. Thermoeconomic cost analysis of CO_2 compression and purification unit in oxy-combustion power plants. Energ Convers Manage 2015;106(2015):53–60.
[19] Seltzer A, Fan Z, Hack H. Oxyfuel coal combustion power plant system optimization. In: 7th annual COAL-GEN conference, Milwaukee, WI; 2007.
[20] Zebian H, Gazzino M, Mitsos A. Multi-variable optimization of pressurized oxy-coal combustion. Energy 2012;38(1):37–57.
[21] Zanganeh KE, Shafeen A. A novel process integration, optimization and design approach for large-scale implementation of oxy-fired coal power plants with CO_2 capture. Int J Greenhouse Gas Control 2007;1(1):47–54.
[22] Singh D, Croiset E, Douglas PL, Douglas MA. Techno-economic study of CO_2 capture from an existing coal-fired power plant: MEA scrubbing vs. O_2/CO_2 recycle combustion. Energ Convers Manage 2003;44(19):3073–91.
[23] Varagani RK, Châtel-Pélage F, Pranda P, Rostam-Abadi M, Lu Y, Bose AC. Performance simulation and cost assessment of oxy-combustion process for CO_2 capture from coal-fired power plants. In: The fourth annual conference on carbon sequestration, May; 2005.
[24] Xiong J, Zhao H, Zheng C, Liu Z, Zeng L, Liu H, et al. An economic feasibility study of O_2/CO_2 recycle combustion technology based on existing coal-fired power plants in China. Fuel 2009;88(6):1135–42.
[25] Xiong J, Zhao H, Zheng C. Techno-economic evaluation of oxy-combustion coal-fired power plants. Chinese Sci Bull 2011;56(31):3333–45.
[26] Strube R, Manfrida G. CO_2 capture in coal-fired power plants—impact on plant performance. Int J Greenhouse Gas Control 2011;5(4):710–26.

[27] Andersson K, Johnsson F. Process evaluation of an 865 MWe lignite fired O_2/CO_2 power plant. Energ Convers Manage 2006;47(18–19):3487–98.

[28] Electric Power Planning and design Institute for China Power Engineering Consulting Group Corporation. Reference cost indexes in quota design for coal-fired projects (2009 levels). Beijing: China Electric Power Press; 2010 [in Chinese].

[29] Andersson K, Johnsson F, Strömberg L. Large scale CO_2 capture: applying the concept of O_2/CO_2 combustion to commercial process data. VGB Powertech 2003;83(10). 29–33+4.

[30] Marion J. Engineering feasibility of CO_2 capture on an existing US coal-fired power plant. Fuel Energy Abstracts 2002;43(3):217.

[31] Kanniche M, Gros-Bonnivard R, Jaud P, Valle-Marcos J, Amann JM, Bouallou C. Pre-combustion, post-combustion and oxy-combustion in thermal power plant for CO_2 capture. Appl Therm Eng 2010;30(1):53–62.

Chapter 12

Control Concepts, Dynamic Behavior and Mode Transition Strategy for Oxy-fuel Combustion Systems

Bo Jin*, Wei Luo†, Haibo Zhao‡, Zhaohui Liu‡
*Hunan University, Hunan, China
†Wuhan Second Ship Design and Research Institute, Hubei, China
‡Huazhong University of Science and Technology, Wuhan, China

12.1 INTRODUCTION

Oxy-fuel combustion uses a blend of oxygen (from the cryogenic air separation unit, ASU) and recycled flue gas (RFG) rather than air to combust with fuel, and the produced flue gas enters the CO_2 compression and purification unit (CPU) to obtain high purity CO_2 products for other applications. In order to match the heat transfer profiles of air combustion and oxy-fuel combustion and thereby minimize retrofitting costs, flue gas is recycled into the burner to moderate the combustion process. This process results in some variations: the atmosphere in the combustion process changes from N_2/O_2 to CO_2/O_2, the plant structure becomes more complex, the flue gas recirculation increases control and operational flexibility, the dual combustion modes (air and oxy) switch smoothly back and forth, and the integrated operations among subsystems influence each other [1]. Therefore, in this chapter, dynamic simulations, control concepts, dynamic behavior, and mode transitions for differently scaled oxy-fuel combustion systems are introduced and explained. Dynamic models for pilot facilities and conceptual commercial-scale plants are established in Sec. 12.2 to lay the foundation for the later complicated analysis. Then the control system design, the dynamic response identification, and the mode switching strategy are analyzed in Sections 12.3–12.5. Finally, these aspects are summarized with the intention of enhancing further development of oxy-fuel combustion.

Oxy-fuel Combustion. http://dx.doi.org/10.1016/B978-0-12-812145-0.00012-8

12.2 DYNAMIC MODELING AND SIMULATION

Different from the steady-state simulation discussed in Chapter 11, dynamic modeling and simulation identifies real-time process responses to different operating tasks (operating command, disturbances, start-up, emergence condition, etc.) and provides useful guidelines throughout a plant's lifespan [2]. At the design stage, dynamic modeling and simulation can be used to confirm the rationality of the process structure and the feasibility of the operating strategy. Then it enables engineers to design and compare different control strategies, helps operators gain operating skills under a "what if" scenario, and helps workers understand the process [3]. Moreover, for an existing system, it can help balance steady-state optimization and dynamic operability [4]. Thus dynamic modeling and simulation should be applied to oxy-fuel combustion systems in order to identify their dynamic behavior and the appropriate operating strategy.

In fact, several studies have been conducted to implement the dynamic modeling of oxy-fuel combustion power plants. Postler et al. [5] built a dynamic model of 250 MW oxy-fuel combustion system and tested it under a load change and maloperation processes. Kuczynski et al. [6] adopted different software to model three subsystems and stated that the key points reside in the boundary assumptions, the communication among subsystems, and the determination of independent systems. Jin et al. [1,7–12] conducted a series of studies on dynamic modeling, control, and analysis for a conceptual 600 MW_e oxy-fuel power plant. Luo et al. [13,14] established the dynamic model for a pilot facility and studied the influences of operating parameters on system real-time performance. Apart from the preceding cases with pulverized coal-fired (PC) furnaces, similar research [15–17] has also been conducted on the fluidized bed types. To illustrate the dynamic modeling procedures, dynamic models for a pilot facility (3 MW_{th}), a commercial scale plant (600 MW_e), and subsystems (ASU and CPU) are presented in this section.

12.2.1 Pilot Facility

A real prototype presented in Chapter 10 was chosen for the case study. As shown in Fig. 12.1, a dynamic model of a 3 MW_{th} oxy-fuel boiler island system was built in the Aspen Plus Dynamics software [14], which included the boiler and air quality control system (AQCS) models. Using this software, the basic procedure included building a steady-state model, finishing dynamic preparations [18,19], and converting the steady-state model into a dynamic model. In this dynamic process model, the following assumptions were considered. (1) RStoic reactor model would represent the coal combustion and pollutant formation processes. (2) HeatX block would be applied to simulate the heat transfer in the boiler, while the unit devices in the AQCS would be simulated by Sep blocking. (3) Compr and Valve would be employed to characterize the compressor and damper, while the FSplit and Mixer model would be used for splitting and mixing processes. (4) Moreover, the related coefficients

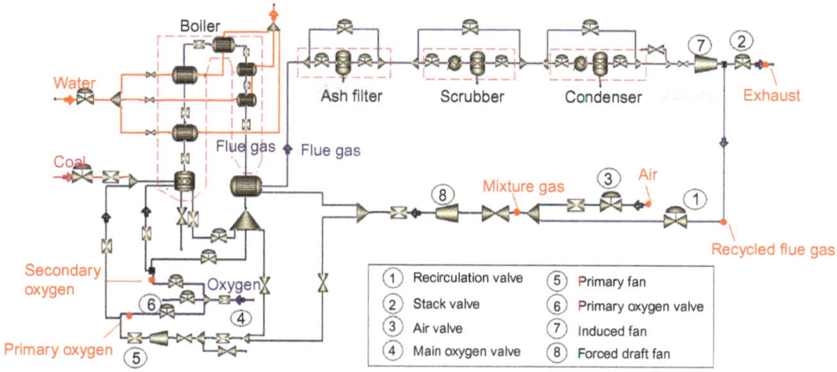

FIG. 12.1 Dynamic model of the 3 MW_{th} oxy-fuel boiler island systems. *(Reprinted with permission from Luo W, Wang Q, Liu Z, Zheng C. Dynamic simulation of the transition process in a 3 MWth oxy-fuel test facility. Energy Proc 2014; 63:6281–88. Copyright (2014) Elsevier.)*

(conversion rate, heat transfer coefficient, separation efficient, etc.) would be adjusted on the basis of the experimental data. (5) The geometry size for equipment would come from the design data of the pilot facility.

Four time periods (time delay $\mathbf{t}_1$, time constant $\mathbf{t}_2$, fluid flow time $\mathbf{t}_3$, and heat transfer time $\mathbf{t}_4$) were chosen to validate the established dynamic model [13]. As shown in Fig. 12.2, heat load rate step change processing was selected to validate the model. During this process, $\mathbf{t}_1$ and $\mathbf{t}_2$ were 0.5 and 2.5 min, respectively, which was consistent with that in the real test. Choosing two temperature measuring points (T_1 and T_5), we found that $\mathbf{t}_3$ and $\mathbf{t}_4$ for the simulation closely matched those of the test data, and this agreement indicated that the dynamic model could be further used to identify its dynamic responses.

12.2.2 Commercial-Scale Plant

Based on the previous studies [20–24], a 600 MW_e oxy-fuel PC power plant was used as an example [8]. The dynamic model shown in Fig. 12.3 comprises only the ASU, boiler island, and CPU, because steam cycle technology is mature and needs almost no retrofitting if an identical temperature profile is obtained. Because the ASU and CPU models are explained in the next sections, here only the details for building the boiler island model are discussed. The main burner zone (MBZ) and the over-fired zone were simulated by the RGibbs reactor and only gas phase reaction was thought to reflect macro-combustion characteristics. Water walls (WW1 and WW2), superheaters (SHPA, SHPL, and FSH), reheaters (FRH and RH), spray water attemperators (AT1, AT2, and AT3), an economizer (ECO), and an air heater (AH) were applied to represent the heat transfers between the water-side and the gas-side. In the AQCS model, simplified separation models were used to simulate selective catalytic reduction (SCR), electrostatic precipitator (ESP), flue gas desulfurization (FGD), and flue gas cooler (FGC). Their separation efficiencies were set at 90%, 100%, 98%,

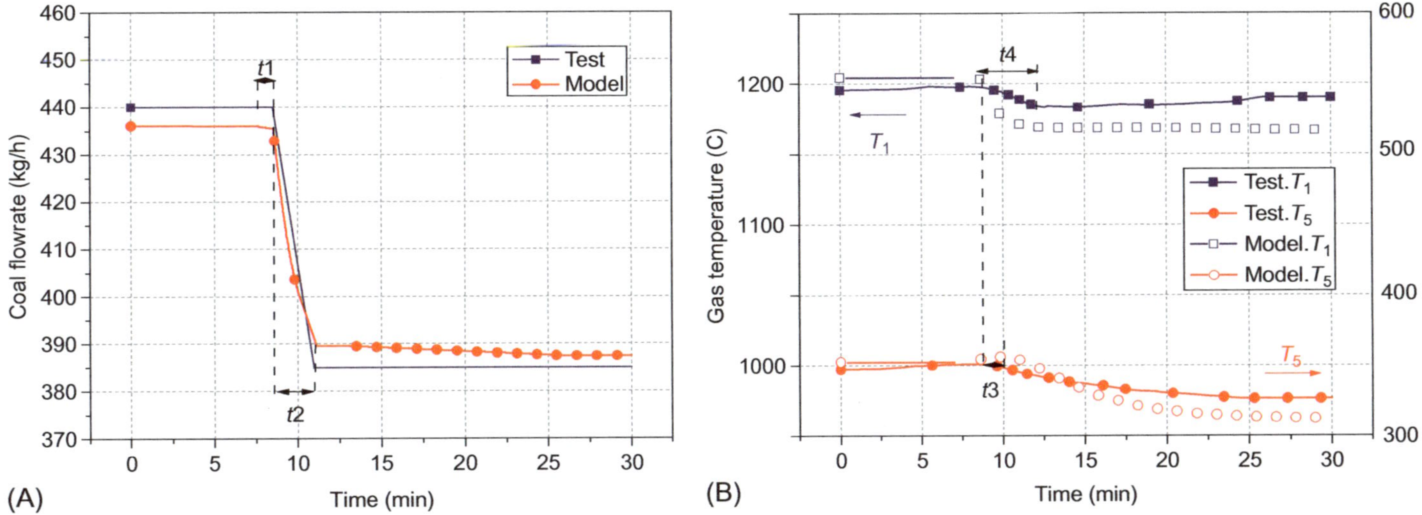

FIG. 12.2 The comparison of dynamic responses in the boiler between simulation and test: (A) coal flow rate step change and (B) gas temperature. *(Reprinted with permission from Luo W, Wang Q, Huang X, Liu Z, Zheng C. Dynamic simulation and transient analysis of a 3 MWth oxy-fuel combustion system. Int J Greenhouse Gas Control 2015;35:138–49. Copyright (2015) Elsevier.)*

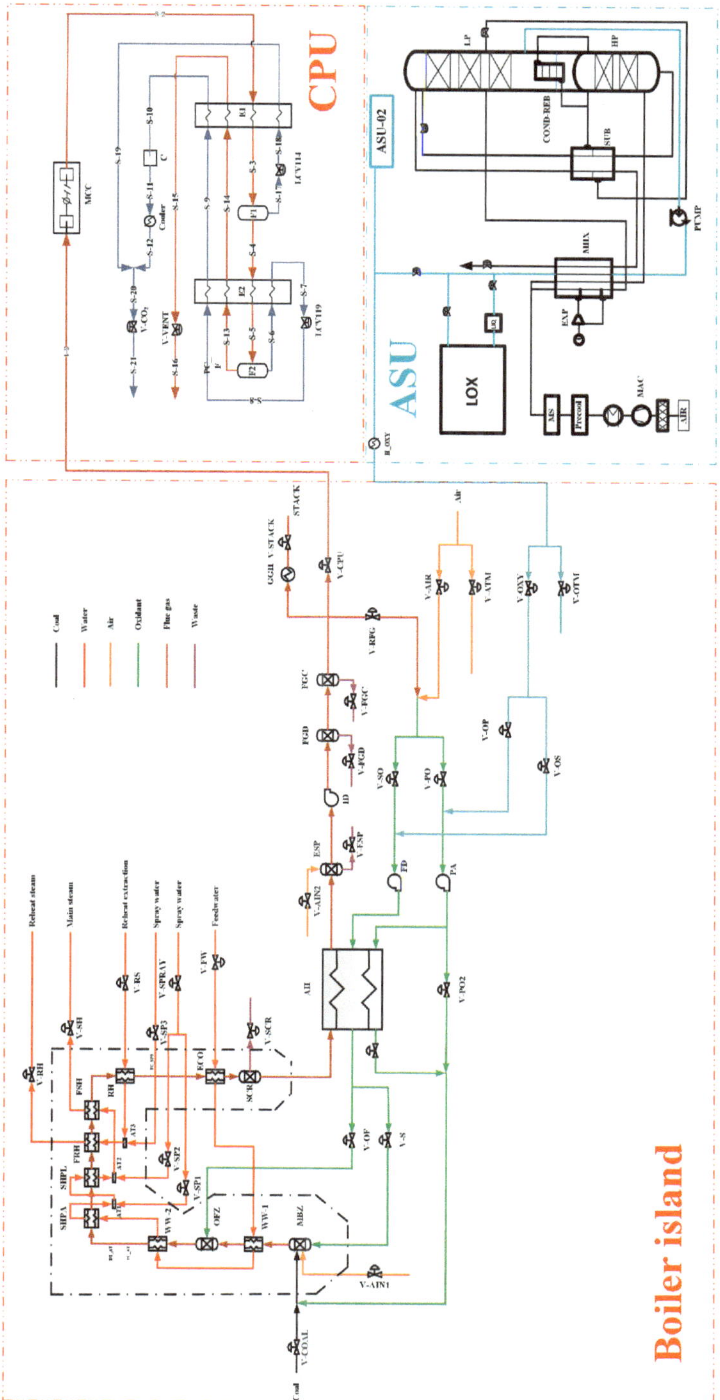

FIG. 12.3 Dynamic model of the analyzed oxy-fuel PC power plant. *(Reprinted with permission from Jin B, Zhao H, Zheng C. Dynamic simulation and control design for pulverized-coal-fired oxy-combustion power plants, in clean coal technology and sustainable development. Springer; 2016. p. 325–33. Copyright (2016) Springer Science + Business Media Singapore and Tsinghua University Press.)*

and 100% to remove NOx, ash, SOx, and water, respectively. To drive the gas flows, induced draft fan (ID), primary air/oxidant fan (PA/PO), and forced draft fan (FD) were modeled by integrating Compr and HeatX blocks. It should be noted that the dynamic modeling procedure was identical with the preceding case but that the equipment sizing was estimated from the steady-state simulation results [7,8,18] and should be updated if actual data is available.

12.2.3 Subsystems

Because the ASU and CPU were added, their dynamic modeling and simulation were also investigated. Because of the consensus that cryogenic ASU is seen as the only available commercial technology to supply oxygen for oxy-combustion application [25], it was selected to identify the control and operation performances. Fig. 12.4 shows a dynamic model of a double-column ASU with a liquid oxygen storage drum (LOX) [9]. The following assumptions were made to finish the dynamic model. (1) Main air compressor (MAC) simulated by MCompr block would boost the air to 5.3 bar, while precool unit (PU) and molecular sieve (MS) would not be considered because the simulated air consisted only of N_2, O_2, and Ar. (2) MHeatX block would be employed to represent main heat exchanger (MHX), while expander (EXP) would be simulated

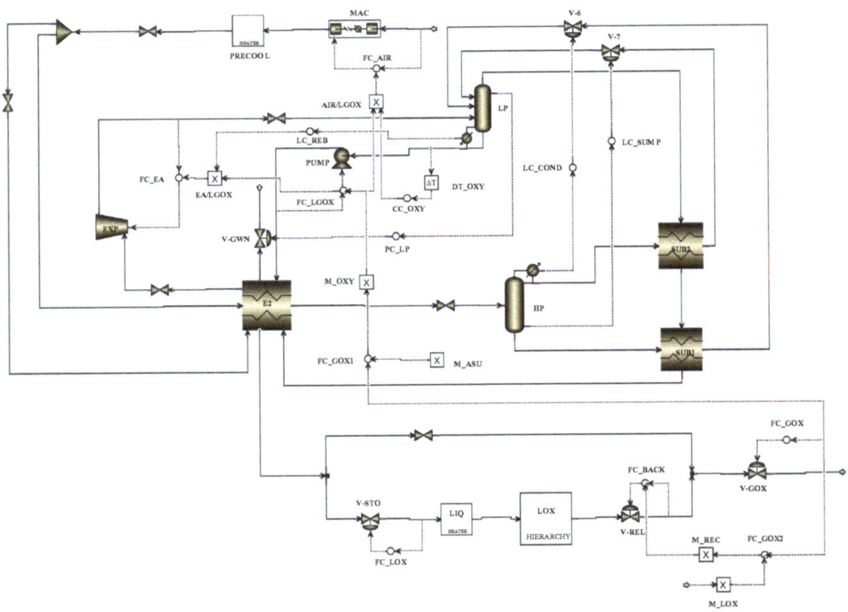

FIG. 12.4 Dynamic model configured with control structure for ASU: black lines for streams and blue lines for control signals. *(Reprinted with permission from Jin B, Su M, Zhao H, Zheng C. Plantwide control and operating strategy for air separation unit in oxy-combustion power plants. Energ Convers Manage 2015;106(2015):782–92. Copyright (2015) Elsevier.)*

by Compr block. (3) Two RadFrac blocks would be used to model high pressure column (HP), low pressure column (LP), and condenser-reboiler (COND-REB) to produce the required oxygen products. (4) HeatX and pump blocks would be applied to simulate the subcooler (SUB) and liquid oxygen pump (PUMP), respectively. (5) Full heat integration and oxygen product quality would be achieved by constraint equation and control loop. (6) Identical equipment sizing method adopted in boiler island model would be used to obtain the dynamic inputs.

For the CPU, a double flash drum type [23] was adopted. Chansomwong et al. [26] established the dynamic mathematical model and identified its transient response to some operating disturbances. Jin et al. [10] used multivariable optimization to obtain optimal operating conditions, built a dynamic model, and obtained the real-time behavior of the flue gas flow rate and composition change processes. In the present case, a dynamic model for this kind of CPU was constructed. As shown in Fig. 12.5, the main components included a multistage CO_2 compressor (MCC), heat exchangers (E1 and E2), and two flash separators (F1 and F2), which were modeled by MCompr, MHeatX, and flash blocks to adjust system pressure, reflect heat transfer, and represent CO_2 separation process, respectively.

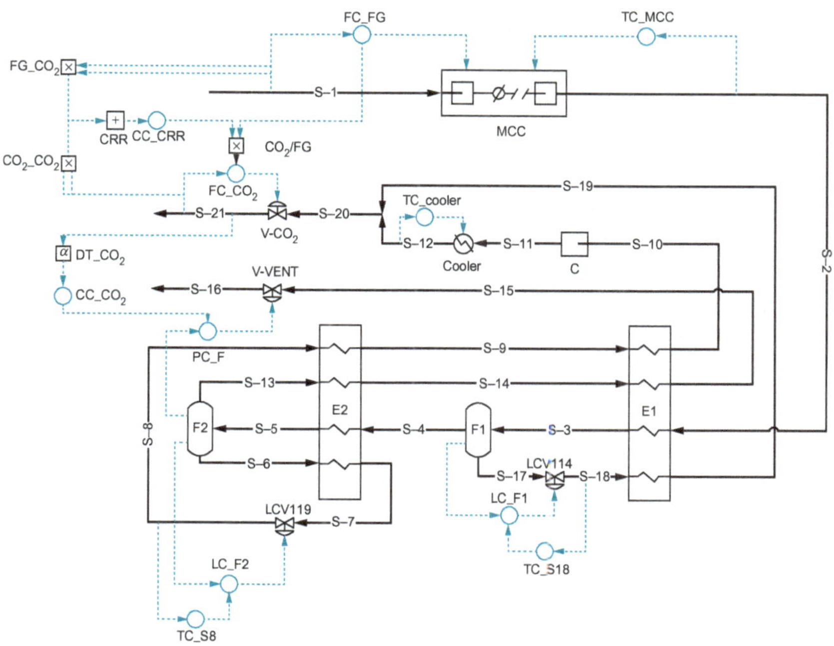

FIG. 12.5 Dynamic model of CPU installed with double temperature control system (DTC). *(Reprinted with permission from Jin B, Zhao H, Zheng C. Optimization and control for CO_2 compression and purification unit in oxy-combustion power plants. Energy 2015;83(2015):416–30. Copyright (2015) Elsevier.)*

12.3 CONTROL CONCEPTS

After completing the dynamic model, the next key step was to design the control system in order to further identify real-time performance. The designed control system was aimed at avoiding the possibility of unexpected emergencies, adjusting the actuators automatically, and ensuring safe, reliable, and effective operations. However, because of the variations in plant arrangements and combustion agents, a traditional control strategy was not suitable for oxy-fuel combustion. Much research has been done in an attempt to handle these issues. In terms of stable combustion, the first method proposed [27] dealt with the manipulation of the oxygen supply to meet boiler load demands and with the adjustment of the RFG flow rate to satisfy boiler heat absorption duties; it was proposed to control the total oxygen content in the oxidant gas entering the boiler (BI_O_2). The second method [15] involved controlling flue gas oxygen content (FG_O_2) or/and BI_O_2 to ensure stable combustion, since flue gas recirculation increases control flexibility and dependence on the measurement of the oxygen content [28]. The secondary oxygen flow or total oxygen flow was used to moderate FG_O_2, while the control of BI_O_2 was realized by regulating the primary and secondary oxygen flows. Another method [29] that manipulated RFG to guarantee the BI_O_2 in a preset range (25–30 wt.%) could also be considered. In addition, a nonlinear model predictive control (NMPC) structure could be used to maintain the water and pressure levels in the drums, steam temperature in the secondary superheater, and FG_O_2 for a water-tube boiler using the oxy-fuel approach [30]. These efforts indicate that it is essential to design an appropriate control system for oxy-fuel combustion systems. Therefore control strategies for a 3 MW_{th} facility and a conceptual 600 MW_e plant are discussed.

12.3.1 Pilot Facility

In the pilot facility, a control system was installed and manipulated by the engineers to achieve certain operating tasks. In the mode switching process, for example, there were issues relating to the design of a suitable control system [14]. Because it introduced disturbances, manipulating the valves and fans to implement a mode transition process presented the first issue. Then a control method was needed to achieve a similar temperature profile between air combustion and oxy-fuel combustion. Finally, because flue gas greatly affects the CPU's operation, the flue gas flow rate and composition needed to be kept stable within acceptable ranges. Therefore the control strategy shown in Table 12.1 was proposed to achieve the mode switching process. In this control method, some key variables were kept within a reasonable range, while the selection of control types for actuators depended on the active function. Because of the role that the primary oxidant gas (PO) played, its flow rate and oxygen concentration (PO_O_2) was kept stable and lower than 18 mol.%. FG_O_2 was set at 3 mol.% to organize reasonable combustion, while BI_O_2 was maintained at around 27 mol.% [5–7] to obtain a heat transfer profile similar to that of air

TABLE 12.1 Control Method for Implementing Mode Switching Process

Measured Variable	Value	Primary Manipulated Devices	Type
BI_O_2	27 mol.%	Recirculation valve	Manual
		Stack valve	Manual
FG_O_2	3 mol.%	Air valve	Manual
		Main oxygen valve	Manual
PO flowrate	Stable	PA	Automatic
PO_O_2	<18 mol.%	Primary oxygen valve	Manual
FD	At a certain range	ID	Automatic

Reprinted with permission from Luo W, Wang Q, Liu Z, Zheng C. Dynamic simulation of the transition process in a 3 MW_{th} oxy-fuel test facility. Energy Proc 2014; 63:6281–88. Copyright (2014) Elsevier.

combustion. The control of BI_O_2 was realized by manually manipulating the recirculation valve, stack valve, air valve, and oxygen valves, while the PA and ID regulations were set automatically such that they would quickly respond if safety issues arose.

12.3.2 Commercial Plant

A "top-down analysis, bottom-up design" method [31] was used to design the closed-loop control system for the conceptual oxy-fuel plant. Top-down analysis aims to define control goals, identify controlled and manipulated variables, and determine the rate of production; a bottom-up design focuses on designing the regulatory control layer and supervisory control layer. First, necessary control objectives were considered for all the subsystems: (1) The ASU needed to satisfy the internal and external requirements. The former means that ASU would maintain the operating parameters (liquid air and oxygen levels, operating pressure, and expander speed) within safe ranges. The latter is mainly derived from the operating demands of the oxy-fuel plant. Namely, the oxygen product quality (purity, flow rate, pressure) satisfies the supply demand for the boiler, the operation handles the high ramp rate, and the control system bears the flexible strategies. (2) In the boiler island system, different control targets would be specified in both the water-side and the flue gas-side. For the water-side, the main steam temperature and reheat steam temperature would be kept within a safe range to avoid going beyond the materials' limitations. With respect to flue gas side, the wet flue gas O_2 content (WFG_O_2) would be maintained around a suitable value to achieve stable and effective combustion, while a negative pressure operation would be selected. Moreover, the control system needed to bear engineering demands such as mode switching, load change, operational disturbances, and emergence conditions. (3) As the final processing unit, the CPU would obtain at least 96 mol.% CO_2 product purity and a 90% or higher CO_2 recovery rate. To keep the pipe from blocking, a stream temperature at lower reaches of throttle valveLCV119 (T_LCV119) would be above the CO_2 freezing point. Then the choice of inputs and outputs would relate to the subsystems' features. As shown in Table 12.2, through a degree of freedom analysis, it was possible to find desirable control pairings (i.e., manipulated and controlled variables). Because feed water represented the load demand from the latter subsystem and connected the boiler island and steam cycle, its flow rate was determined as the product rate.

Finally, control layers were designed and configured in the dynamic model to form a closed-loop control structure for the analyzed oxy-fuel combustion system. As illustrated in Fig. 12.6, a simple PID-based control system [1] was proposed and discussed. (1) As shown in Fig. 12.4, the feedforward-feedback control strategy [9] adopted in the ASU was used to meet the demands for the oxygen supply and flexible operations. Its regulatory control layer configured flow and pressure control loops to control the flow rates and operating pressure, while two cascade control loops (namely, CC_OXY and LC_REB) were employed in the supervisory control layer to approach the desirable oxygen product purity and liquid oxygen level. (2) A flue gas O_2 control strategy [8] was used in the boiler island system to organize the internal furnace combustion,

TABLE 12.2 Primary Inputs and Outputs for Each Subsystem

Input	Output	Input	Output
ASU		Reheat steam	Main steam temperature
Air flow rate	Oxygen product purity	RFG flow rate	Reheat steam temperature
Expanded air flow rate	Liquid oxygen level	CPU	
Liquid oxygen	LP operating pressure	Flue gas flow rate	CO_2 product purity
Oxygen products	HP sump level	MCC exit pressure	CO_2 recovery rate
Boiler island		F1 temperature	T_LCV119
Coal flow rate	WFG_O_2	F2 temperature	
Feed water flow rate	Furnace exit pressure		

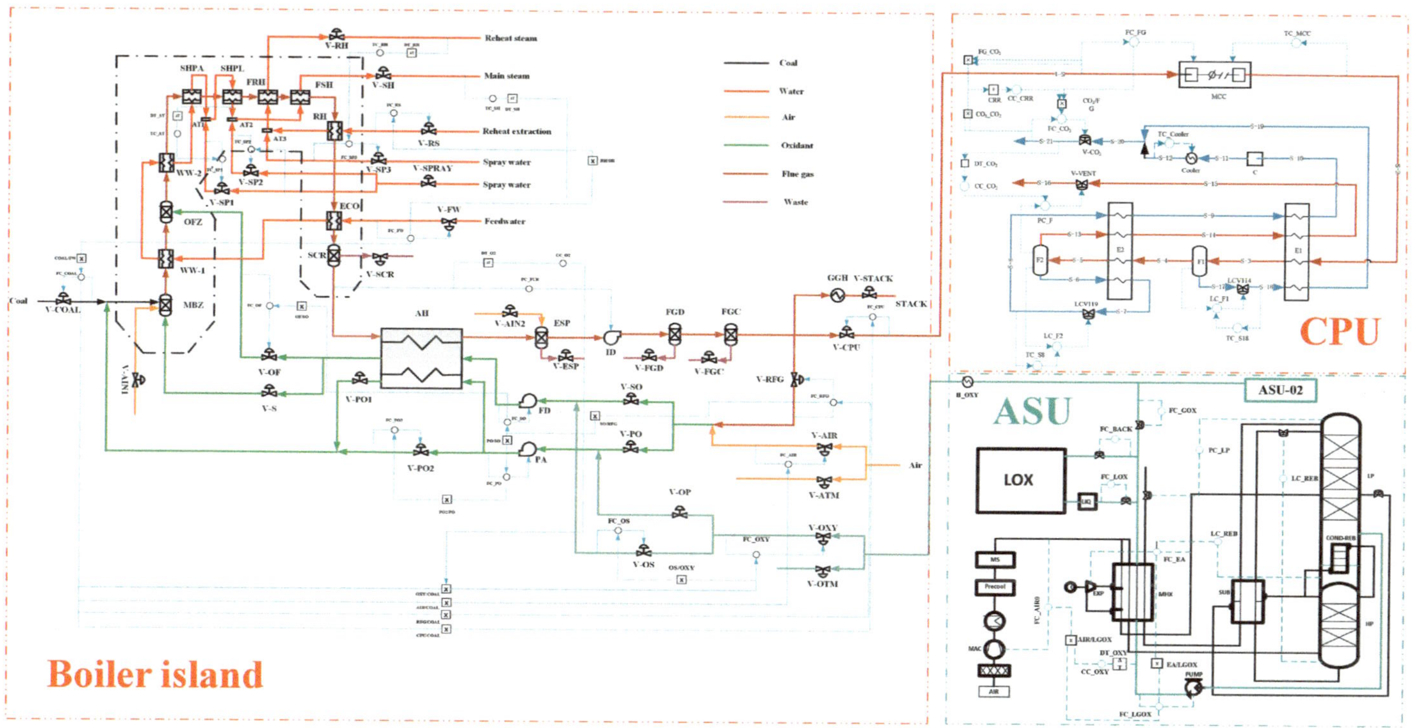

FIG. 12.6 Dynamic model for oxy-fuel combustion power plants with the proposed control system. *(Reprinted with permission from Jin B, Zhao H, Zheng C. Dynamic simulation and control design for pulverized-coal-fired oxy-combustion power plants, in clean coal technology and sustainable development. Springer; 2016. p. 325–33. Copyright (2016) Springer Science + Business Media Singapore and Tsinghua University Press.)*

consisting of steam flow rate controls, firing rate controls, furnace pressure control (PC_FUR), gas distribution controls, ratio controls, steam temperature controls (TC_SH and TC_RH), and flue gas O_2 content control (CC_O_2). The first four control loops were installed in the regulatory control layer, while the supervisory control layer contained the latter three control loops. The realization of TC_SH and TC_RH came from the adjustments of AT1, AT2, and AT3, while CC_O_2 denoted that WFG_O_2 was controlled by regulating the total oxygen flow from the ASU. (3) A DTC [10] was used in the CPU to gain the desirable CO_2 product purity, maintain the CO_2 recovery rate, and avoid stream temperatures below the CO_2 triple-phase temperature point. In Fig. 12.5, the double temperature represents the operating temperature of two flash separators. Control of the CO_2 product purity and CO_2 recovery rate was achieved via two cascade control loops by sending the corresponding control actions to the actuators.

12.4 DYNAMIC BEHAVIOR

A dynamic model configured with a control system can be used to execute the dynamic simulations under most common operating scenarios, and the identified dynamic process responses would provide important information for engineers and operators. The validated dynamic model of a pilot facility was first applied to identify the influence of disturbances on the system operation, and then dynamic tests for a commercial plant were conducted to manifest the dynamic behavior of the ASU, boiler island, and CPU.

12.4.1 Pilot Facility

Fig. 12.7 shows the system dynamic performance when four operating disturbances were tested. For simplicity, only the dynamic response of the heat duties are presented to illustrate the dynamic characteristics. The detailed information can be found in reference [13]. When the fuel flow rate got a step decrease, the heat duties ($Q1$, $Q2$, and $Q3$) for three tube banks all decreased. Because of its close position to the disturbance, $Q1$ changed more sharply. During oxygen flow rate step change processing, the sensitivity of the heat duties was similar to that in the fuel flow rate case. The smallest change occurred for $Q3$, which was attributed to it having the weakest heat transfer capacity. The radiation-dominated heat transfer contributed to the decrease of $Q1$. A very slight variation was observed for heat duties when the feed water flow rate varied, which may have been because a change in the feed water flow rate has no effect on the heat transfer if it reaches its limitation [13]. In these three cases, a comparison of the real-time responses between air mode and oxy mode indicated that the operation and control approaches for two combustion modes may be distinctive in terms of dealing with disturbances. On the other hand, the recycle ratio case could be tested only in oxy mode, and it was found that $Q3$ increased while $Q1$ and $Q2$ decreased.

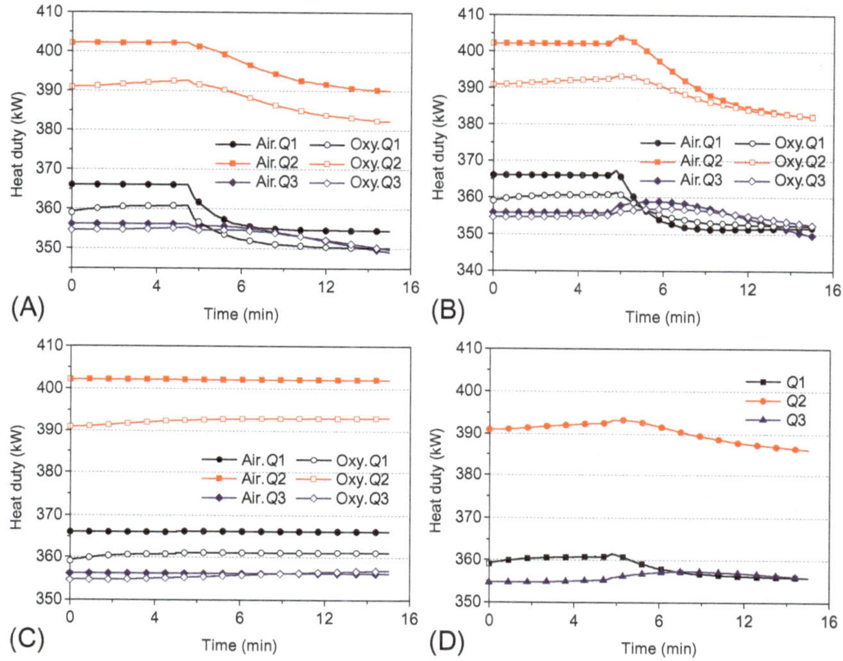

FIG. 12.7 Dynamic process responses to different disturbances in the step change method: (A) fuel flow rate, (B) oxygen supplying flow rate, (C) feed water flow rate, and (D) recycle ratio. *(Reprinted with permission from Luo W, Wang Q, Huang X, Liu Z, Zheng C. Dynamic simulation and transient analysis of a 3MWth oxy-fuel combustion system. Int J Greenhouse Gas Control 2015;35:138–49. Copyright (2015) Elsevier.)*

12.4.2 Commercial Plant

For each subsystem, dynamic simulations under different operating disturbances and strategies were conducted to clarify their own characteristics. The oxy-fuel boiler island, as the core component, was first tested in three operating cases (the changes of load, oxygen supply purity, and air in-leakage) to garner the dynamic responses as the outputs of the ASU and the inputs of the CPU. Then the ASU was dynamically operated under a high ramp rate and flexible operations to supply oxygen smoothly and economically, while the change of flue gas flow rate and composition was acted as dynamic disturbances for the CPU to produce CO_2 steadily and effectively. Finally, the dynamic simulation results provided the basic data for sophisticated analysis of the real-time exergy property and the impact of controls in system operations.

(1) Dynamic behavior of ASU

The ASU needed to bear a fast ramping process in order to meet the varied load demands of power plants and to achieve economic operations and reduce operating costs. When integrated into oxy-fuel plants, three operating schemes were proposed to achieve these two requirements [9]. The first pattern was validated by dynamic tests under a 5%/min ramp rate process and achieved stable operation with the targeted oxygen product purity and supplying rate. An oversized ASU with LOX was designed to confirm the second pattern, that is, that an oxy-fuel boiler island always operates in full load while the ASU takes the strategy of a peak and off-peak (POP) operation. Fig. 12.8 shows the dynamic responses during this operating scheme. Two control strategies, ASU-following and LOX-following, were put forward to complete this process when three flow controls (FC_LGOX,

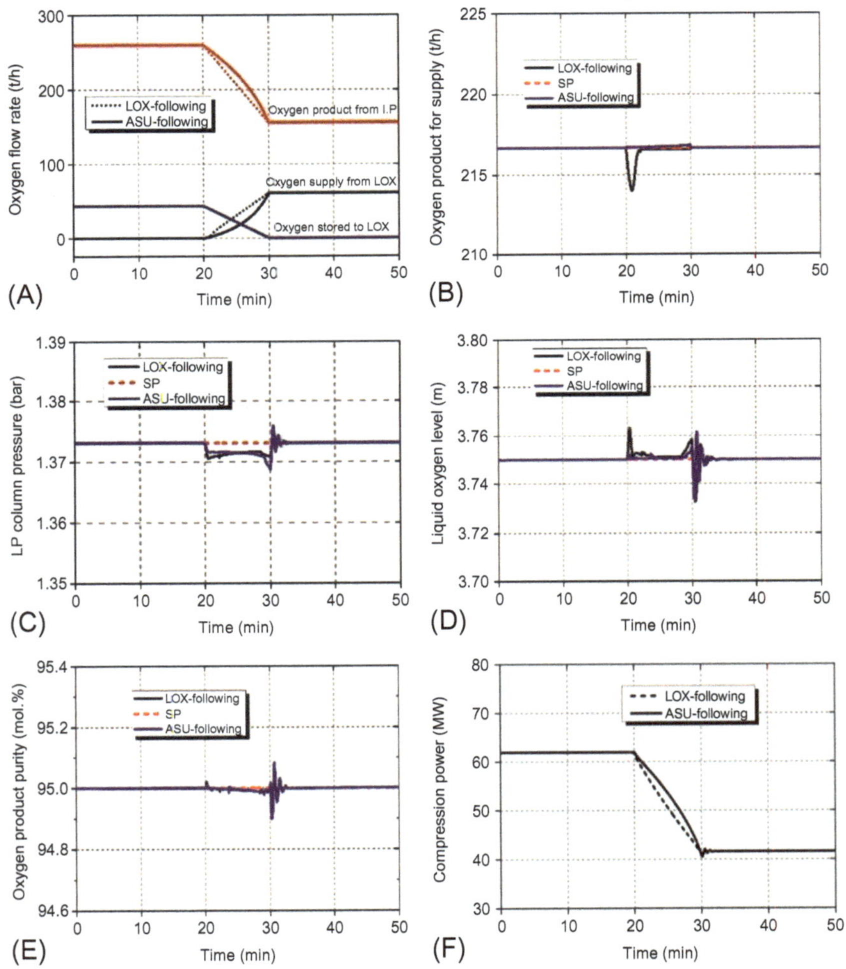

FIG. 12.8 Dynamic performance for ASU under flexible operation: (A) flow rate change, (B) oxygen product for supply, (C) LP operating pressure, (D) liquid oxygen level, (E) oxygen product purity, and (F) compression power consumption. *(Reprinted with permission from Jin B, Su M, Zhao H, Zheng C. Plantwide control and operating strategy for air separation unit in oxy-combustion power plants. Energ Convers Manage 2015;106(2015):782–92. Copyright (2015) Elsevier.)*

FC_BACK, and FC_LOX) were regulated while the oxygen supplying rate was kept constant. The difference between the two control methods concentrated mainly on what kinds of oxygen flows were pre-specified and remotely controlled. The former sets the oxygen supplying rate from LOX in advance and controls oxygen products from LP remotely, while the latter sets oxygen products from LP at a preset rate and the LOX operation is moderated to follow the oxygen supplying goal. Nevertheless, their dynamic behavior was similar and supportable for economic operations, because 32.85% energy consumption was reduced and operating parameters (oxygen supply rate, oxygen product purity, liquid oxygen level, etc.) were maintained around their setpoints. For the third pattern, a downsized ASU with LOX was used and an energy storage method was applied to achieve a flexible and economic operation. Dynamic simulation results showed that only the LOX-following strategy worked and that compression power decreased about 27.71%. Compared to a pure feedback control system, a very slight difference was observed for their dynamic process responses. If the dynamic exergy method is applied to compare the preceding two control systems, it might be concluded that feedforward-feedback control is more favorable than pure feedback control because it achieves a greater savings in energy during operating conditions.

(2) Dynamic behavior of oxy-fuel boiler island

The oxy-fuel boiler island needed to achieve stable combustion, obtain desirable steam quality, and reject unexpected disturbances. As shown in Fig. 12.9, robust dynamic responses for all control objectives were obtained during the load change process [8]. The WFG_O_2 was maintained around 3.364 mol.%, the negative operating mode was realized, and the steam temperature fluctuated within reasonable ranges. When oxygen supply purity increased to 0.2%/min, it contributed to a 4.62% decrease in the oxygen supplying rate and a 1.09% increment in the BI_O_2. Fortunately, the WFG_O_2, the main steam temperature, and the reheat steam temperature all stayed around their setpoints, although some fluctuations were observed. An air in-leakage step increase led to increments in the main steam temperature and the reheat steam temperature on the water-side and a decrement in the oxygen supplying rate and BI_O_2, which may have been caused by the introduction of external oxygen into the oxy-fuel boiler island system. When the present control system was compared to an alternative control structure that set BI_O_2 as the control target, very similar dynamic behavior was obtained. Their difference was in the oxygen concentration selected as the control objective to reflect combustion conditions. Based on the aspects of mode switching and variations in the flue gas sent to the CPU, the flue gas O_2 control proved to be more suitable for oxy-fuel combustion application than for that of oxidant O_2 control. The dynamic exergy method was applied to the oxy-fuel boiler island system to optimize system operations and identify the control impacts [12]. The flow rate change process was

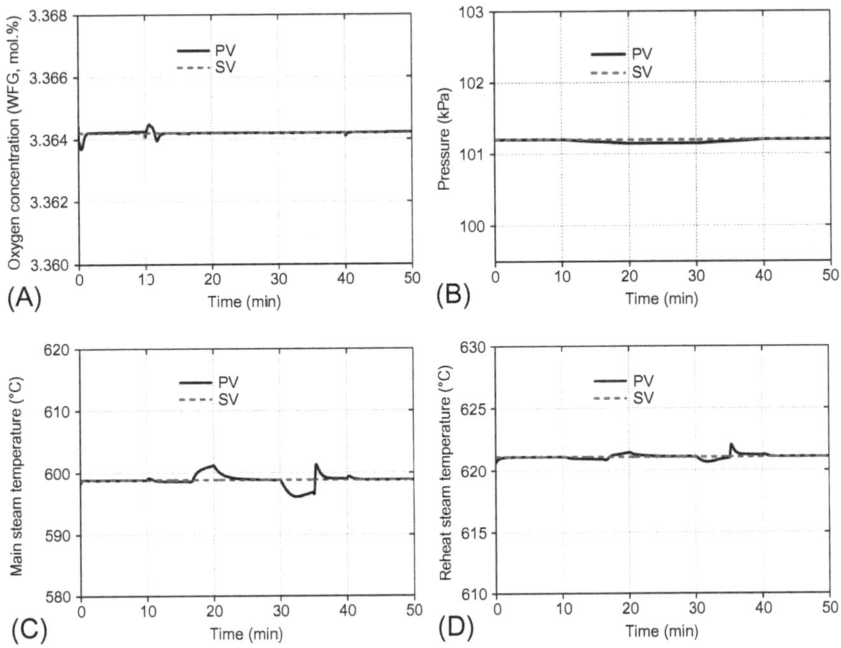

FIG. 12.9 Dynamic responses during load change process: (A) WFG_O_2, (B) furnace operating pressure, (C) main steam temperature, and (D) reheat steam temperature. *(Reprinted with permission from Jin B, Zhao H, Zheng C. Dynamic modeling and control for pulverized-coal-fired oxy-combustion boiler island. Int J Greenhouse Gas Control 2014;30(2014):97–117. Copyright (2014) Elsevier.)*

more energy-efficient than the other cases, while the oxygen supplying rate change had the greatest influence on system operations and required more control efforts. Although the control system had no effect on identifying the sensitivity of the operating parameters, it might save or consume energy while accomplishing operating commands. More importantly, control penalties and costs for control systems, layers, and loops were distinctive and related to their functions during dynamic operation.

(3) Dynamic behavior of CPU

The CPU needed to resist frequent variations in the oxy-fuel boiler, produce high purity CO_2 products, and ensure a high CO_2 recovery rate. When the load ramped down, the system operating pressure, CO_2 product purity, and stream S-8 temperature followed the evolution of the load change [10]. The stream S-18 temperature and specific energy consumption varied in the opposite direction, while the CO_2 recovery rate showed a stepwise decrease first and then returned to the setpoint. As shown in Fig. 12.10, different real-time behavior was presented when the flue gas composition changed. The CO_2 recovery rate appeared with an almost opposite variation than the CO_2 recovery rate, even though the latter was below 90% at the 28th

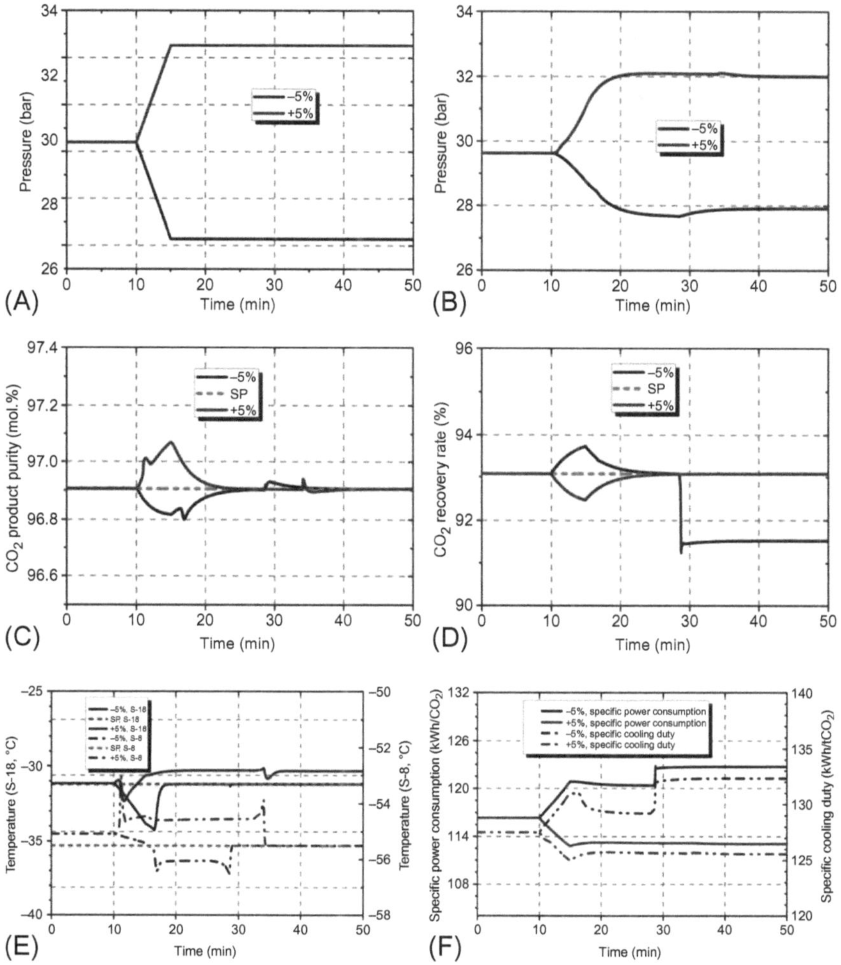

FIG. 12.10 Dynamic responses of process variables under flue gas composition change process: (A) CO_2 concentration with $\pm 5\%$ ramp change, (B) system pressure, (C) CO_2 product purity, (D) CO_2 recovery rate, (E) stream temperature, and (F) energy consumption. *(Reprinted with permission from Jin B, Zhao H, Zheng C. Optimization and control for CO2 compression and purification unit in oxy-combustion power plants. Energy 2015;83(2015):416–30. Copyright (2015) Elsevier.)*

minute during a 5% decreasing process. The specific energy consumption negatively correlated with the flue gas composition, while the stream temperature was always above the CO_2 triple point to avoid CO_2 solidification. When the present case was compared to CPU with a single temperature control (STC), it was found that a DTC system might be more suitable for a CPU because a DTC system is better able to keep control objectives close to their setpoints, which consumes less energy. The current study's dynamic exergy evaluation results [11] indicated that the dynamic exergy property was related to the operation type, that the system operation was more sensitive to flue gas composition changes, that the control system helped save energy during the flue gas composition reduction process, and that more control efforts need to be taken to gain the desirable CO_2 quality.

12.5 MODE SWITCHING STRATEGY

Finding a feasible mode switching strategy may be one of most important factors for attaining commercial demonstration oxy-fuel combustion plants [32], because dual combustion modes are often considered during its routine operations. Plant is first started up from air mode, then is switched to oxy-fuel mode at a certain operating load, and finally ends in air mode [33]. Currently, two mode transition approaches have been investigated [15]. One approach adjusted all oxidant flows simultaneously, including primary air, secondary air, primary RFG, secondary RFG, PO, and SO. The second approach performed oxidant flows sequentially so that oxygen and RFG flows were first manipulated while the air was gradually decreased. The proper switching method depends on the plant type, control strategy, and operating target. In this section, a simultaneous approach is adopted to illustrate and identify the dynamic behavior of a pilot facility and a commercial plant.

12.5.1 Pilot Facility

To implement mode switching from air mode to oxy mode, a series of operating commands need to be done correctly. When flue gas begins to be recycled, the stack valve and air valve are gradually closed, and the main oxygen valve is opened. During this process, stable combustion should be maintained and certain operating parameters (oxygen concentration, furnace pressure, CO_2 concentration) should be measured to achieve the control objectives. Fig. 12.11 shows the dynamic behavior of the pilot facility during the mode switching process. Fig. 12.11A and B indicate that the dynamic simulation results were consistent with that of experimental tests and that the dynamic model would be further applied to predict real-time system performance. The variation of the heat duties in the tube banks showed that the heat transfer condition will undergo slight changes during the mode switching process. An 80 mol.% above CO_2 concentration in the flue gas demonstrated that the control method used efficiently accomplished mode switching.

12.5.2 Commercial Plant

As shown in Fig. 12.12, the transition from oxy mode to air mode was investigated to identify the corresponding dynamic behavior of the studied commercial plant. A staged operating strategy [7] and a partial control strategy [8] were chosen to complete this process. In the former, the oxygen and RFG streams were manipulated with

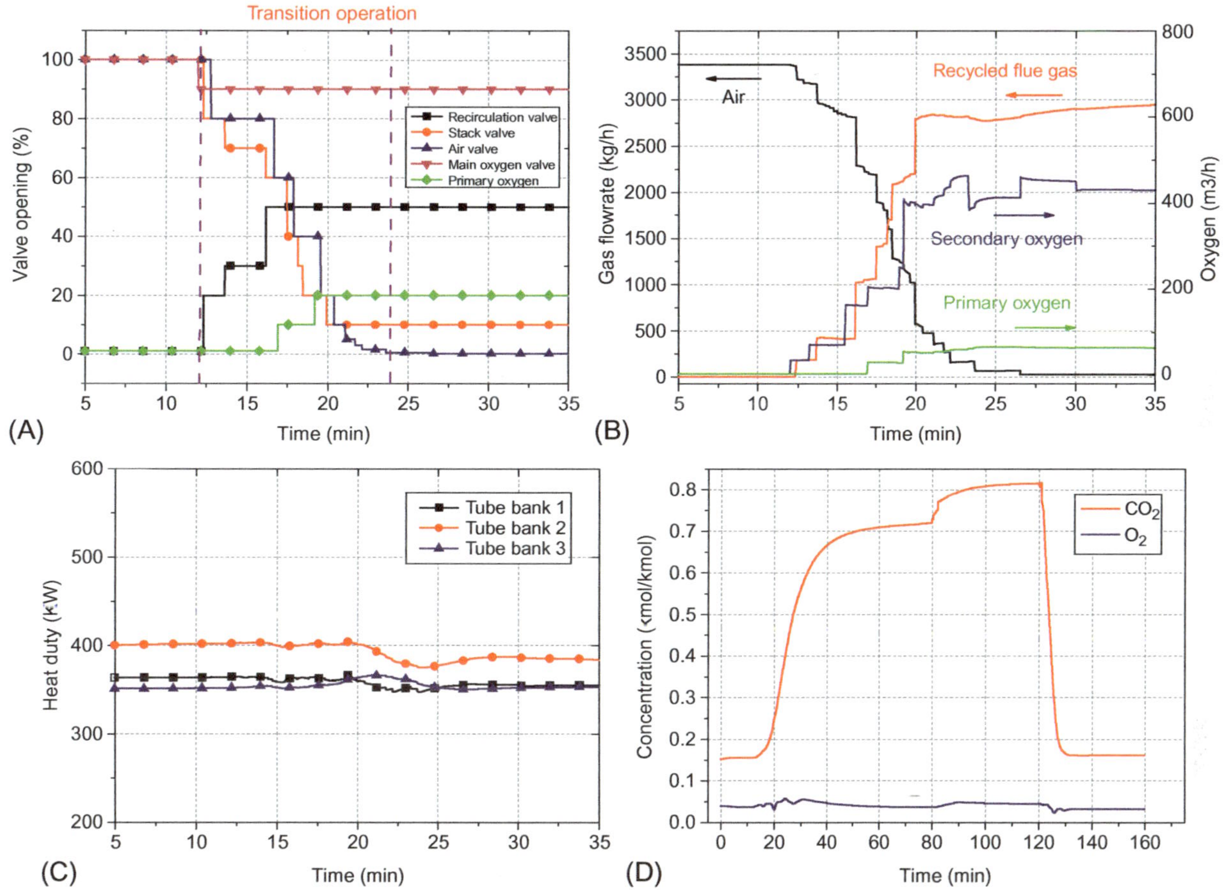

FIG. 12.11 Dynamic simulation responses during mode switching process: (A) valve opening, (B) gas flow, (C) heat duty, and (D) CO_2 and O_2 concentration. *(Reprinted with permission from Luo W, Wang Q, Liu Z, Zheng C. Dynamic simulation of the transition process in a 3 MWth oxy-fuel test facility. Energy Proc 2014; 63:6281–88. Copyright (2014) Elsevier.)*

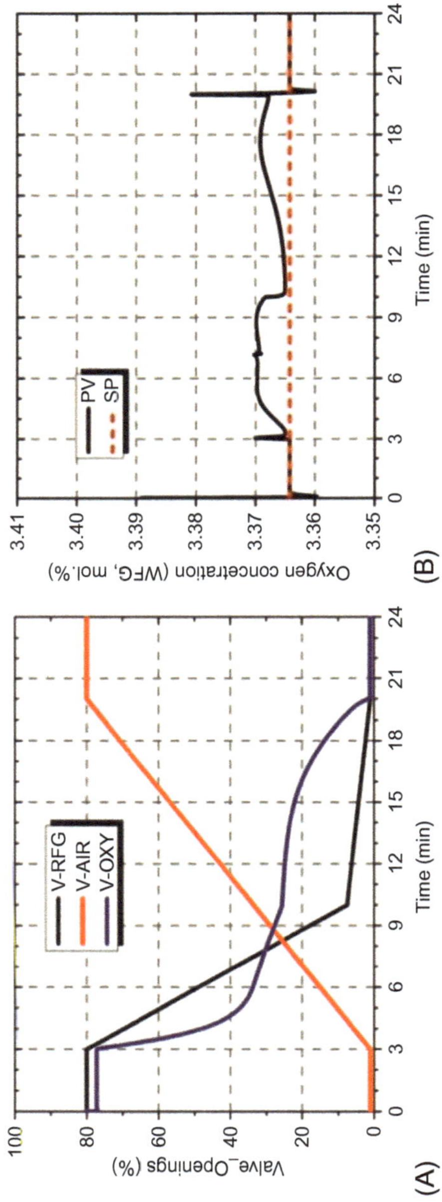

FIG. 12.12 Dynamic process responses during mode switching process: (A) valve openings, (B) WFG_O_2,

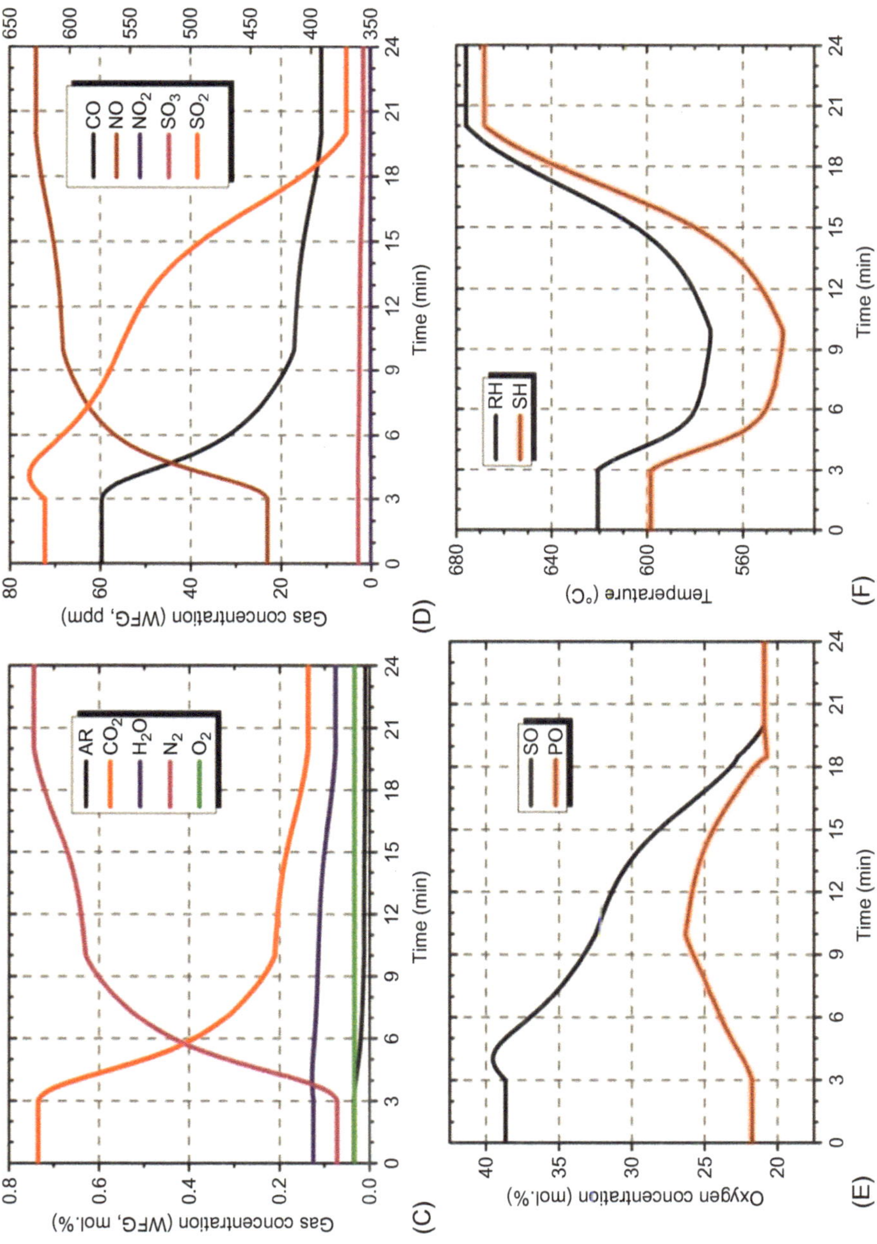

FIG. 12.12, CONT'D (C) WFG composition responses, (D) PO_O_2 and SO_O_2 responses, and (E) steam temperature. *(Reprinted with permission from Jin B, Zhao H, Zheng C. Dynamic modeling and control for pulverized-coal-fired oxy-combustion boiler island. Int J Greenhouse Gas Control 2014;30(2014):97–117. Copyright (2014) Elsevier.)*

different rates of change in corresponding valve positions (i.e., V-RFG and V-OXY) during different switching stages (as shown in Fig.12.12A) while air stream with a constant rate of change of corresponding valve position (i.e., V-AIR). The latter set the fuel flow control and CC_O_2 automatically, configured other control loops in manual mode, and defined some operating tasks. Fig. 12.12A shows the implementation of the stage operating strategy when the maximum valve opening was 80%. Although WFG_O_2 appeared with some fluctuations and always above the setpoint during mode switching, it performed robustly, which indicated that favorable combustion was maintained. The variation of the flue gas composition indicated that the mode switching was successful and emerged in a similar S-shape [34] fashion. Because of some manual adjustments, the evolution of the oxygen concentrations in the oxidant flows and steam temperature exceeded desirable conditions. Nevertheless, it provides important guidelines that can help engineers understand this process and how to operate it so that these adverse situations can be avoided.

12.6 SUMMARY

This chapter presented a comprehensive review of dynamic modeling, control performance, and mode switching strategy for oxy-fuel combustion power plants. The chapter provided deep insight into how to identify the dynamic behavior and operating performance of pilot-scale and commercial-scale oxy-fuel combustion systems under common operating conditions, which will help engineers gain skills useful in the operation of real plants. Although great efforts have been made, oxy-fuel combustion still faces great challenges in terms of developing a highly robust and reliable control system, of achieving energy efficiency and economically feasible operations, of optimizing the effects of process control on system thermodynamic performance, and of reducing the environmental impact.

REFERENCES

[1] Jin B, Zhao H, Zheng C. Dynamic simulation and control design for pulverized-coal-fired oxy-combustion power plants, in clean coal technology and sustainable development. Springer; 2016. p. 325–33.

[2] Svrcek WY, Mahoney DP, Young BR. A real-time approach to process control. New York: John Wiley & Sons; 2006.

[3] Pottmann M, Engl G, Stahl B, Ritter R. Dynamic simulation of oxyfuel CO_2 processing plants. Energy Proc 2011;4(2011):951–7.

[4] Luyben WL. Process modeling, simulation and control for chemical engineers. New York: McGraw-Hill Higher Education; 1989.

[5] Postler R, Epple B, Kluger F, Mönkert P, Heinz G. Dynamic process simulation model of an oxyfuel 250 MWel demonstration power plant. In: 2nd oxyfuel combustion conference; 2011.

[6] Kuczynski K, Fitzgerald F, Adams D, Glover F, White V, Chalmers H, et al. Dynamic modelling of oxyfuel power plant. Energy Proc 2011;4(2011):2541–7.

[7] Jin B, Zhao H, Zheng C. Dynamic simulation for mode switching strategy in a conceptual 600MWe oxy-combustion pulverized-coal-fired boiler. Fuel 2014;137(2014):135–44.

[8] Jin B, Zhao H, Zheng C. Dynamic modeling and control for pulverized-coal-fired oxy-combustion boiler island. Int J Greenhouse Gas Control 2014;30(2014):97–117.
[9] Jin B, Su M, Zhao H, Zheng C. Plantwide control and operating strategy for air separation unit in oxy-combustion power plants. Energ Convers Manage 2015;106(2015):782–92.
[10] Jin B, Zhao H, Zheng C. Optimization and control for CO_2 compression and purification unit in oxy-combustion power plants. Energy 2015;83(2015):416–30.
[11] Jin B, Zhao H, Zheng C. Dynamic exergy method and its application for CO_2 compression and purification unit in oxy-combustion power plants. Chem Eng Sci 2016;144(2016):336–45.
[12] Jin B, Zhao H, Zheng C, Liang Z. Dynamic exergy method for evaluating the control and operation of oxy-combustion boiler island systems. Environ Sci Technol 2017;51:725–32.
[13] Luo W, Wang Q, Huang X, Liu Z, Zheng C. Dynamic simulation and transient analysis of a 3 MW_{th} oxy-fuel combustion system. Int J Greenhouse Gas Control 2015;35:138–49.
[14] Luo W, Wang Q, Liu Z, Zheng C. Dynamic simulation of the transition process in a 3 MW_{th} oxy-fuel test facility. Energy Proc 2014;63:6281–8.
[15] Hultgren M, Ikonen E, Kovács J. Oxidant control and air-oxy switching concepts for CFB furnace operation. Comput Chem Eng 2014;61(2014):203–19.
[16] Lappalainen J, Tourunen A, Mikkonen H, Hänninen M, Kovács J. Modelling and dynamic simulation of a supercritical, oxy combustion circulating fluidized bed power plant concept—firing mode switching case. Int J Greenhouse Gas Control 2014;28(2014):11–24.
[17] Niva L, Ikonen E, Kovács J. Self-optimizing control structure design in oxy-fuel circulating fluidized bed combustion. Int J Greenhouse Gas Control 2015;43(2015):93–107.
[18] Luyben W. Plantwide dynamic simulators in chemical processing and control. New York: Chemical Industries. CRC; 2002. p. 448.
[19] Aspen Technology Inc. Aspen Dynamics™ 11.1 User Guide; 2001.
[20] Xiong J, Zhao H, Chen M, Zheng C. Simulation study of an 800 MWe oxy-combustion pulverized-coal-fired power plant. Energy Fuels 2011;25(5) 2405–15.
[21] Xiong J, Zhao H, Zheng C. Exergy analysis of a 600 MWe oxy-combustion pulverized-coal-fired power plant. Energy Fuels 2011;25(8):3854–64.
[22] Xiong J, Zhao H, Zheng C. Techno-economic evaluation of oxy-combustion coal-fired power plants. Chinese Sci Bull 2011;56(31):3333–45.
[23] Dillon DJ, White V, Allam RJ, Wall RA, Gibbins J. Oxy-fuel combustion processes for CO_2 capture from power plant. IEA greenhouse gas R&D programme; 2005.
[24] Ciferno J. Pulverized coal oxycombustion power plants. Final Report; 2007.
[25] Darde A, Prabhakar R, Tranier J-P, Perrin N. Air separation and flue gas compression and purification units for oxy-coal combustion systems. Energy Proc 2009;1(1):527–34.
[26] Chansomwong A, Zanganeh K, Shafeen A, Douglas P, Croiset E, Ricardez-Sandoval L. Dynamic modelling of a CO_2 capture and purification unit for an oxy-coal-fired power plant. Int J Greenhouse Gas Control 2014;22(2014):111–22.
[27] Yamada T, Fujimori T, Takano S. Method and apparatus for controlling combustion in oxygen fired boiler. Google Patents; 2013.
[28] McDonald D, Zadiraka A. Control of pulverized coal oxy-combustion systems. In: 17th annual joint ISA POWID/EPRI controls & instrumentation conference, Pittsburgh, PA; 2007.
[29] Terushita S, Yamada T, Watanabe S, Uchida T, Misawa N Method and apparatus of controlling combustion in oxyfuel combustion boiler. Google Patents; 2008.
[30] Haryanto A, Hong K-S, Jeon CH. Nonlinear model predictive control of an oxy-fuel combustion boiler for regulating the excess oxygen percentage in the flue gas. In: SICE annual conference. IEEE; 2008.
[31] Skogestad S. Control structure design for complete chemical plants. Comput Chem Eng 2004;28(1):219–34.

[32] Wall T, Stanger R, Santos S. Demonstrations of coal-fired oxy-fuel technology for carbon capture and storage and issues with commercial deployment. Int J Greenhouse Gas Control 2011;5:S5–S15.

[33] Stone BB, McDonald DK, Zadiraka AJ. Oxy-combustion coal fired boiler and method of transitioning between air and oxygen firing. Google Patents; 2009.

[34] Weigl S. Modellierung und experimentelle Untersuchungen zum Oxyfuel-Prozess an einer 50 kW-Staubfeuerungs-Versuchsanlage. Dresden, Techn. Univ., Diss.; 2009.

Chapter 13

Oxygen Production for Oxy-fuel Combustion

Cong Luo, Jing Liu
Huazhong University of Science and Technology, Wuhan, China

13.1 INTRODUCTION OF CRYOGENIC AIR SEPARATION UNIT

The air separation unit (ASU) provides the oxygen for the combustion process. Air separation at a level required for a full-scale power plant can be performed by using existing commercially available cryogenic technology. Fig. 13.1 illustrates the five major unit operations required to cryogenically separate air into useful products. An air pretreatment section downstream from the air compression and after cooling removes process contaminants, including water, carbon dioxide, and hydrocarbons. The air is then cooled to cryogenic temperatures and distilled into oxygen, nitrogen, and, optionally, argon streams. Warming these product streams against the incoming air feed conserves refrigeration, with any deficit made up by expanding a small portion of pressurized air or nitrogen. Numerous configurations of heat exchange and distillation equipment can separate air into the required product streams. These process alternatives are selected based on the purity and number of product streams required, trade-offs between capital costs and power consumption, and the degree of integration between the ASU and other facility units. Characterization of the types of cryogenic processes used for air separation, hereinafter referred to as cycles, can be based on the method of pressurizing the product streams or on the air feed pressure to the ASU.

Pressure swing adsorption (PSA), vacuum swing adsorption (VSA), and membrane separation technologies are well suited for small-scale oxygen production (<200 t/d), whereas cryogenic plants can be designed for large-scale oxygen production (Fig. 13.2). Industrial sites exist with cumulative oxygen production of 30,000 t/d; currently, the largest oxygen production plant in the world is operated at a capacity of 4000 t/d. Existing technologies would allow capacities up to 5000 t/d. Design studies for 7000 t/d by a single train have been performed. Nevertheless, once CO_2 transport networks have been developed, decentralized power generation from oxy-fuel combustion could be deployed, in which case PSA or VSA technologies could prove viable. As a consequence,

Oxy-fuel Combustion. http://dx.doi.org/10.1016/B978-0-12-812145-0.00013-X

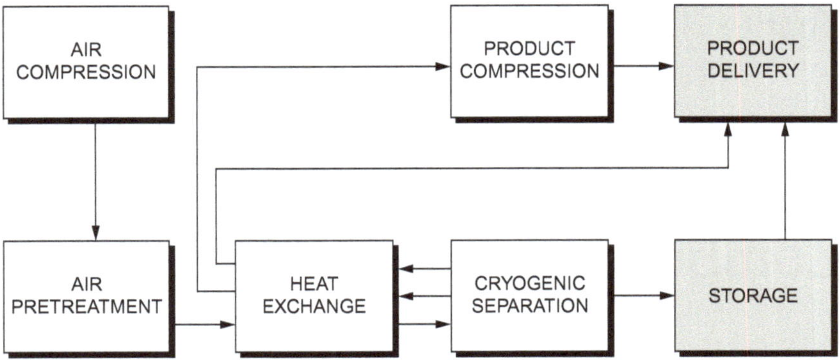

FIG. 13.1 Unit operations for a cryogenic air separation process.

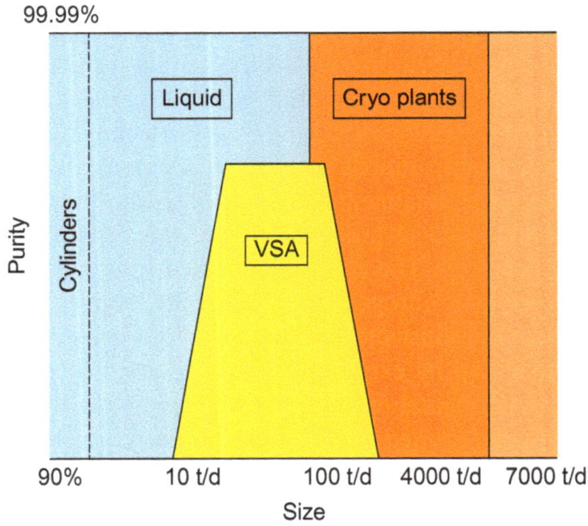

FIG. 13.2 Oxygen production technology selection as a function of quantity.

cryogenic distillation is currently the only proven option for an oxy-fuel combustion power plant on an industrial scale.

Large volumes of high purity products are required for oxy-fuel combustion, so distillation at cryogenic conditions is the first choice. The first steps in a cryogenic ASU are to compress the feed air and remove water, carbon dioxide, and other hydrocarbon contaminants from it. The resulting mixture consists of nitrogen, oxygen, argon, and some traces of other noble gases. Next, the cleaned air is cooled to cryogenic conditions in the main heat exchanger (MHX) and subsequently fed to the distillation unit. The distillation unit separates the air into an oxygen stream and a nitrogen stream; sometimes an argon stream is produced as well. The product streams are passed again through the MHX and compressed or pumped to pressures that are required downstream of the ASU.

A wide variety of process designs exist for cryogenic separation of air. The main differences between the designs relate to the method used for providing refrigeration, to the method used for pressurizing the products, to the operating pressures, and to the column configuration in the distillation section. The type of process that is chosen depends usually on the feed and product specifications of the specific application. For a process that is not producing argon, the most common configuration contains two columns. In order to improve the market competitiveness of oxy-fuel combustion, it is necessary to reduce the energy consumption of the ASU. To meet the unique demand of oxy-fuel combustion, improvement from double columns to three columns is significant.

As shown in Fig. 13.3, because the oxygen concentration and the pressure of oxy-fuel combustion are not high, the reflux ratio of the low pressure column (LPC, 1.3 bar) is relatively superfluous. Thus, a medium pressure column (MPC, 2.8 bar) is added to the system, and some of the air is compressed into a high pressure column (HPC, 4.8 bar). Because the air flowing into the HPC is reduced, the ASU's consumption of power to produce oxygen decreases, thereby reducing the compressor's overall power consumption.

Compared with the traditional double-column process, the three-column air separation process can significantly reduce the oxygen consumption of products; when the oxygen concentration is 97%, the destruction of three columns and double columns amounts to 4.12 and 4.67 kJ/mol air. This means that the three-column design reduces energy by 12%, as well as 7% of the network input requirements. The actual energy efficiencies of these two designs are 35% and 38%, respectively, as a result of adding the third column.

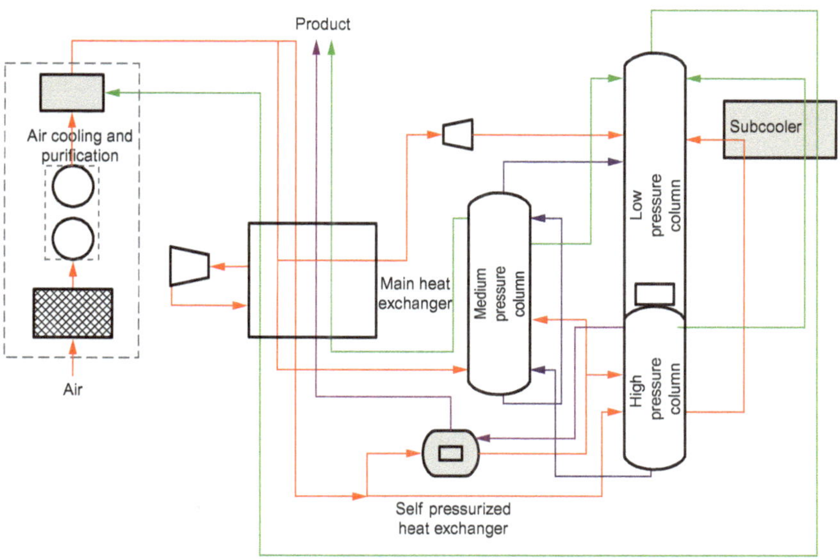

FIG. 13.3 A schematic process flow chart of the three-column ASU design.

Producing oxygen with this technology results in an energy penalty that, together with the energy demand for CO_2 compression and conditioning, causes a greater efficiency penalty than a power plant without CO_2 capture does. However, no other mature technologies exist that can be applied to the necessary size of industrial scale operation.

Cryogenic distillation consumes a great amount of energy in the form of electricity. Preliminary calculations indicate that this operation alone requires about 60% of the power consumed for carbon capture and reduces the overall efficiency of a power plant by about 7–9 percentage points. The availability of an oxygen plant is crucial to a power plant's operation. Suppliers report that typical oxygen plant availability is from 98% to 99%, with only a limited amount of time needed for maintenance. Another important requirement is the ability to follow the load changes of a power plant. According to Xu et al. [1] the maximum ramp rate for an ASU is 3%/min. The boiler, however, can generally be operated at a ramp of up to 6%/min. At the same time, ASUs are able to operate at only 60%–100% load and show rather poor efficiencies below 80%, which creates some operational difficulties that need to be further investigated.

A liquid oxygen storage approach could provide a solution, thus, oxygen demand and production was provided by the storage during the load changes of the boiler. The storage facility also makes it possible to shift to air-firing mode within a relatively short time without risking boiler trips in cases of unexpected ASU shutdowns. Even without load changing challenges, it may be beneficial to decouple the ASU and power plant by introducing oxygen storage. In that way, the liquid oxygen could provide a means of energy storage in high electricity-production periods via, for example, windmills. The safety issues related to liquid oxygen storage at power plants should be assessed thoroughly, however, in order to evaluate its applicability. Preliminary studies by Xu et al. [1] suggested the installation of a tank with a 500 tonne capacity for liquid oxygen at 2.5 bar (about −175°C), corresponding to approximately 1.5 h of operation at their 413 MW_e, net, oxy-fuel reference plant.

Further studies have focused on reducing the energy consumption of the ASU by using a chemical looping process for oxygen production as an alternative to conventional cryogenic distillation methods. Current investigations into chemical looping technologies are at the conceptual or laboratory-scale stage. Studies on the possible use of the chemical looping process for oxygen production have shown that it limits the drop in overall net plant efficiency by 5.2%. Typical oxygen carriers like perovskite and CuO that have high permeation rates at the conditions of interest are seen as new prospects for oxygen production. The following sections discuss the mechanism and development of two kinds of chemical looping processes for O_2/CO_2 production.

13.2 CHEMICAL LOOPING PROCESS OF PEROVSKITES FOR HIGH TEMPERATURE O_2/CO_2 PRODUCTION

13.2.1 Introduction of Perovskites for O_2/CO_2 Production

Chemical looping air separation for O_2/CO_2 production is a new technology for air separation. The process can be cost-effective, depending on the cyclic reduction and oxidation of an oxygen carrier. Because of their feasible application in photocatalysis, oxygen-permeable membranes, energy storage, gas sensors, superconductivity, and magnetoresistance materials, perovskite-type metal oxides have received considerable attention. Perovskite-type oxides are capable of acting as oxygen carriers providing pure O_2 or O_2/CO_2-mixed gas streams for oxy-fuel combustion. Such oxygen production comprises two main steps as shown in Fig. 13.4: oxygen adsorption and oxygen desorption. In the first step, air is used as feed gas to saturate the perovskite oxygen carrier with O_2; in the second step, CO_2 is used as sweep gas to desorb O_2 from the perovskite and produce an O_2-enriched CO_2 flue gas stream. The reversible adsorption/desorption processes based on the perovskite-type oxygen carriers are described as follows:

$$2ABO_{3-\delta} + 2CO_2 \Leftrightarrow 2ACO_3 + B_2O_3 - \frac{1-2\delta}{2} O_2 \quad (13.1)$$

Perovskite-type oxides are generally described by the stoichiometric formula ABO_3, where A can be an alkaline earth or a lanthanide metal and B is a transition metal. Their compositions can be varied over a wide range by partial substitution of cations in positions A and B, yielding compounds of the formula $(A_xA_{1-x})(B_yB_{1-y})O_{3-\delta}$. Several perovskite-type oxides offer potential application for O_2 production via high temperature air separation. The sorption and desorption mechanism is based on the reversible defect reaction

$$1/2O_2(g) + V_O^* \Leftrightarrow O^* + 2h \quad (13.2)$$

where V_O^* is the oxygen vacancy, and O^* and h denote lattice oxygen and electronic-hole, respectively. The main feature of this type of perovskite oxygen carriers is that they can adsorb a large quantity of O_2 from air, but no other gases.

Though several types of perovskite oxygen carriers have been investigated for oxygen production application, the major problem encountered with these perovskite materials is their relatively low level of oxygen desorption. Yang et al. [2] illustrated a process that used perovskite oxygen carriers to produce O_2-enriched CO_2 stream. The perovskite-type sorbents $La_{0.1}Sr_{0.9}Co_{0.5}Fe_{0.5}O_{3-\delta}$ (LSCF) and $Sr_{0.5}Ca_{0.5}Co_{0.5}F_{e0.5}O_{3-\delta}$ (SCCF) were studied as materials to obtain an O_2/CO_2 gas stream. The processes based on such materials were reversible and can be expressed as

$$La_{0.1}Sr_{0.9}Co_{0.5}Fe_{0.5}O_{2.6} + 0.9CO_2 \Leftrightarrow 0.9SrCO_3 + 0.05La_2O_3 + 0.5CoO + 0.25Fe_2O_3 + 0.15O_2 \quad (13.3)$$

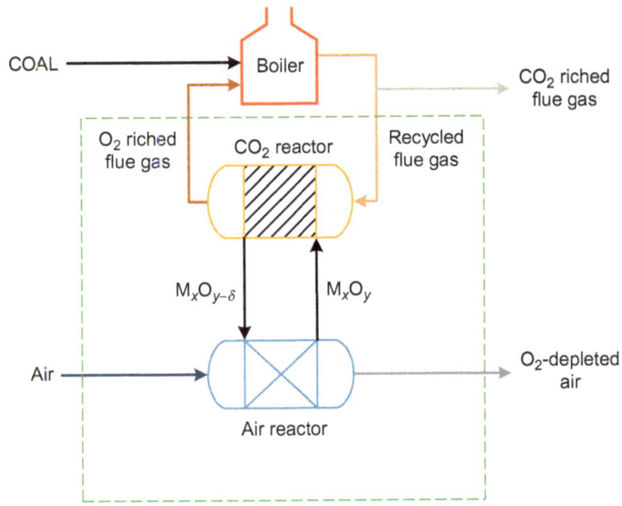

FIG. 13.4 Simplified schematic of O_2/CO_2 production of perovskite materials for oxy-fuel combustion. *(Reproduced from Shen Q, Zheng Y, Luo C, Ding N, Zheng C-G, Thern M. Effect of A/B-site substitution on oxygen production performance of strontium cobalt based perovskites for CO_2 capture application, RSC Adv 2015;5:39785–90, with permission from the Royal Society of Chemistry.)*

$$\begin{aligned} &Sr_{0.5}Ca_{0.5}Fe_{0.5}O_{2.47} + CO_2 \Leftrightarrow 0.5SrCO_3 + 0.5CaCO_3 \\ &+0.5CoO + 0.25Fe_2O_3 + 0.11O_2 \end{aligned} \tag{13.4}$$

According to their research, the oxygen production rate for $La_{0.1}Sr_{0.9}Co_{0.5}Fe_{0.5}O_{3-\delta}$ was 0.272 mL/min (3.6 g sample). Compared with $La_{0.1}Sr_{0.9}Co_{0.5}Fe_{0.5}O_{3-\delta}$, $Sr_{0.5}Ca_{0.5}Co_{0.5}Fe_{0.5}O_{3-\delta}$ had a faster oxygen desorption rate and a higher oxygen desorption capacity. Kusaba et al. [3] also studied a strontium cobaltite-based perovskite oxygen carrier and found that $SrCo_{0.4}Fe_{0.6}O_{3-\delta}$ had the best oxygen adsorption/desorption properties under temperatures between 300°C and 500°C. The oxygen production rate for $SrCo_{0.4}Fe_{0.6}O_{3-\delta}$ was 25 m^3 (STP)/ton of oxide/h. Recently, our group researched the O_2/CO_2 production performances of a series of $SrCo_{1-x}Fe_xO_{3-\delta}$ ($x = 0.2, 0.4, 0.6, 0.8$) perovskites and found that the oxygen production performance of $SrCo_{1-x}Fe_xO_{3-\delta}$ was improved by co-doping. $SrCo_{0.8}Fe_{0.2}O_{3-\delta}$ showed the best oxygen production property among the $SrCo_{1-x}Fe_xO_{3-\delta}$ ($x = 0.2, 0.4, 0.6, 0.8$).

Obviously, in oxygen production by a perovskite oxygen carrier system, advances of perovskite with a high amount of oxygen desorption and a strong cyclic performance are crucial to the industrial application of this promising alternative for the production of oxygen enriched stream. The performance of perovskite oxygen carriers is closely associated with the related preparation methods.

13.2.2 Characteristics and Performance of Ba-Sr-Co-Fe-Based Perovskites

Cobalt-based $Ba_{1-x}Sr_xCo_{0.8}Fe_{0.2}O_{3-\delta}$ perovskite oxides have attracted great interest in research because of their extremely high oxygen permeability and favorable structural stability. Several methods have been used to synthesize perovskite-type oxide powders, such as the combined EDTA-citrate method, the solid-state reaction method, wet chemical synthesis, the sol-gel method, and so on. Among these methods, the combined EDTA-citrate method is one of the most widely used. In this method, all the reactants are mixed in the solution at a molecular level, which leads to a homogeneous product [4]. $Ba_{1-x}Sr_xCo_{0.8}Fe_{0.2}O_{3-\delta}$ ($x=0$, 0.5, 1) powders were synthesized by the two methods, the EDTA method and the MWA-EDTA method for comparative analysis. Fig. 13.5 shows the XRD patterns of as-synthesized samples. The XRD characterization showed that pure perovskite structures were indeed formed after a calcining process at high temperature. The major diffraction peaks of the powders were matched with the theoretical ones. The strongest five diffraction peaks according to the 2θ around 22.3°, 32.18°, 38.2°, 46.14°, and 57.37° were characteristic peaks for perovskites.

Fig. 13.6 compares the initial morphology of three different perovskite powders prepared by the two methods. The crystallite size of the powders synthesized by EDTA was larger, with evident crystallite agglomeration and lower porosity, leading to the formation of a relatively condensed and compact surface. In comparison, powders synthesized by the MWA-EDTA method comprised numerous relatively ordered spherical nanocrystallines with a highly porous surface. Because pore distribution reflects the change in surface area values, it can be concluded that the synthesis method affected the morphology and the crystallite size of the prepared perovskite powders. Because surface area and porosity are very important for gas-solid reactions, the MWA-EDTA method is expected to improve the contact of gas and perovskites and the performance of perovskite for O_2-enriched CO_2 stream production.

Fig. 13.7 compares the morphologies of fresh $BaCoO_{3-\delta}$ powders, samples after different and multiple cycles' desorption, and the reverted products of $BaCoO_{3-\delta}$ after 22 test cycles [5]. The fresh $BaCoO_{3-\delta}$ powder's particles consisted of numerous spherical nanocrystallines, and the nanosize grains merged to form a porous and rough structure. When the fresh sample underwent desorption processes in 10 cycles, the particles were composed of irregular agglomerations, leading to the formation of a relatively condensed and compact surface. After more test cycles, the agglomeration of particles became severe. Fig. 13.7D shows the reverted products of perovskite after 22 test cycles. As shown in Fig. 13.7D, the morphologies were similar to those in the fresh sample (Fig. 13.7A) and presented an irregular particle shape and a porous surface. The morphology and microstructure of fresh $BaCoO_{3-\delta}$ samples and samples after multiple cycles' oxygen adsorption/desorption tests shown by SEM images also

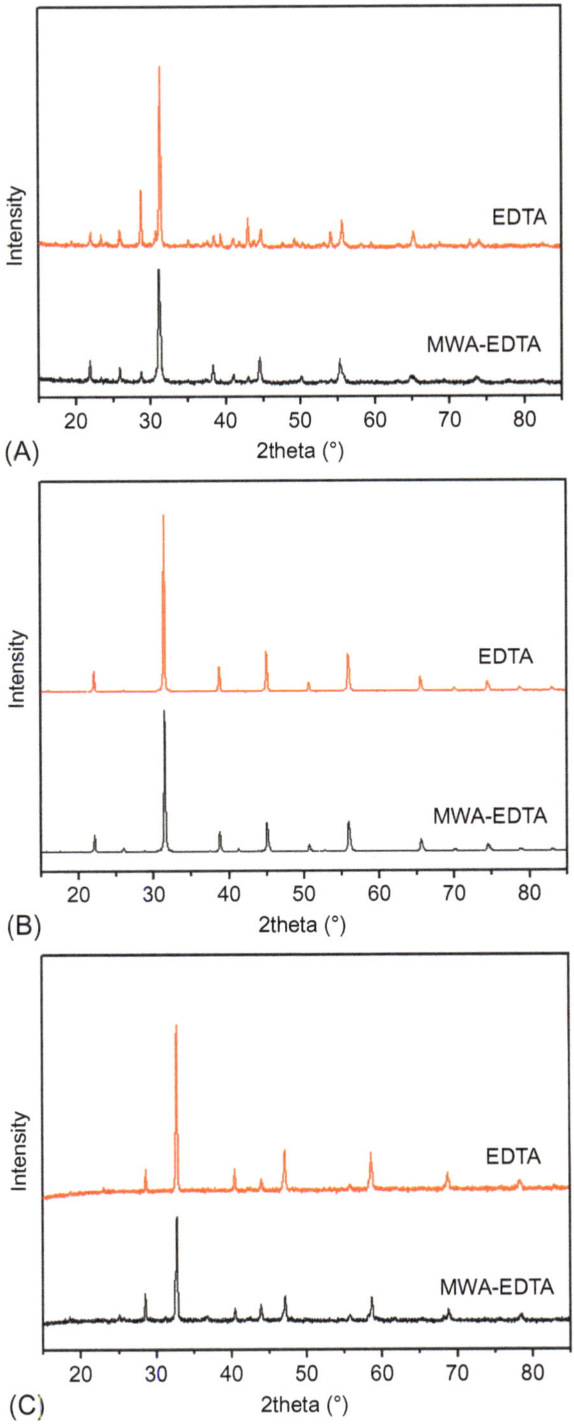

FIG. 13.5 XRD patterns of $Ba_{1-x}Sr_xCo_{0.8}Fe_{0.2}O_{3-\delta}$ (x=0, 0.5, 1) powders prepared by EDTA method and MWA-EDTA method: (A) $BaCo_{0.8}Fe_{0.2}O_{3-\delta}$ (BCF), (B) $Ba_{0.5}Sr_{0.5}Co_{0.8}Fe_{0.2}O_{3-\delta}$ (BSCF), (C) $SrCo_{0.8}Fe_{0.2}O_{3-\delta}$ (SCF). The microwave time is 20 min. *(Reproduced from Shen Q, Zheng Y, Luo C, Zheng C. Development and characterization of* $Ba_{1-x}Sr_xCo_{0.8}Fe_{0.2}O3-\delta$ *perovskite for oxygen production in oxyfuel combustion system. Chem Eng J 2014;255:462–70, with permission from Elsevier.)*

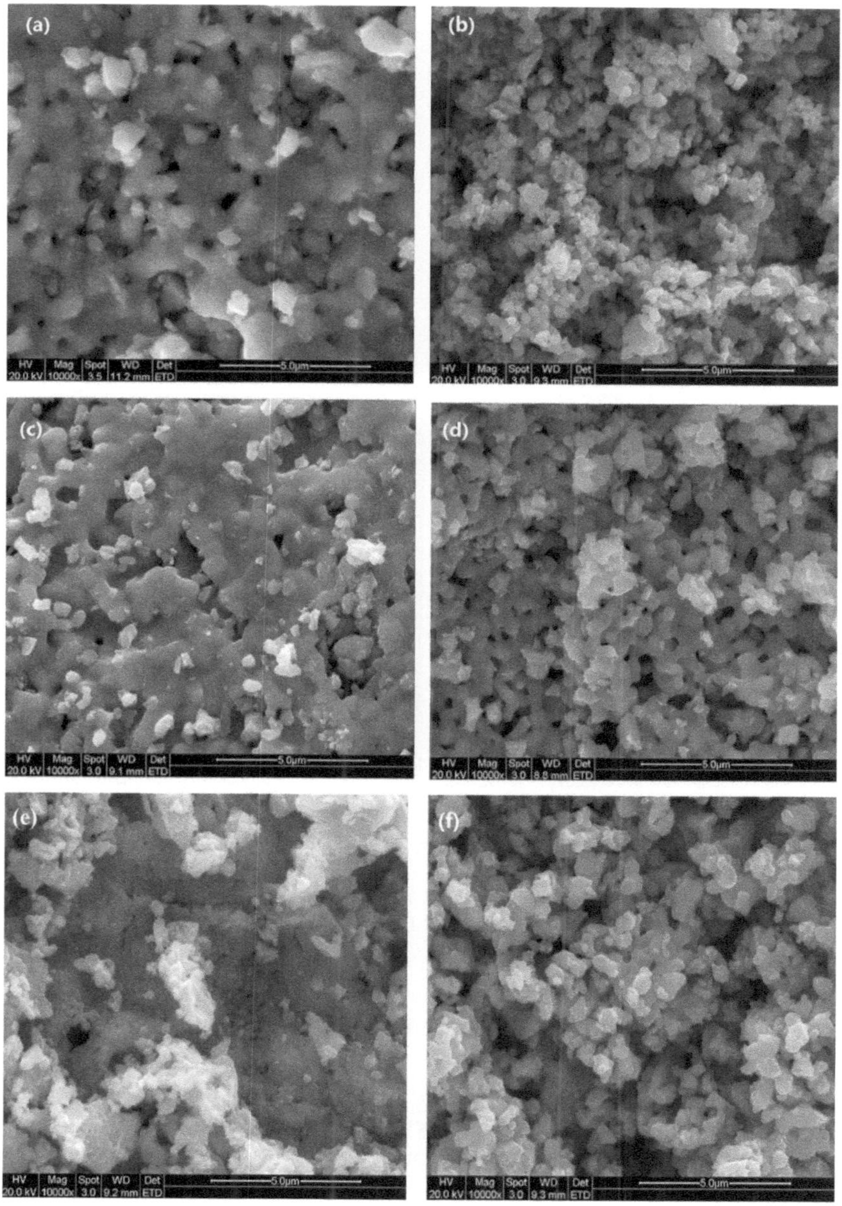

FIG. 13.6 SEM images of perovskite samples of (A) BCF, (C) BSCF, and (E) SCF, as synthesized by EDTA method; and (B) BCF, (D) BSCF, and (F) SCF, as synthesized by MWA-EDTA method. The microwave time was 20 min. *(Reproduced from Shen Q, Zheng Y, Luo C, Ding N, Zheng C-G, Thern M. Effect of A/B-site substitution on oxygen production performance of strontium cobalt based perovskites for CO_2 capture application, RSC Adv 2015;5:39785–90, with permission from the Royal Society of Chemistry.)*

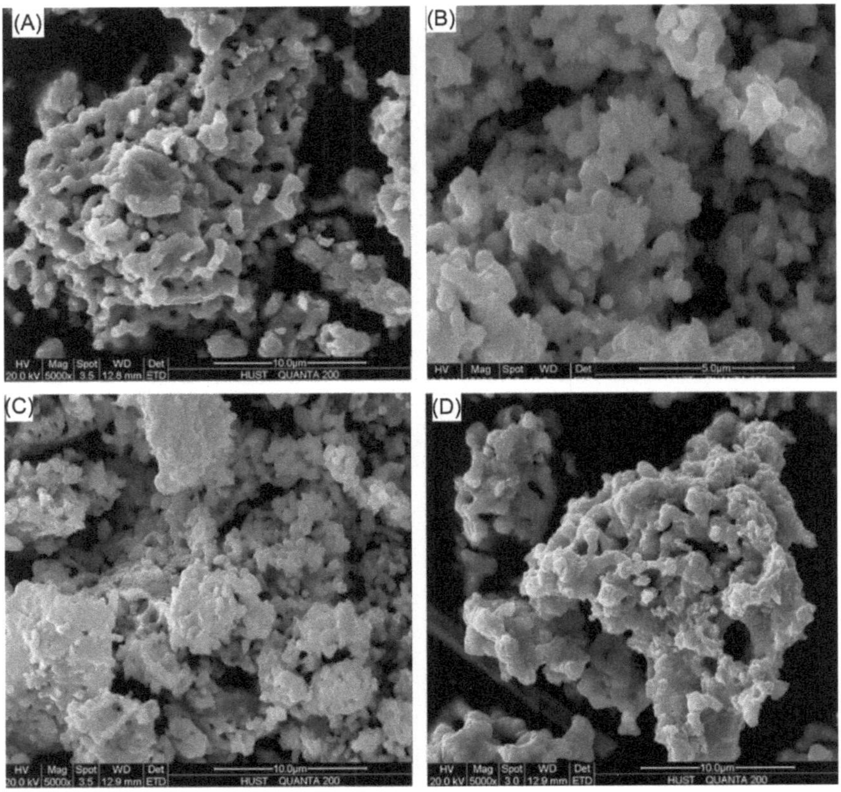

FIG. 13.7 SEM images of $BaCoO_{3-\delta}$ powders: (A) fresh sample, (B) sample after 10 cycles' oxygen desorption, (C) sample after 22 cycles' oxygen desorption, and (D) reverted products after 22 cycles. *(Reproduced from Shen Q, Zheng Y, Li S, Ding N, Xu Y, Zheng C, Thern M. Optimize process parameters of microwave-assisted EDTA method using orthogonal experiment for novel BaCoO3−δ perovskite. J Alloys Compd 2016;658:125–31, with permission from Elsevier.)*

reflected the change in grain size and surface area values. The fresh $BaCoO_{3-\delta}$ powders had a grain size of 42.2 nm and BET surface area of 38.4 m^2/g; after multiple cycles, the grain size increased to 56.1 nm, and the BET surface area of sample decreased to 24.5 m^2/g.

For all the samples, fixed-bed experiments were conducted to study the oxygen production performance. For typical oxygen adsorption and desorption processes, 0.5 g of the samples packed in the quartz reactor were exposed to an air flow of 100 mL/min at 850°C for 10 min for the oxygen adsorption step; then for the oxygen desorption step, the sweep gas from air was switched to a CO_2 flow of 50 mL/min. Fig. 13.8 compares the oxygen desorption amount of $Ba_{1-x}Sr_xCo_{0.8}Fe_{0.2}O_{3-\delta}$ (x=0, 0.5, 1). It is clear that samples prepared by MWA-EDTA showed better oxygen production properties than the samples synthesized by EDTA. Furthermore, the oxygen desorption amount of $Ba_{1-x}Sr_xCo_{0.8}Fe_{0.2}O_{3-\delta}$ (x=0, 0.5, 1) perovskites increased in the order of BCF > BSCF > SCF.

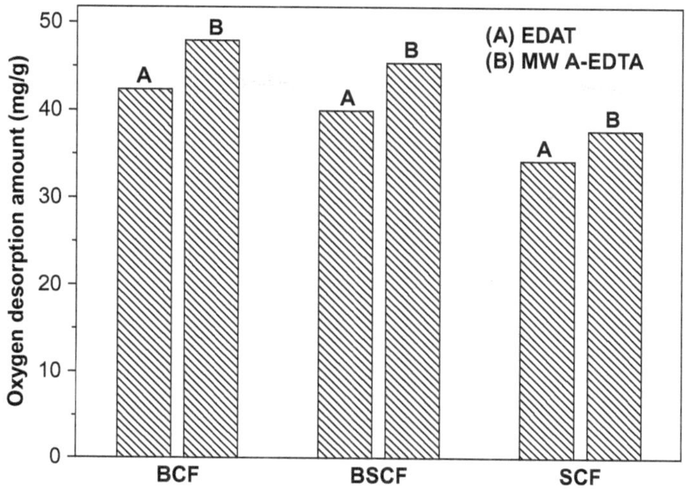

FIG. 13.8 Comparison of oxygen desorption amount of $Ba_{1-x}Sr_xCo_{0.8}Fe_{0.2}O_{3-\delta}$ ($x=0, 0.5, 1$) synthesized by different synthesis methods: (A) EDTA, (B) MWA-EDTA. The adsorption/desorption temperatures were both 850°C. *(Reproduced from Shen Q, Zheng Y, Luo C, Zheng C. Development and characterization of* $Ba_{1-x}Sr_xCo_{0.8}Fe_{0.2}O_{3-\delta}$ *perovskite for oxygen production in oxyfuel combustion system. Chem Eng J 2014;255:462–70, with permission from Elsevier.)*

Substitution of Sr^{2+} by Ba^{2+} increased the oxygen desorption amount. Among the three samples, BCF prepared by MWA-EDTA displayed the most oxygen production. The higher desorption amount may have resulted from the larger ionic radius; the ionic radius was in the order Ba (1.75Å) > Sr (1.58Å), and the larger ionic radius of Ba may have resulted in the increase of the lattice volume, thus leading to the oxygen ions' transition in the crystal.

BCF synthesized by MWA-EDTA under optimal synthesis conditions was chosen to investigate the effects of the operation temperature on the fixed-bed adsorption/desorption process for oxygen production. Fig. 13.9 shows the oxygen desorption curves of BCF at different adsorption/desorption temperatures (400°C, 500°C, 600°C, 700°C, 800°C). As shown in Fig. 13.9, the oxygen desorption amount of BCF increased with the increase of the reaction temperature. The oxygen desorption amount of BCF was negligible at temperatures equal to or lower than 600°C. However, when the temperature continued increasing from 600°C to 800°C, the oxygen desorption amounts sharply increased. It was concluded that adsorption and desorption temperatures play a crucial role in the oxygen production process for BCF powder. Higher adsorption and desorption temperatures led to a shorter oxygen desorption time and to a production of oxygen. This temperature dependency represents the well-known kinetics law for a gas-solid reaction, in which the rate increases with temperature dependence and the equilibrium state is the temperature. Thus this study showed that the optimal temperatures for the adsorption and desorption process of prepared BCF powder were 800°C and 900°C.

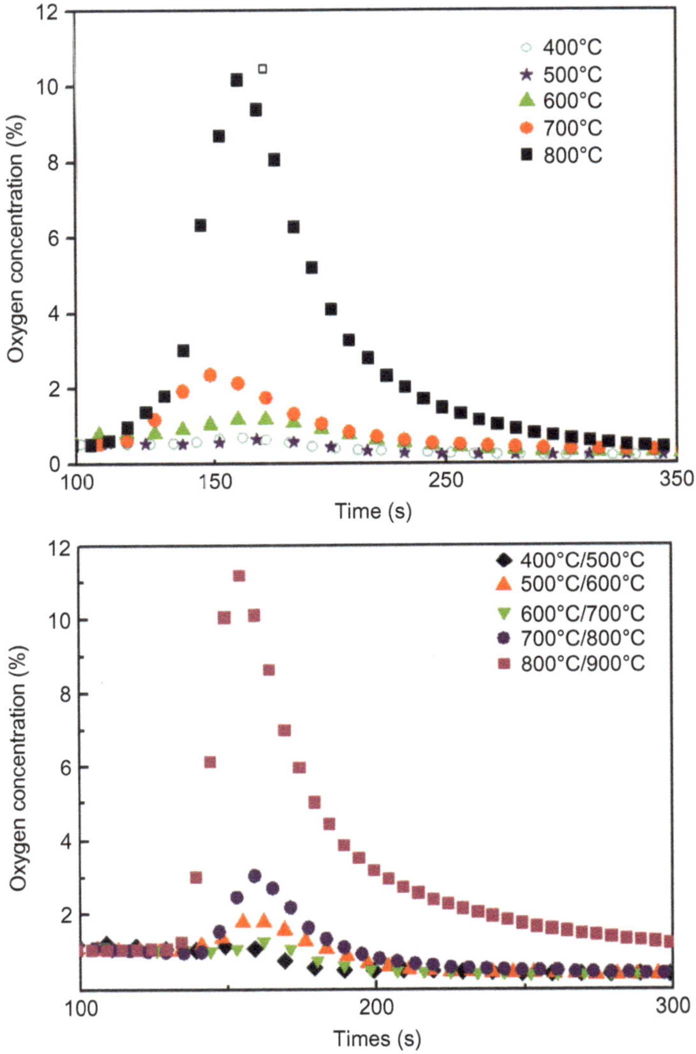

FIG. 13.9 Oxygen desorption curves of BCF at different sets of absorption/desorption temperatures. *(Reproduced from Shen Q, Zheng Y, Luo C, Zheng C. Development and characterization of $Ba_{1-x}Sr_xCo_{0.8}Fe_{0.2}O_{3-\delta}$ perovskite for oxygen production in oxyfuel combustion system. Chem Eng J 2014;255:462–70, with permission from Elsevier.)*

In addition to their excellent oxygen production performance and great potential as oxygen carriers to provide stable O_2/CO_2 cycled gas for oxy-combustion, perovskite-type oxygen carriers require long-term cyclic use. Fig. 13.10 presents the cyclic performance of BCF with the optimum operating adsorption and desorption temperatures (800°C and 900°C, respectively). For each cycle, the sample was exposed to an air flow of 100 mL/min for 10 min,

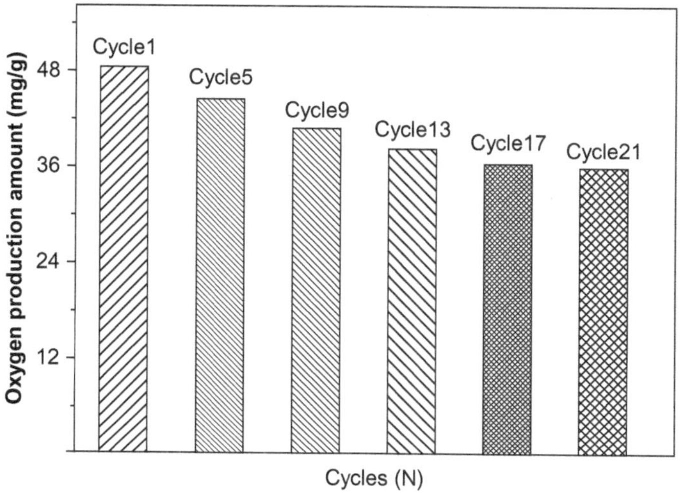

FIG. 13.10 Oxygen production amount of $BaCoFeO_{3-\delta}$ with the number of cycles (adsorption proceeded at 800°C for 10 min; desorption proceeded at 900°C). *(Reproduced from Shen Q, Zheng Y, Luo C, Zheng C. Development and characterization of $Ba_{1-x}Sr_xCo_{0.8}Fe_{0.2}O_{3-\delta}$ perovskite for oxygen production in oxyfuel combustion system. Chem Eng J 2014;255:462–70, with permission from Elsevier.)*

then switch the sweep gas from air to a 50 mL/min CO_2 flow, and the desorption step was stopped when the O_2 concentration dropped to nearly zero. The oxygen desorption amounts were checked for 21 continuous cycles. It was evident that the oxygen desorption capacity did not show an obvious decrease during the 21 cycles. The oxygen production amount of BCF could still achieve a 38.21 mg/g sample after 21 cycles.

Fig. 13.11 shows the cyclic behavior of a $BaCoO_3$ perovskite oxygen carrier prepared under optimal synthesis conditions via a microwave-assisted EDTA method based on the preceding analysis. It demonstrated that the oxygen desorption capacity did not show an obvious decrease during the 22 cycles. The oxygen production amount of $BaCoO_{3-\delta}$ could still achieve a 52.1 mg/g sample after 22 cycles.

13.2.3 Effect of A/B-Site Substitution on Oxygen Production Performance

Perovskite-type metal oxides have the general formula ABO_3, where A-site can be an alkali, alkaline earth, rare earth, or other large ion, and B-site can be a transitional metal ion. If A- and/or B-sites are substituted with lower valent cations, in order to maintain charge neutrality, some oxygen ions are removed from the lattice thereby creating oxygen vacancies [6]. Perovskite metal oxides have excellent mixed ionic-electronic conductivity and exhibit extremely high selectivity for oxygen because of the existence of oxygen vacancies. Perovskites

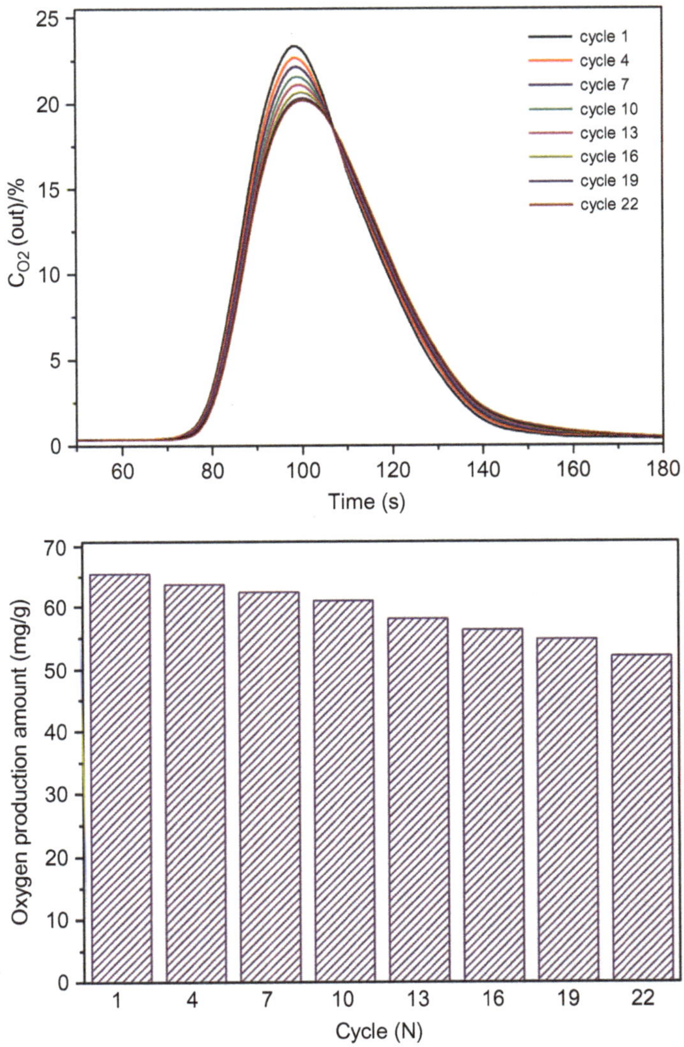

FIG. 13.11 Oxygen production amount of $BaCoO_{3-\delta}$ with the number of cycles. *(Reproduced from Shen Q, Zheng Y, Li S, Ding N, Xu Y, Zheng C, Thern M. Optimize process parameters of microwave-assisted EDTA method using orthogonal experiment for novel BaCoO3−δ perovskite. J Alloys Compd 2016;658:125–31, with permission from Elsevier.)*

are highly selective for oxygen over nitrogen or other non-oxygen-containing gases, which makes them ideal candidates for oxygen permeable membranes.

Effects of A-site substitution on oxygen adsorption/desorption performance[7,8]: Fig. 13.12 compares the oxygen adsorption breakthrough curves of different A-site total/partially substituted $SrCo_{0.8}Fe_{0.2}O_{3-\delta}$ perovskite samples denoted as $SrCo_{0.8}Fe_{0.2}O_{3-\delta}$ (SCF), $BaCo_{0.8}Fe_{0.2}O_{3-\delta}$ (BCF), $Ba_{0.5}Sr_{0.5}Co_{0.8}Fe_{0.2}O_{3-\delta}$ (BSCF), $Sr_{0.5}Ca_{0.5}Co_{0.8}Fe_{0.2}O_{3-\delta}$ (SCCF), and $MgCo_{0.8}$

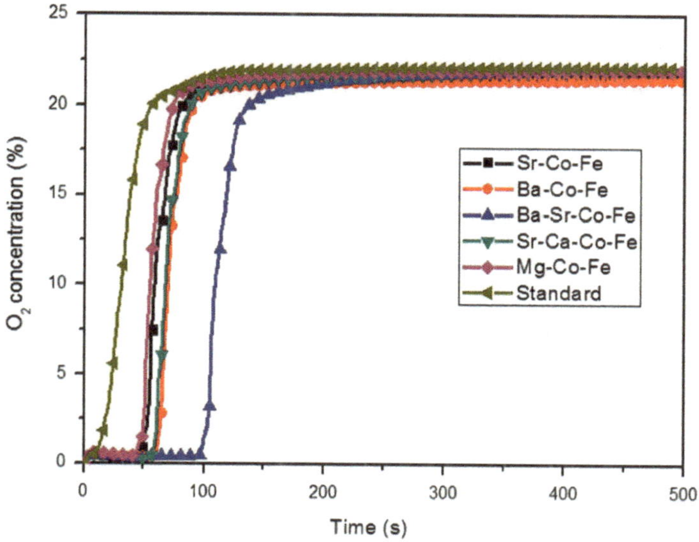

FIG. 13.12 Oxygen adsorption curves for A (A = Sr, Ba, Mg, Ca) $Co_{0.8}Fe_{0.2}O_{3-\delta}$. *(Reproduced from Shen Q, Zheng Y, Luo C, Ding N, Zheng C-G, Thern M. Effect of A/B-site substitution on oxygen production performance of strontium cobalt based perovskites for CO_2 capture application, RSC Adv 2015;5:39785–90, with permission from the Royal Society of Chemistry.)*

$Fe_{0.2}O_{3-\delta}$ (MCF) at the air flow rate of 200 mL/min^{-1} at 850°C. A standard oxygen adsorption breakthrough curve was introduced for comparison. The results showed that the breakthrough curves were similar in shape with a sharp increase after a breakthrough time (50–100 s) and that soon afterward all the curves were characterized by a long tail. For example, for BSCF, no outgoing oxygen was detected by the oxygen sensor before 100 s, and after 150 s, the outlet oxygen concentration was equal to the feed gas concentration (20.9%). Fig. 13.12 indicates that the MCF sample had the shortest breakthrough time (oxygen adsorption capacity) and that the BSCF sample had the longest breakthrough time.

Fig. 13.13 depicts the comparison of oxygen desorption curves of SCF with different A-site total/partial substitution. It is clear that A-site total/partial substitution significantly influenced the oxygen desorption properties of SCF. Fig. 13.13 shows the oxygen desorption amount of each desorption process for the different samples. The results showed that the oxygen desorption amount was in the order BCF > BSCF > SCF > SCCF > MCF. Fig. 13.13 also shows that the slope of the curves of BCF and BSCF was greater than for others, which demonstrated that the oxygen desorption rates of BCF and BSCF were faster than the rest. The outlet oxygen concentration of BCF reached a maximum value of 25% after 150 s. On the other hand, the substitution of Sr^{2+} with Ca^{2+}/Mg^{2+} reduced the oxygen desorption amount for $SrCo_{0.8}Fe_{0.2}O_{3-\delta}$. The lower desorption amount may have resulted from the smaller ionic radius, as the ionic radius was in the order Ba (1.75 Å) > Sr (1.58 Å) > Ca (1.48 Å) > Mg (1.03 Å);

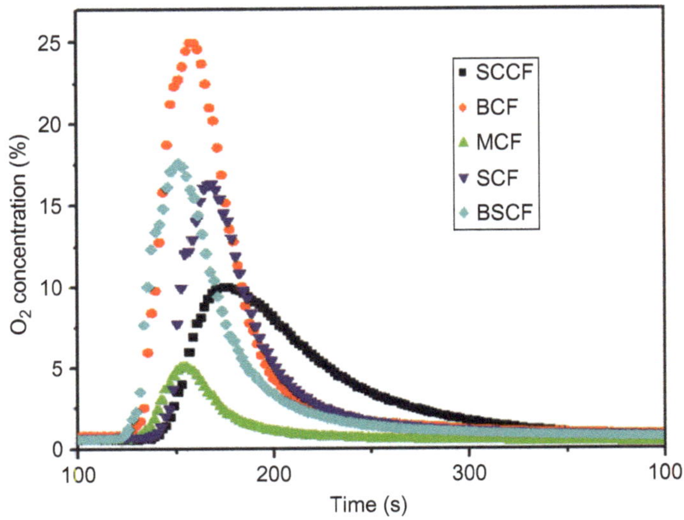

FIG. 13.13 Oxygen desorption curves for A (A = Sr, Ba, Mg, Ca) $Co_{0.8}Fe_{0.2}O_{3-\delta}$. *(Reproduced from Shen Q, Zheng Y, Luo C, Ding N, Zheng C-G, Thern M. Effect of A/B-site substitution on oxygen production performance of strontium cobalt based perovskites for CO_2 capture application, RSC Adv 2015;5:39785–90, with permission from the Royal Society of Chemistry.)*

the larger ionic radius of Ba may have resulted in the increase in the lattice volume, thus contributing to the oxygen ions' transition in the crystal. Therefore, though high temperature oxygen (β-oxygen) desorption is usually related to B-site substitution, it is also affected by A-site substitution. Moreover, for a fixed B-site composition, A-site ionics with the same valency but different ionic radius affect the oxygen desorption property.

Effects of B-site substitution on oxygen adsorption/desorption performance: The effects of B-site substitution by different transition metal ions on oxygen adsorption/desorption performance of $SrCo_{0.8}B_{0.2}O_3$ (B = Ni, Zn, Zr, Cr, Cu, Fe) were studied. The equilibrium weights (mg) of $SrCo_{0.8}B_{0.2}O_{3-\delta}$ (B = Fe, Ni, Cu, Zn, Cr, Zr) at different conditions are listed in Table 13.1. Fig. 13.14 compares the oxygen adsorption capacity, which was calculated from data in Table 13.1. The results show that the substitution of Fe by Zr, Cr, and Cu in $SrCo_{0.8}B_{0.2}O_{3-\delta}$ increased the oxygen adsorption capacity and that the adsorption capacity decreased in the order $SrCo_{0.8}Zr_{0.2}O_{3-\delta}$ (Sr-Co-Zr) > $SrCo_{0.8}Cr_{0.2}O_{3-\delta}$ (Sr-Co-Cr) > $SrCo_{0.8}Cu_{0.2}O_{3-\delta}$ (Sr-Co-Cu) > $SrCo_{0.8}Fe_{0.2}O_{3-\delta}$ (Sr-Co-Fe) > $SrCo_{0.8}Ni_{0.2}O_{3-\delta}$ (Sr-Co-Ni) > $SrCo_{0.8}Zn_{0.2}O_{3-\delta}$ (Sr-Co-Zn).

Generally speaking, with the substitution of B-site by higher valent cations, such as Zr, the Cr results in a lower adsorption capacity because the charge neutrality rule restrains the formation of an oxygen vacancy. Because this contradicts the results of this study, it may appear that, except for oxygen vacancy of perovskite, the oxygen adsorption capacity also relates to the surface adsorption activity.

TABLE 13.1 Equilibrium weight (mg) of $SrCo_{0.8}B_{0.2}O_{3-\delta}$ (B = Fe, Ni, Cu, Zn, Cr, Zr) at initial conditions and 850°C in air and N_2, respectively

Samples	Sr-Co-Ni	Sr-Co-Zn	Sr-Co-Zr	Sr-Co-Cr	Sr-Co-Cu	Sr-Co-Fe
W_{air} (mg)	10.500	10.400	8.969	9.319	9.372	9.655
W_{N2} (mg)	10.465	10.370	8.907	9.259	9.314	9.615
W_0 (mg)	10.67	10.64	9.789	9.646	9.563	9.867

Reproduced from Shen Q, Zheng Y, Luo C, Ding N, Zheng C-G, Thern M. Effect of A/B-site substitution on oxygen production performance of strontium cobalt based perovskites for CO_2 capture application, RSC Adv 2015;5:39785–90, with permission from the Royal Society of Chemistry.

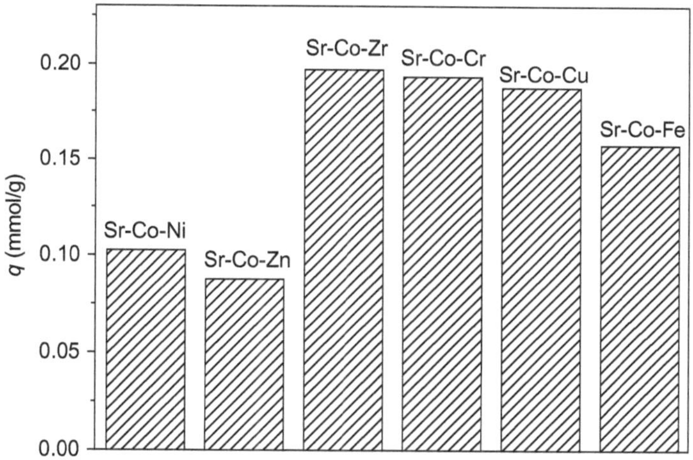

FIG. 13.14 Comparison of oxygen adsorption capacity for $SrCo_{0.8}B_{0.2}O_{3-\delta}$ (B = Fe, Ni, Cu, Zn, Cr, Zr). *(Reproduced from Shen Q, Zheng Y, Luo C, Ding N, Zheng C-G, Thern M. Effect of A/B-site substitution on oxygen production performance of strontium cobalt based perovskites for CO_2 capture application, RSC Adv 2015;5:39785–90, with permission from the Royal Society of Chemistry.)*

Fig. 13.15 shows a comparison of the cyclic oxygen desorption performance for $SrCo_{0.8}B_{0.2}O_{3-\delta}$ (B = Fe, Ni, Cu, Zn, Cr, Zr), which indicated that substitution of Fe in a B-site with different transition metal ions significantly affected the oxygen desorption performance. The substitution of Fe by Zr, Ni, Zn, and Cr in $SrCo_{0.8}Fe_{0.2}O_{3-\delta}$ enhanced the oxygen desorption amount. On the other hand, substitution of Fe by Cu reduced the oxygen desorption amount, which resulted in the lowest cyclic ability for $SrCo_{0.8}Cu_{0.2}O_{3-\delta}$ among the samples, as shown in Fig. 13.15. These results indicated that substitution of Fe with Zr and Cr in $SrCo_{0.8}Fe_{0.2}O_{3-\delta}$ enhanced both the oxygen adsorption capacity and the desorption amount and that substitution with Zn and Ni increased the oxygen desorption

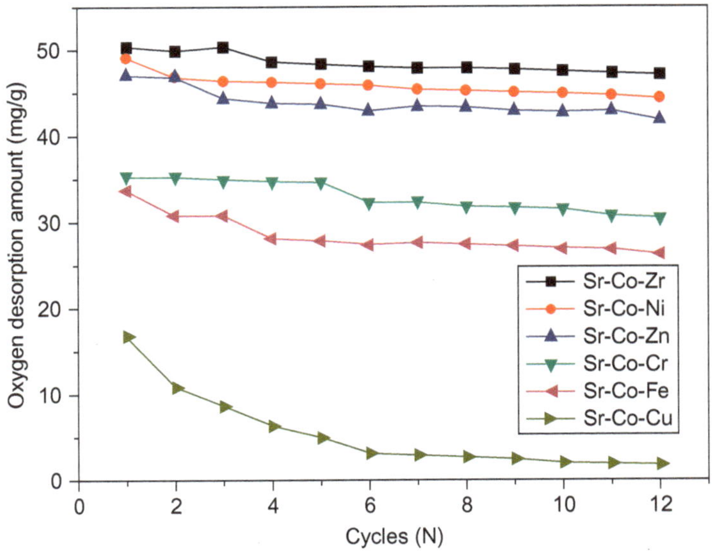

FIG. 13.15 Comparison of oxygen desorption amount for $SrCo_{0.8}B_{0.2}O_{3-\delta}$ (B = Fe, Ni, Cu, Zn, Cr, Zr) with the number of cycles. *(Reproduced from Shen Q, Zheng Y, Luo C, Ding N, Zheng C-G, Thern M. Effect of A/B-site substitution on oxygen production performance of strontium cobalt based perovskites for CO_2 capture application, RSC Adv 2015;5:39785–90, with permission from the Royal Society of Chemistry.)*

amount, but adversely reduced the oxygen adsorption capacity. Considering both adsorption/desorption, $SrCo_{0.8}Zr_{0.2}O_{3-\delta}$ performed best followed by $SrCo_{0.8}Cr_{0.2}O_{3-\delta}$ among the different compositions of $SrCo_{0.8}B_{0.2}O_{3-\delta}$ with different B-site transition metal iron substitutions.

In summary, the oxygen desorption amount of different A-site-substituted perovskites decreased in the order of $BaCo_{0.8}Fe_{0.2}O_{3-\delta} > Ba_{0.5}Sr_{0.5}Co_{0.8}Fe_{0.2}O_{3-\delta} > SrCo_{0.8}Fe_{0.2}O_{3-\delta} > Sr_{0.5}Ca_{0.5}Co_{0.8}Fe_{0.2}O_{3-\delta} > MgCo_{0.8}Fe_{0.2}O_{3-\delta}$. The lower desorption amount was related to the smaller ionic radius, and the larger ionic radius of Ba resulted in an increase in the lattice volume, thus contributing to the oxygen ions' transition in the crystal. The study showed that B-site substitution by different transition metal ions can significantly modify the oxygen adsorption capacity and the oxygen desorption performance of $SrCo_{0.8}Fe_{0.2}O_{3-\delta}$. Fe substituted by Zr, Cr, and Cu in $SrCo_{0.8}B_{0.2}O_{3-\delta}$ increased the oxygen adsorption capacity. Therefore it may be that, except for oxygen vacancy of perovskite, the oxygen adsorption capacity is also related to the surface adsorption activity. Moreover, substitution of a Fe ion in $SrCo_{0.8}Fe_{0.2}O_{3-\delta}$ with Zr, Ni, Zn, and Cr ions enhanced the oxygen desorption amount. However, substitution of a Fe ion with Cu reduced the oxygen desorption amount. Considering both adsorption/desorption, the performance of $SrCo_{0.8}Zr_{0.2}O_{3-\delta}$ was best followed by $SrCo_{0.8}Cr_{0.2}O_{3-\delta}$ among the different compositions of $SrCo_{0.8}B_{0.2}O_{3-\delta}$ with different B-site transition metal iron substitutions.

13.3 CHEMICAL LOOPING PROCESS OF CU-BASED CARRIER FOR HIGH TEMPERATURE O_2/CO_2 PRODUCTION

Along with the oxy-fuel combustion process, the chemical looping air separation (CLAS) process based on metallic oxide carriers has also been developed for air separation and O_2/CO_2 mixed gas production, including the two-step redox cycle. In the oxidation process, the oxygen carrier reacts with O_2 in the air in the oxidation reactor. In the reduction process, CO_2 is then applied as the sweep gas to desorb O_2 from the oxygen carrier to obtain O_2/CO_2 mixed gas. The metallic oxide systems, such as CuO/Cu_2O, Co_3O_4/CoO, and Mn_2O_3/Mn_3O_4, have proved to be feasible in O_2 sorption/desorption reversible processes [9].

13.3.1 Zr-Doped Cu-Based Oxygen Carrier for O_2/CO_2 Mixed Gas Production

The Cu-based oxygen carrier is inexpensive and has a higher oxygen-carrying capacity than the Co-based oxygen carrier. The higher oxygen-carrying capacity can reduce the size of reactors, thereby reducing the cost of air separation [10]. However, the Cu-based oxygen carriers have some challenges that need to be solved, including a high agglomeration trend and low mechanical stability [11]. The addition of a dopant is considered as an available method for increasing the stability of the Cu-based oxygen carrier. Zr doping can improve the reactivity and oxygen storage capacity of catalysts and oxygen storage materials. Therefore the Zr-doped oxygen carrier is a promising carrier material for the production of O_2/CO_2 mixtures.

A series of Zr-doped Cu-based oxygen carriers were prepared by the sol-gel combustion method, and the undoped carrier consisted of 60 wt.% CuO and 40 wt.% $CuAl_2O_4$ [12]. The series of Zr-doped Cu-based oxygen carriers (Zr mass fraction: 0, 2, 4, 8, 10 wt.%) were named CuAl, $CuAlZr_2$, $CuAlZr_4$, $CuAlZr_8$, and $CuAlZr_{10}$, respectively. The phase structure, the surface topography, the O_2 absorption/desorption, and the cyclic reaction of the Zr-doped Cu-based oxygen carrier were investigated [12]. Then O_2 sorption/desorption experiments were performed at 880°C in a fixed-bed reactor using pure CO_2 as the sweep gas, and the air was used for reoxidation.

The powder XRD analysis was applied to identify the chemical composition and crystalline phases of the Cu-based oxygen carriers. The XRD patterns of the fresh and cycled oxygen carriers are shown in Fig. 13.16. The XRD measurements of the unreacted oxygen carriers revealed that CuO and CuA12O4 were the major phases in all the fresh samples. The spinel phase ($CuAl_2O_4$) was an inert carrier that could influence the oxygen sorption and desorption kinetics of the carrier material, depending on its amount, size, and dispersion [13]. With the increase of the Zr content, the additional reflections became evident, indicating

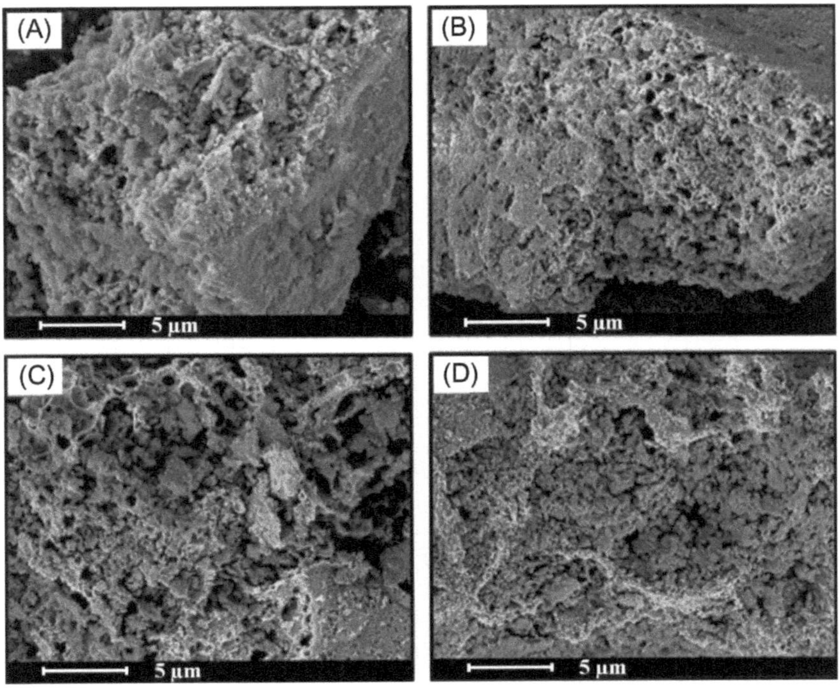

FIG. 13.16 SEM micrographs of the fresh oxygen carriers (A: CuAl, B: $CuAlZr_2$, C: $CuAlZr_4$, D: $CuAlZr_8$).

the formation of an impurity phase. The by-phase in CuAlZr8 was identified to be ZrO_2. The XRD patterns of the cycled oxygen carriers were similar to those of the unreacted samples, indicating that the oxygen carriers synthesized were recycled. No carbonate of copper was detected, suggesting that the Cu-based oxygen carrier did not react with CO_2 during the desorption process, which is important for oxy-fuel combustion when the recycled flue gas is supplied as sweep gas.

The scanning electron micrographs of the fresh oxygen carriers with various doping contents indicated that the fresh oxygen carriers had a fine-grained microstructure structure, and no abnormal grain growth and sintering behavior were observed. With an increase in the Zr doping content, the oxygen carrier particles revealed a narrower size range and a smaller average size. Also, the pore amount of the Zr-doped Cu-based oxygen carrier was more than that of the undoped sample. The maximum pore amount of the synthesized oxygen carriers was observed inside the CuAlZr8 particle, while its pore size was smaller. The porous structure resulted in a larger specific area in the oxygen carriers, which was beneficial to the chemical reactivity of the oxygen carriers. The preceding experimental results indicated that the grain size of the Cu-based oxygen carrier was minimized, that grain size uniformity was enhanced, and that the pore amount was increased via Zr doping. In addition, the high-resolution transmission electron microscopy (HRTEM) results showed that more lattice imperfections were found in CuAlZr8. The microstructure of the Cu-based oxygen carrier was modified with Zr doping, for example, the by the generation of a lattice defect and the increase of the oxygen's mobility in the bulk, which were beneficial to the improvement of oxygen absorption/desorption kinetics [12].

In order to investigate the kinetics of O_2 sorption and desorption, the experiments were performed at 880°C in a fixed-bed reactor using pure CO_2 as the sweep gas, and the air was used for reoxidation. Fig. 13.17A shows O_2 desorption curves of Cu-based oxygen carriers with different contents of Zr doping. It can be seen that the O_2 fraction in the outlet mixture gas increased with increasing Zr doping content. Especially, the oxygen release capacity increased significantly after the Zr doping. At the same time, the oxygen release time of all samples was similar, which indicated that the Zr-doped carriers had high desorption reaction rates. This may be attributed to the following: (1) The Zr doping led to a porous structure and larger specific area of the carrier particles, which were validated through morphological and BET surface analyses. (2) More oxygen vacancies were formed because the lattice defects increased with the Zr doping, thus more oxygen was generated during the desorption step. (3) More lattices were in the activated state owing to the occurrence of lattice distortion with the formation of a solid solution of Zr in the Cu-based oxygen carrier. It should be noted that the oxygen release capacity of $CuAlZr_{10}$ increased very little compared with CuAlZr8, which was mainly due to the densification of the oxygen carrier with the 10 wt.% Zr doping content [12].

The O_2 sorption results are shown in Fig. 13.17B. The oxygen sorption curve can be divided into three stages: (1) the pre-breakthrough stage during which the O_2 sorption rate was the fastest; (2) the breakthrough stage, in which the O_2 content increased with the decrease of the O_2 sorption efficiency; and (3) the post-breakthrough stage. There was an obvious improvement in the oxygen sorption rate of the Zr-doped carrier in comparison with the undoped Cu-based oxygen carrier. Moreover, the O_2 sorption rates of $CuAlZr_2$ and $CuAlZr_4$ were much higher than those of $CuAlZr_8$ and $CuAlZr_{10}$. A possible explanation is that the ZrO_2 in $CuAlZr_8$ and $CuAlZr_{10}$ increased the activation energy during the oxidation process of Cu_2O by O_2 [14]. Based on the preceding analysis, 8 wt.% was the optimal Zr doping ratio in the experiments.

Next experiments on the reversibility of reaction $2Cu_2O+O_2=4CuO$ and the stability in the cyclic operation of $CuAlZr_8$ with five cycles were carried out via a fixed-bed process at 880°C [12]. In the third cycle, the oxygen release amount of the oxygen carrier increased a little, which may have resulted from the formation of the larger inner pores caused by the thermal shock and chemical reaction [15]. Then the oxygen release capacity remained stable. The scanning electron micrograph of the $CuAlZr_8$ after five cycles showed that the $CuAlZr_8$ still had a porous structure and that no abnormal grain growth, though a slight sintering, occurred.

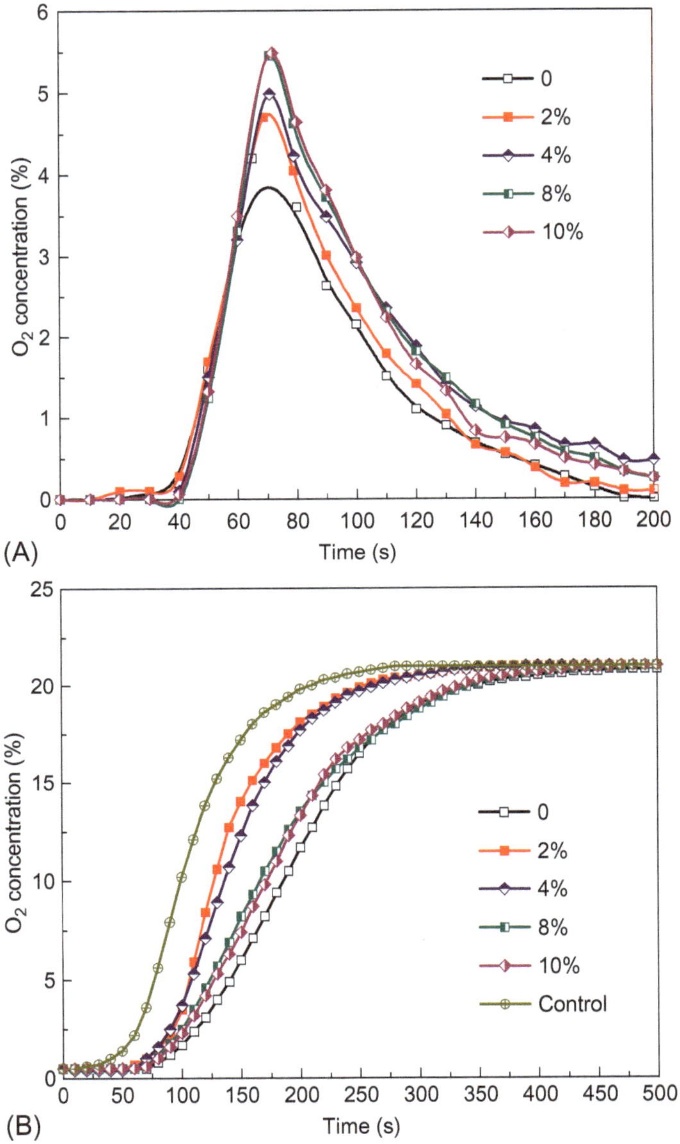

FIG. 13.17 O_2 desorption and sorption curves of Cu-based oxygen carriers with different Zr doping content: (A) O_2 desorption and (B) O_2 sorption.

13.3.2 Influencing Mechanism of Zr Doping on O_2 Release of CuO

Density functional theory (DFT) is suited for investigating the electronic structures, adsorption, and reaction mechanisms. In order to understand the effect of Zr on O_2 release mechanisms of CuO, the oxygen vacancy formation energies, the energy barriers for O_2 desorption and O anion diffusion, and the band gap of Zr-doped CuO and pure CuO were calculated using the DFT.

The oxygen vacancy formation energy (E_{vac}) is defined according to

$$E_{vac} = E_{def} + 0.5E_{O_2} - E_{per} \quad (13.5)$$

where E_{per} is the energy for CuO supercell, E_{def} is the energy for CuO supercell with one oxygen vacancy, and E_{O2} is the energy of an optimized oxygen in the gas phase with ground state. A positive value of E_{vac} means that energy is needed to generate an oxygen vacancy.

The energy barrier (E_b) is defined as

$$E_b = E_{TS} - E_{IS} \quad (13.6)$$

where E_{TS} is the energy of the transition state (TS), and E_{IS} is the energy of the initial state (IS).

The O_2 formation and desorption pathway is shown in Fig. 13.18A, and the O anion diffusion process is shown in Fig. 13.18B.

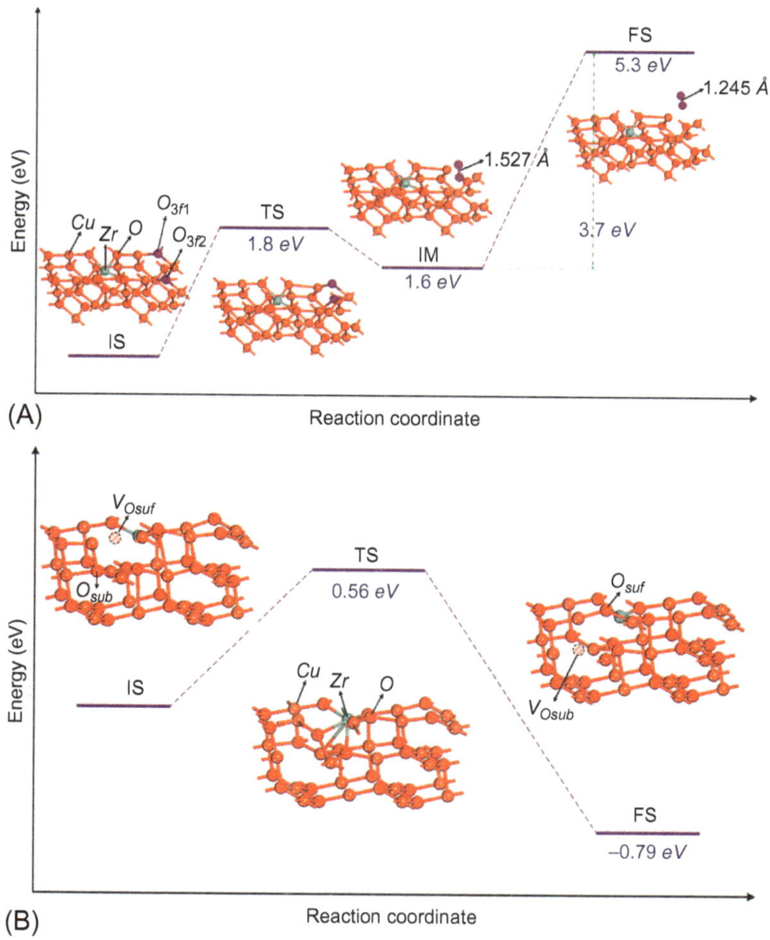

FIG. 13.18 Structures of the initial state (IS), transition state (TS), intermediate state (IM), and final state (FS) of Zr-doped CuO supercell bulk for O_2 release process and the corresponding potential energy diagram: (A) O_2 formation and desorption and (B) O anion diffusion.

In the O_2 formation process, an O_2 complex formed in the intermediate state (IM) structure with a O–O bond length of 1.527 Å. The energy barrier (E_{b1}) was 1.8 eV, which was lower than that of the pure CuO (2.3 eV [16]). In the O_2 desorption process, O_2 was desorbed from the surface, and the O–O bond length was 1.245 Å, which was similar to the bond length of the free O_2 molecule (1.240 Å). The energy barrier of Zr-doped CuO for the O anion diffusion process (E_{b2} = 0.56 eV) was also lower than that of the pure CuO (0.88 eV [16]). These results showed that the Zr-doped CuO released oxygen more easily than the pure CuO. Moreover, the Zr-doped CuO had a lower vacancy formation energy of 6.6 eV compared with the pure CuO. The results indicated that oxygen vacancy formation in the Zr-doped CuO was more favorable than that in pure CuO. After the vacancy formation, the reactive sites were present and oxygen vacancy migration became possible [17].

13.4 SUMMARY

The production of high concentration oxygen is the basis for oxy-fuel combustion technology. In a practical oxy-fuel combustion system, part of the flue gas containing a number of CO_2 was recycled back into the combustion chamber to provide the dilution effect of the absent nitrogen (N_2) and to moderate the furnace temperatures. Conventional O_2 production utilizing the cryogenic distillation process is highly expensive. The chemical looping air separation process for O_2/CO_2 production is receiving more attention because of its potential to mitigate the cost of O_2 production, thus helping to promote this clean energy technology. Typical oxygen carrier materials like $Ba_{1-x}Sr_xCo_{0.8}Fe_{0.2}O_{3-\delta}$ perovskite (BSCF) and CuO doped with ZrO_2 that have high oxygen desorption rates in the interested conditions have been identified as suitable materials for chemical looping air separation. The substitution of Fe in $SrCo_{0.8}B_{0.2}O_{3-\delta}$ perovskite with Zr, Cr, and Cu increases the oxygen adsorption capacity and oxygen desorption amount. The doping of Zr in Cu-based oxygen carriers can increase the chemical stability, the mechanical strength, and the reactivity of the oxygen carrier.

Progress has clearly been made in the preparation and testing of new carrier materials for O_2/CO_2 production; however, further modifications and developments in batch production techniques to accelerate the deployment of cost-effective oxygen carrier materials to the market are needed. Despite the challenges surrounding oxygen carriers for O_2/CO_2 production, chemical looping air separation (CLAS) can play a central role in solving the O_2/CO_2 production problems for oxy-fuel combustion system.

REFERENCES

[1] Xu B, Stobbs RA, White V, Wall RA, Gibbins J, Iijima M. Future CO_2 capture technology options for the Canadian market. Technical Report Report No. COAL R309 BERR/Pub URN 07/1251, Doosan Babcock Energy Limited; 2007.

[2] Yang Q, Lin YS. Improved sorbent for high-temperature production of oxygen-enriched carbon dioxide stream. Ind Eng Chem Res 2007;46:6025–31.
[3] Kusaba H, Sakai G, Shimanoe K, Miura N, Yamazoe N. Oxygen-sorptive and -desorptive properties of perovskite-related oxides under temperature-swing conditions for oxygen enrichment. Solid State Ionics 2002;152–153:689–94.
[4] Shen Q, Zheng Y, Luo C, Zheng C. Development and characterization of $Ba_{1-x}Sr_xCo_{0.8}Fe_{0.2}O_{3-\delta}$ perovskite for oxygen production in oxyfuel combustion system. Chem Eng J 2014;255:462–70.
[5] Shen Q, Zheng Y, Li S, Ding H, Xu Y, Zheng C, Thern M. Optimize process parameters of microwave-assisted EDTA method using orthogonal experiment for novel $BaCoO_{3-\delta}$ perovskite. J Alloys Compd 2016;658:125–31.
[6] Ortiz-Vitoriano N, Bernuy-López C, Ruiz de Larramendi I, Knibbe R, Thydén K, Hauch A, Holtappels P, Rojo T. Optimizing solid oxide fuel cell cathode processing route for intermediate temperature operation. Appl Energy 2013;104:984–91.
[7] Shen Q, Zheng Y, Luo C, Zheng C, Thern M. Effect of A/B-site substitution on oxygen production performance of strontium cobalt based perovskites for CO_2 capture application. RSC Adv 2015;5:39785–90.
[8] Shen Q, Zheng Y, Luo C, Zheng C. Characteristics of $SrCo_{1-x}Fe_xO_{3-\delta}$ perovskite powders with improved O 2/CO_2 production performance for oxy-fuel combustion. Bull Kor Chem Soc 2014;35:1613.
[9] Moghtaderi B. Application of chemical looping concept for air separation at high temperatures. Energy Fuel 2010;24:190–8.
[10] Alipour M, Nychka JA, Gupta R. Evaluation of copper-aluminum oxides as sorbents for high-temperature air separation. Energy Fuel 2014;28:319–28.
[11] Wang K, Yu Q, Qin Q. Reduction kinetics of Cu-based oxygen carriers for chemical looping air separation. Energy Fuel 2013;27:5466–74.
[12] Wang M, Liu J, Hu J, Liu F. O_2-CO_2 mixed gas production using a Zr-doped Cu-based oxygen carrier. Ind Eng Chem Res 2015;54:9805–12.
[13] Arjmand M, Leion H, Mattisson T, Lyngfelt A. ZrO_2-supported CuO oxygen carriers for chemical-looping with oxygen uncoupling (CLOU). Energy Procedia 2013;37:550–9.
[14] Tan Q, Qin W, Chen Q, Dong C, Li W, Yang Y. Synergetic effect of ZrO_2 on the oxidation-reduction reaction of Fe_2O_3 during chemical looping combustion. Appl Surf Sci 2012;258:10022–7.
[15] Zhao H-b, Liu L-m, Di X, Zheng C-g, Liu G-j, Jiang L-l. $NiO/NiAl_2O_4$ oxygen carriers prepared by sol-gel for chemical-looping combustion fueled by gas. J Fuel Chem Technol 2008;36:261–6.
[16] Zhang Y, Zhao H, Guo L, Zheng C. Decomposition mechanisms of Cu-based oxygen carriers for chemical looping with oxygen uncoupling based on density functional theory calculations. Combust Flame 2015;162:1265–74.
[17] Nolan M, Fearon JE, Watson GW. Oxygen vacancy formation and migration in ceria. Solid State Ionics 2006;177:3069–74.

FURTHER READING

[1] Li Z-s, Zhang T, Cai N-s. Experimental study of O_2-CO_2 production for the oxyfuel combustion using a Co-based oxygen carrier. Ind Eng Chem Res 2008;47:7147–53.
[2] Wang M, Liu J, Shen F, Cheng H, Dai J, Long Y. Theoretical study of stability and reaction mechanism of CuO supported on ZrO_2 during chemical looping combustion. Appl Surf Sci 2016;367:485–92.

Chapter 14

MILD Oxy-fuel Combustion

Pengfei Li*, Fan Hu*, Biao Xiong*, Jianpeng Zhang†, Jianchun Mi†
**Huazhong University of Science and Technology, Wuhan, China*
†Peking University, Beijing, China

14.1 INTRODUCTION

Research and development (R&D) [1–3] of oxy-fuel combustion technology [4,5] in demonstration and large pilot projects has made significant progress in carbon capture and storage (CCS) technology. However, this technology still faces a number of challenges. At least three aspects of oxy-fuel combustion need to be addressed [1]: (1) a way to further increase the system's thermal performance, (2) a way to further reduce the NO_x formation caused by the leakage of air or nitrogen in the fuel, and (3) the challenge of combustion stability caused by the lower adiabatic flame temperature, the delayed ignition, and the lower burning rate in a CO_2-diluted environment compared to a N_2 environment. The most commonly used way to enhance thermal performance is to increase the flame temperature so as to increase the combustion intensity [6]. Nevertheless, this approach often comes at the price of high pollutant emissions (e.g., NO_x). Although the NO_x emission can be controlled by fuel or air staging or other low-NO_x combustion technologies, the flame temperature also decreases to an undesirable level. It is difficult to simultaneously fulfill the requirements of high thermal performance and low pollution in oxy-fuel combustion processes. One approach to solving this problem is to use moderate or intense low oxygen dilution (MILD) combustion.

MILD combustion is a clean combustion technology [7–11] in which extremely low NO_x emissions and uniform thermal distribution can be simultaneously achieved. In the early 1990s, three groups—the International Flame Research Foundation (IFRF) [12–14], Wünning and Wünning [9], and Nippon Furnace Kogyo Kaisha, Ltd. (NFK-Japan) [8]—each independently developed this technology. Wünning and Wünning [9] referred to this technology as flameless oxidation (FLOX), and the NFK-Japan and the IFRF referred to it as high temperature air combustion (HTAC). Later Cavaliere and de Joannon et al. [7] used the acronym MILD for this technology. Now the term MILD

Oxy-fuel Combustion. http://dx.doi.org/10.1016/B978-0-12-812145-0.00014-1

combustion is well recognized by the international combustion community for this technology.

The key advantages of MILD combustion are its (1) extremely low pollutant emissions (i.e., NO_x and soot), (2) uniform thermal distribution, (3) low temperature fluctuation, (4) stable combustion operating conditions without flame stabilization problems, (5) low combustion noise, and (6) safe operation because no high temperature flame is attached to the burner exit. Therefore this technology has attracted significant attention of both the fundamental research and industrial combustion communities. Under MILD combustion conditions, reactants are strongly diluted with the exhaust gas because of the large-scale in-furnace recirculation. The chemical reaction occurs in the entire furnace over a large reaction volume. There is no visible flame front under MILD combustion conditions, thus the temperature peak is avoided. Consequently, semiuniform temperature distribution is obtained, and NO_x emissions (especially thermal NO_x route) are largely suppressed. Therefore combining oxy-fuel combustion with MILD combustion to achieve MILD oxy combustion is more advantageous than using conventional oxy-fuel combustion.

Although a significant amount of work has been conducted on MILD combustion and oxy-fuel combustion, R&D on their combination (i.e., MILD oxy combustion) is quite limited. A brief literature review of MILD oxy combustion of both gaseous and solid fuels is given next.

For gaseous fuels, the IFRF experimentally and numerically investigated MILD oxygen-natural gas combustion from 1995 to 1999 [15,16]. The characteristics of the combustion were discussed, such as the flame structure, heat transfer, and NO_x emissions. The eddy dissipation concept with a full equilibrium chemistry procedure was found to obtain good predictions. Blasiak et al. [17] used MILD oxygen-natural gas combustion in the thermal treatment processes of wastes and in the recovery of zinc veering feed in a rotary kiln. Their experiments revealed that MILD oxy combustion not only increased the productivity of the rotary kiln but also reduced fuel consumption and NO_x emissions. Krishnamurthy et al. [18] investigated the characteristics of MILD oxygen-propane combustion in a 200 kW furnace. They found that a slight asymmetric injection of the high velocity oxygen jet led to a large recirculation pattern, resulting in no visible flame and reduced soot and NO_x formations relative to conventional oxy combustion. Li et al. [19] investigated the characteristics of MILD oxy combustion using three gaseous fuels (natural gas, liquefied petroleum gas, and ethylene) at a 13 kW firing rate. They found that MILD oxy combustion could be established as long as the inert gas dilution or equivalence ratio was sufficiently high. The N_2O-intermediate route was found to play a crucial role in producing NO. The NO-reburning chemistry is also important and should not be ignored under the MILD condition. Zhang et al. [20] found that the requirement for establishing MILD combustion was less stringent with dilution by CO_2 rather than by N_2. Zhang et al. [20] and Tu et al. [21] found that CO_2 dilution substantially lowered the rise in temperature, delayed the ignition, slowed the overall reaction rate, and thus facilitated the occurrence of MILD combustion. Sabia et al. [22] performed unsteady calculations of a perfectly stirred flow reactor for methane/oxygen mixtures in high CO_2 dilution conditions in order to study the chemical effects of CO_2. Their study emphasized the effect of combustion kinetics on the temperature oscillation dynamic behaviors of the oxidation process. Mardani and Ghomshi [23] carried out a numerical study on MILD oxy combustion. They found that as the initial oxygen level increased, the concentrations of CO, NO, CH_2O, and HCO decreased in the reaction zone. They also found that by reducing the fuel hydrogen content, the uniformity of the temperature field and the volume of the reaction zone increased, while the formation of NO and CO decreased.

For MILD oxy combustion of pulverized coal, Stadler et al. [24,25] experimentally investigated MILD combustion of pulverized coal in three oxidizer atmospheres: air, CO_2/O_2, and O_2-enriched flue gas recirculation. They discovered that with an increase in the oxygen ratio, the NO_x emissions from both air and oxy combustion under both conventional and MILD combustion conditions increased. They also attributed the reduction of NO_x emissions by flue gas recirculation to both the reduction of recirculated NO_x and the reduced conversion of fuel-N to NO. Li et al. [26] investigated MILD oxy combustion of light oil and pulverized coal by experiments in a 300kW pilot-scale furnace. They examined the effects of burner design and dilution level on the combustion appearance of liquid and solid fuels and emissions of CO and NO_x. They found that the NO_x reduction was more efficient under MILD oxy combustion than in the air case. Tu et al. [27] investigated the MILD oxy combustion of pulverized coal in a 0.58MW furnace through computational fluid dynamics simulations. They also found that MILD oxy combustion had a greater potential on the NO_x reduction than MILD air combustion.

The preceding literature review is aimed at discussing and summarizing the fundamental issues and technological development of MILD oxy combustion. The rest of the chapter is organized as follows. The next section analyzes the physical and chemical effects of CO_2 addition on MILD oxy combustion. Then various burner configurations for MILD oxy combustion are presented. Based on the MILD oxy combustion burner, the characteristics of MILD oxy combustion of pulverized coal are discussed in detail. Finally, a summary and conclusions are given.

14.2 PHYSICAL AND CHEMICAL EFFECTS OF CO_2 ADDITION ON MILD OXY COMBUSTION

In this section, physical and chemical effects of CO_2 and N_2 dilutions on MILD combustion are discussed. Table 14.1 summarizes these physical and chemical processes, and the properties and characteristics are discussed next.

TABLE 14.1 Important Physical Properties and Chemical Characteristics for Combustion

Physical Process	Property	Chemical Process	Characteristic
Momentum transfer	Density	Ignition	Ignition delay time
	Kinematic viscosity	Chemical equilibrium	Adiabatic equilibrium temperature
Thermodynamic	Density		Heat release
	Heat capacity		
Heat transfer	Thermal conductivity		

14.2.1 Physical Effects

Density

The effects of density difference on the physical process of combustion when N_2 is replaced by CO_2 (see Table 14.1) are two-fold. On one hand, the CO_2-diluted reactants have greater momentum than those diluted with N_2 under the same equivalence ratio and volumetric flow rate, because the density of CO_2 is greater than that of N_2, as shown in Fig. 14.1A. In Fig. 14.1A, data from Ref. [28] is fitted with the ideal gas law. The injection momentum for the establishment of MILD combustion was critical. Szegö et al. [29,30] found that a certain jet momentum threshold was required to establish MILD combustion in their investigation of the stability limits of their combustion system under various operational conditions. Mi et al. [31] also experimentally and numerically confirmed that MILD combustion could not occur when the injection momentum was below the criterion. Once MILD combustion was established, the influence of the injection momentum became insignificant on the combustion process and NO_x emissions, which implied the unique importance of the injection momentum on the establishment of MILD combustion. From this respect, replacing N_2 with CO_2 is positive for the realization of MILD combustion. On the other hand, the density of the gases affects the thermodynamic process; the corresponding discussion is provided later in the Section "Specific Heat Capacity."

Kinematic Viscosity

In Fig. 14.1B, the kinematic viscosities ν of CO_2 and N_2 are presented, and it is clear that $\nu_{CO2} < \nu_{N2}$ from 300 to 2000 K. Fluid dynamics play a vital role in the establishment of MILD combustion in a real furnace. Previous investigations declared the importance of injection momentum [29–31], velocity [32],

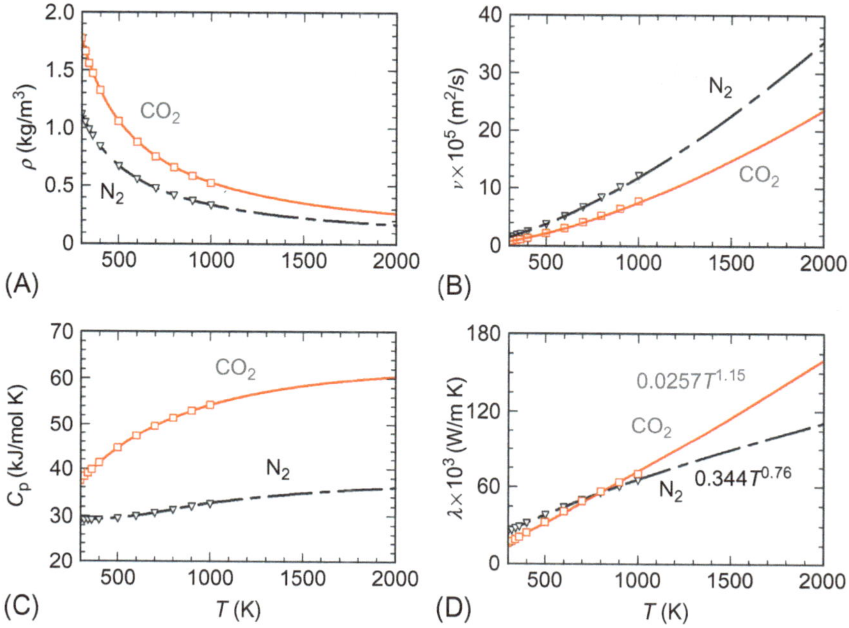

FIG. 14.1 Comparisons of the CO_2 and N_2 physical properties: (A) density, (B) kinematic viscosity, (C) specific heat capacity, and (D) thermal conductivity. Symbols: Data from Ref. [28]. Lines: Curve fitting by (A) ideal gas law, (B) Sutherland's formula, (C) polynomial from CHEMKIN data set, and (D) power law. *(Reproduced with permission from Zhang J, Mi J, Li P, Wang F, Dally BB. MILD combustion of methane diluted by CO2 and N2. Energy Fuel 2015;29(7):4576–85. Copyright 2015 American Chemical Society.)*

inlet Reynolds number [33], and mixing with flue gas or recirculation [9] on the establishment of MILD combustion. Li et al. [33] found that the inlet Reynolds number $Re = uD/\nu$ (u being the inlet velocity, D being the equivalent inlet diameter) needed to be reach in order to produce substantial recirculation in the furnace to dilute the reactants for MILD combustion. The smaller ν_{CO2} compared to ν_{N2} can increase the initial Re under the same inlet conditions and further enhance exhaust gas recirculation in the furnace. Therefore ν_{CO2} is positive for establishing MILD combustion when replacing N_2 with CO_2.

Specific Heat Capacity

As shown in Fig. 14.1C, the constant pressure heat capacity of CO_2 $C_{p,CO2}$ is almost twice that of $C_{p,N2}$, according to the polynomial fitting from the CHEMKIN data set. While their temperatures rise, the dilution gases affect the thermodynamic process during combustion by absorbing heat. The absorbed heat under constant pressure can be calculated as

$$H = \int_{T_1}^{T_2} \rho C_p dT. \tag{14.1}$$

Eq. (14.1) shows that ρ and C_p together determine the heat absorption. The greater these two factors, the greater the amount of heat absorbed as the gas temperatures rise. According to Eq. (14.1) and Fig. 14.1AC, CO_2 absorbs almost twice as much heat as N_2 absorbs for certain rises in temperature. This means CO_2 dilution is positive for lowering the furnace temperature during combustion through heat absorption, and thus it is beneficial for establishing MILD combustion.

Thermal Conductivity

The thermal conductivities of CO_2 and N_2 are presented in Fig. 14.1D, together with their respective power-law fitting according to the data from Ref. [28]. For $T < 850\,K$, λ_{CO2} is less than λ_{N2}, while $\lambda_{CO2} > \lambda_{N2}$ for $T > 850\,K$. With higher thermal conductivity, the heat released by chemical reactions can be transferred more quickly to the gas within the entire furnace, which helps to build a uniform temperature field and to reduce the local temperature rise [34–36]. From this respect, CO_2 is positive for achieving the MILD condition.

In summary, the physical properties of density, kinematic viscosity, heat capacity, and thermal conductivity of CO_2 are more beneficial for the establishment of MILD combustion than those of N_2.

14.2.2 Chemical Effects

Ignition Delay

Ignition significantly influences flame propagation and stability and thus is an important process in combustion [6]. Various criteria, developed in both experiments and numerical simulations, determine ignition delay time, including the interval between reflected-shock passage [37], onset of rapid rise of pressure [38], occurrence of maximum chemiluminescence or intermediate species (e.g., OH, CH_2O) emission [37–41], and maximum temperature increase rate [42] or certain temperature rise [39]. In the present study, the ignition delay time was defined as the period from the beginning of calculation to the occurrence of maximum OH emission during the ignition process, as shown in Fig. 14.2. The GRI-Mech 3.0 mechanism was used in the ignition calculation where $X_{O2} = 0.05, 0.1, 0.2$, $\Phi = 0.8$, and $T = 1345\,K$. Note that the temperature increase rate reached its maximum at the occurrence of the greatest OH emission, which also implied the ignition of the mixture. The present prediction of ignition delay times is illustrated in Fig. 14.3A. Good agreement was obtained between the present prediction and the previous results [43,44], which confirmed the accuracy of the predicted ignition delay times for both CH_4 and CH_4/C_3H_8 blends. Note that the temperature range for the present prediction was from 1100 to 1600 K, which is the typical temperature for MILD combustion. As shown in Fig. 14.2, as the reactants were diluted with more N_2, the ignition was delayed and the final temperature lowered, which corresponded with the conclusion of van Oijen [45].

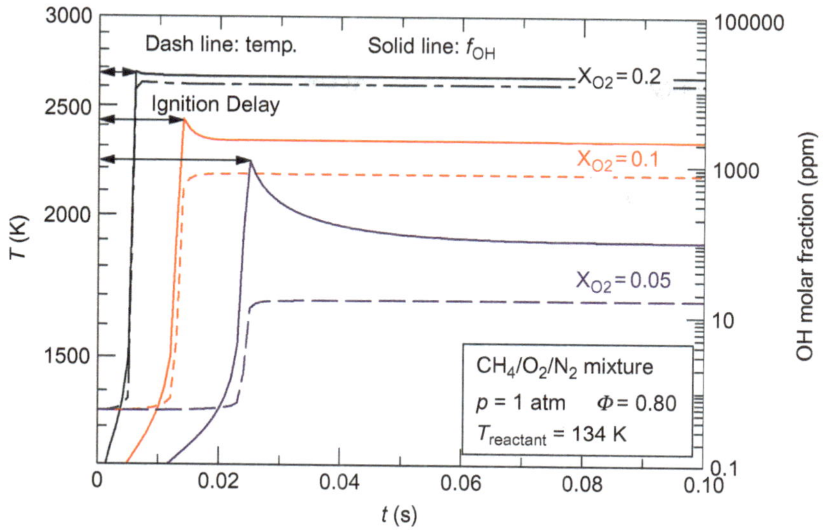

FIG. 14.2 Variations of temperature and OH molar fraction during ignition.

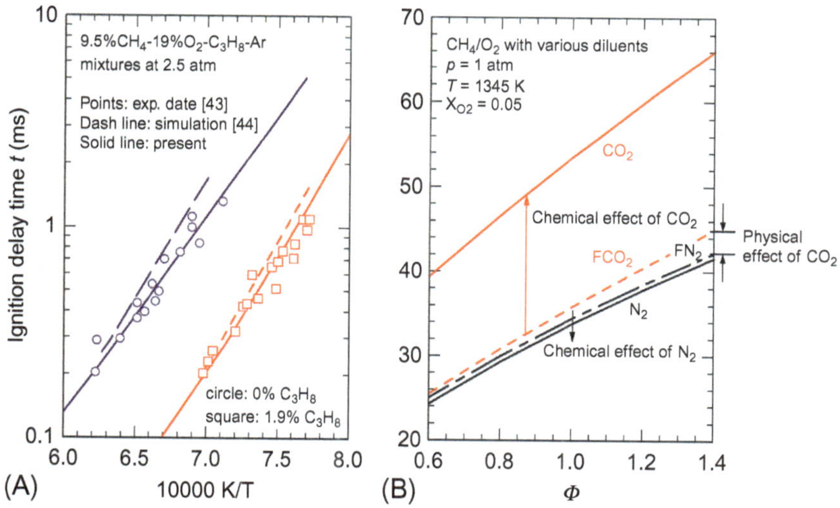

FIG. 14.3 Ignition delay times: (A) verification with experimental data [43] and numerical simulations [44] and (B) present predictions for CH_4/O_2 with various diluents and equivalence ratios under $p=1$ atm, $T=1345$ K, $X_{O2}=0.05$.

In MILD conditions, the flame instability and aerodynamic extinction encountered in traditional combustion can be ignored because MILD combustion is based on fuel self-ignition, which depends on reactant temperatures rather than heat release [7]. Therefore, although delayed ignition may increase the instability of traditional combustion, it on the other hand becomes beneficial for MILD combustion because (1) delayed ignition allows the developing turbulence to interact more with the ignition chemistry [45] and also ensures well-mixed reactants and flue gases, and (2) delayed ignition retards the heat release process of chemical reactions, thus delayed ignition is positive for reducing the local temperature and even the average temperature of a combustion furnace.

In order to examine the physical and chemical effects of N_2 and CO_2 dilutions, two additional fictitious species, FN_2, FCO_2, were included in the present study. FN_2 and FCO_2 had exactly the same physical properties as N_2 and CO_2, respectively, but were not involved in the chemical reactions. FCN_2 represented a species with the density of N_2 and the heat capacity of CO_2, and it was not involved in the chemical reactions, either. The effects on the ignition delay are presented in the following discussion. As shown in Fig. 14.3B, as Φ increased, ignition was delayed, though interestingly the chemical effects of CO_2 and N_2 dilutions on ignition delays were the opposite. Because FN_2 was not involved in the chemical reactions, the discrepancy between the ignition delay times of FN_2 and N_2 dilutions was mainly due to the chemical effect of N_2. As can be seen clearly in Fig. 14.3B, with the involvement of N_2 in the chemical reactions, the ignition accelerated slightly. Guo and Chen [46] also found this phenomenon in their investigation of ignition enhancement of ethylene/air by the addition of NO_x. Also, a comparison of the plots of CO_2 and FCO_2 dilutions showed that the chemical effect of CO_2 significantly prolonged the ignition delay times to at least 1.5 times that with FCO_2 dilution. Because neither FN_2 nor FCO_2 were relevant to the chemical process, the discrepancy of these two plots, as shown in Fig. 14.3B, was induced by the physical property differences between N_2 and CO_2, namely, their difference in density and heat capacity. Similarly, the physical effects of CO_2 also delayed the ignition of the mixture. Therefore, from the respect of ignition, the chemical effects of CO_2 replacement were positive for establishing MILD combustion.

Adiabatic Equilibrium Temperatures

Fig. 14.4 shows the adiabatic equilibrium temperatures $T_{ab\text{-}eq}$ of different diluents based on the calculations of CHEMKIN's well-stirred reactor (WSR) model and GRI-Mech 3.0. The definition of MILD combustion proposed by Cavaliere and de Joannon [7] was used in numerical simulations because (1) this definition is mathematically stated according to the MILD combustion process in a WSR, thus it is more rigorous and convenient than other furnace-based or observation-based definitions [8,9,35,36]; and (2) this definition has been widely applied in combustion regime identifications [47–52] and also in direct numerical simulations (DNS) of MILD combustion [53,54], which have proved its connection to practical application. According to the definition of MILD combustion by Cavaliere and de Joannon [7], and as shown in Fig. 14.4, the combustion conditions under $X_{O2}=0.05$ and $X_{O2}=0.2$ were MILD and high temperature combustion, respectively. (The numerical

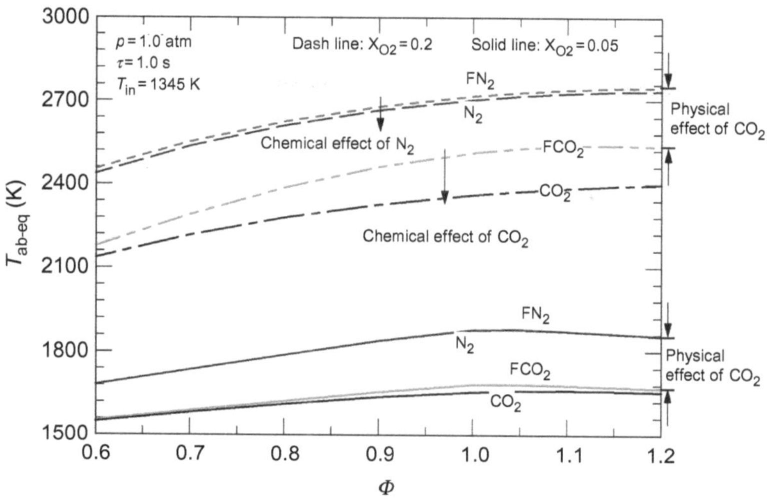

FIG. 14.4 Adiabatic equilibrium temperatures calculated with WSR for CH_4/O_2 mixtures ($X_{O2}=0.05$ and $X_{O2}=0.2$) with different diluents under $p=1$ atm, $T_{in}=1345$ K.

investigation on the establishment of MILD combustion is discussed in detail in Section 14.3.2). As shown in Fig. 14.4, the combination of the physical and chemical effects of CO_2 significantly lowered $T_{ab\text{-}eq}$. For both $X_{O2}=0.05$ and $X_{O2}=0.2$, comparisons between $T_{ab\text{-}eq}$ of FN_2 and FCO_2 implied that the physical effect of CO_2 was important for reducing the furnace temperature. However, the chemical effect of CO_2 became less prominent for reducing the furnace temperature when MILD combustion occurred ($X_{O2}=0.05$). It is also interesting to note that the chemical effect of N_2 was positive for decreasing the final temperature, because the chemical process for N_2 dissociation and NO_x production was endothermic. In summary, based on the perspective of reducing equilibrium temperature, replacing N_2 with CO_2 was beneficial for the occurrence of MILD combustion.

Reaction Heat Releases

The total heat releases Q during the ignition process (from $t=0$ to $t=1$ s) and the corresponding final temperatures with different dilutions under $p=1$ atm, $T=1345$ K, and $\Phi=0.8$ are shown in Fig. 14.5. The total heat releases were also obtained by using calculations based on the WSR model and GRI-Mech 3.0. The relationship of the total heat releases was $Q_{CO_2} < Q_{FCO_2} < Q_{N_2} < Q_{FN_2}$ for both low and high oxygen concentrations, as were the final temperatures; that is, $T_{ad-eq,CO_2} < T_{ad-eq,FCO_2} < T_{ad-eq,N_2} < T_{ad-eq,FN_2}$. Note that the total heat releases of these four diluents were less under $X_{O2}=0.05$ than under $X_{O2}=0.2$, mainly because of the difference in the dilution level. The numbering markers in Fig. 14.5A represent dilution replacements; the corresponding influences are discussed next.

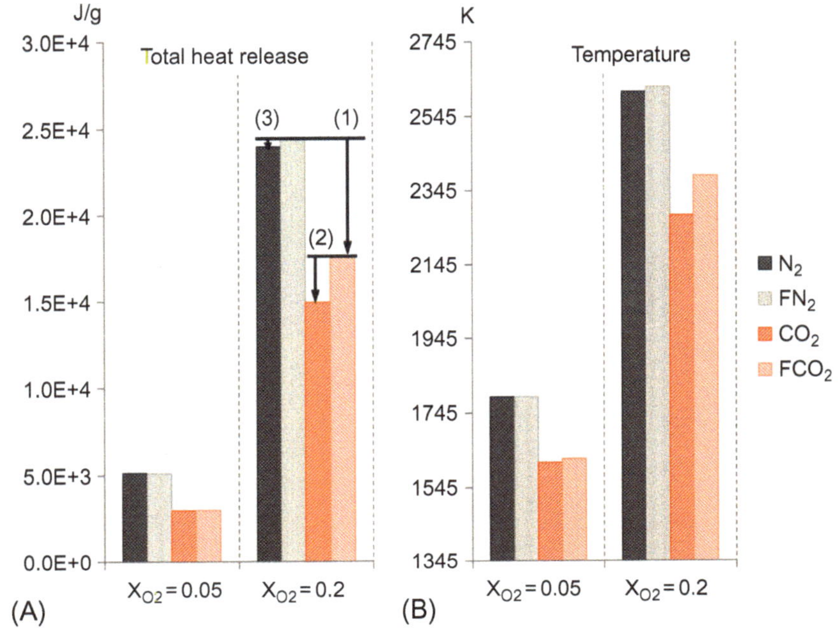

FIG. 14.5 (A) Total heat releases during ignition and (B) final temperatures with different dilutions under low and high O_2 concentration conditions ($X_{O2}=0.05$ and $X_{O2}=0.2$) for $p=1$ atm, $T=1345$ K, and $\Phi=0.8$.

(1) $FN_2 \rightarrow FCO_2$. As previously mentioned, these two fictitious species were not involved in the chemical reactions, thus the heat release difference caused by this replacement was attributed to the differences between the physical properties of N_2 and CO_2. As (1) shows, the reaction heat release was greater with FN_2 dilution, which was because the heat capacity of FCO_2 was higher than that of FN_2. More combustion heat converted to the internal energy of FCO_2, so the total heat release and the final temperature were both reduced.

(2) $FCO_2 \rightarrow CO_2$ and (3) $FN_2 \rightarrow N_2$. After these replacements, the diluents were involved in chemical reactions, and the reaction heat releases were both lowered, as shown in Fig. 14.5A, which implied that the chemical effects of both CO_2 and N_2 were reducing Q because of the energy consumption in the yields of intermediate components. Particularly, more CO was produced through $CO_2+H=CO+OH$ under CO_2 dilution, making the combustion less complete. In response, the reduced heat release caused by the chemical effect of CO_2 was more than that caused by the chemical effect of N_2.

In summary, replacing N_2 with CO_2 suppressed Q and the final temperature because of the differences between the physical properties of CO_2 and N_2. Also, the chemical effect of CO_2 and N_2 reduced Q.

14.3 ESTABLISHMENTS OF MILD COMBUSTION DILUTED BY CO_2 AND N_2

14.3.1 Experimental Observations

To compare the use of MILD combustion under CO_2 and N_2 dilutions, observational experiments were conducted in the present work. For the present observational experiment on firing gaseous fuels, MILD combustion was considered as the flameless combustion, where no visible flame occurs at all. For comparison, during observational experiments, many instantaneous images of the flames were taken under different combustion conditions and under different dilution ratios. The details of the experimental 20 kW furnace at the University of Adelaide are given in Refs. [19,20,29–31,33,55]. Fig. 14.6 presents some typical images of establishing MILD air combustion and MILD oxy combustion in the furnace when firing (A) NG and (B) LPG with an oxygen concentration of $X_{O2}=21\%$ in the oxidant streams and (C) NG diluted by the mixture of N_2 and CO_2 with $X_{O2}=20\%$ and an equivalence ratio $\Phi=0.83$. Note that Φ was varied by changing the flow rate of the oxidant stream, while the fuel supply was kept constant. When burning NG with air (21% O_2+79% N_2) (see Fig. 14.6A), the combustion changed from MILD to a flame condition as Φ was decreased from 1.0 to 0.78, which meant that with N_2 dilution, MILD combustion could be established when Φ was above 0.78. The underlying reason was that, as Φ decreased, the oxygen concentration within the furnace increased, which was opposite the requirement of the low oxygen environment for establishing MILD combustion. Thus decreasing Φ eventually led to the occurrence of the conventional flame condition. The critical equivalence ratio for establishing MILD combustion was denoted by Φ^*, below which flame was visible. With O_2+CO_2 instead of air, Φ^* expanded to 0.71. Similarly, when firing LPG, the replacement of N_2 with CO_2 expanded Φ^* from 0.9 to 0.85 (see Fig. 14.6B). The reduced range of Φ for LPG MILD combustion resulted from its higher heat value compared to that of NG.

The preceding observations showed that the establishment of MILD oxy combustion was less stringent than that of MILD air combustion, because the critical Φ^* was expanded by CO_2 dilution when burning either NG or LPG. To further investigate the importance of CO_2 dilution, oxidant streams consisting of various volume fractions of N_2 and CO_2 were injected into the furnace when firing NG. Fig. 14.6C shows that as the volume fraction of CO_2 in the oxidant streams increased from $X_{CO2}=0\%$ to 20%, sporadic flame fronts disappeared and MILD conditions were established. Thus it was demonstrated that either fully or partially replacing N_2 with CO_2 facilitated the establishment of MILD combustion. A chart of the combustion state against Φ and X_{CO2} is shown in Fig. 14.6D, which shows better variations of Φ^* for the three cases. In summary, the present experimental investigation confirms that MILD combustion is more likely to occur with dilution of CO_2 than N_2 for different gaseous fuels.

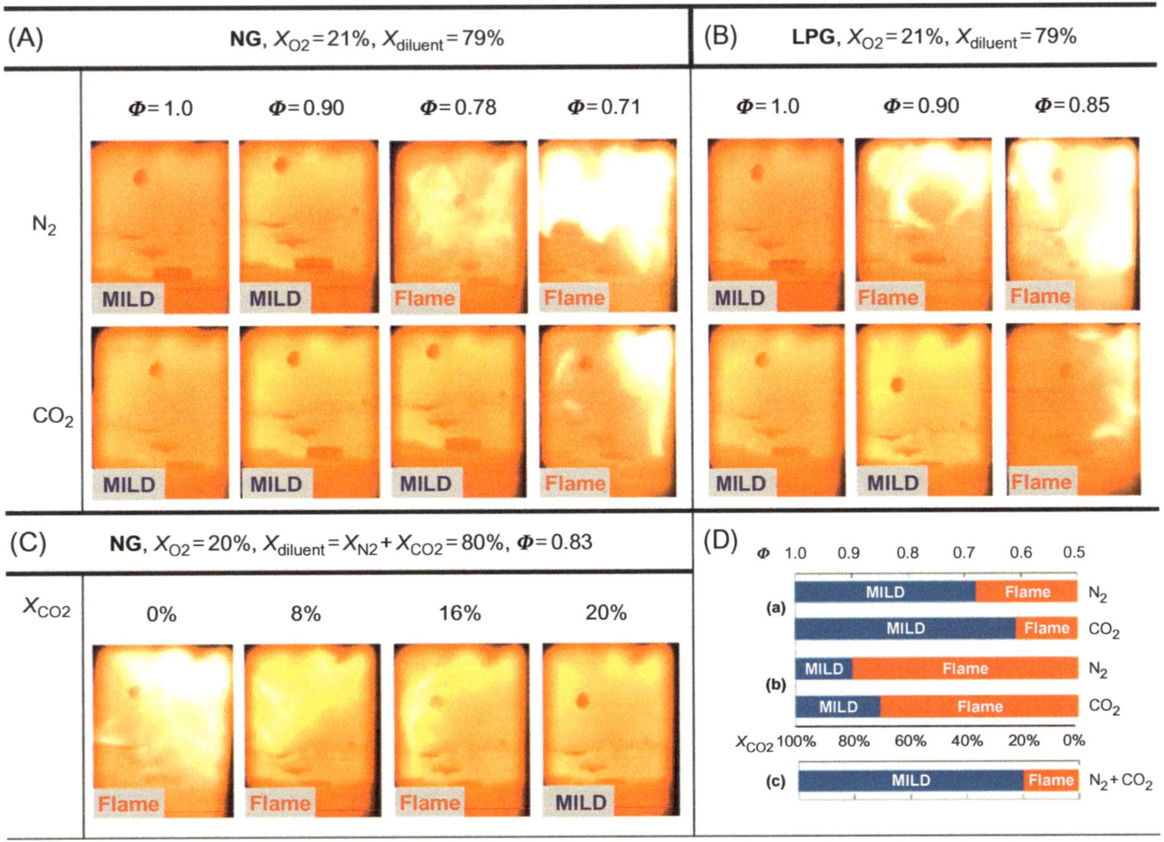

FIG. 14.6 Typical images of establishing MILD combustion with N_2 and CO_2 dilutions for different fuels: (A) NG, (B) LPG, (C) NG with N_2/CO_2 blended dilution at Φ=0.83, (D) the stability limits for MILD combustion. *(Reproduced with permission from Zhang J, Mi J, Li P, Wang F, Dally BB. MILD combustion of methane diluted by CO2 and N2. Energy Fuel 2015;29(7):4576–85. Copyright 2015 American Chemical Society.)*

14.3.2 Numerical Investigation

The combustion of CH_4 in a WSR was considered to numerically investigate the establishment of MILD conditions with different temperatures, equivalence ratios, and dilution levels. According to the definition of MILD combustion by Cavaliere and de Joannon [7], the numerical definition of MILD combustion requires the maximum temperature rise

$$\Delta T = T_{WSR} - T_{in} \langle T_{si}, \quad \text{and} \quad T_{in} \rangle T_{si},$$

where T_{WSR} is the working temperature of a WSR at the end of the residence time τ, T_{in} is the temperature of inlet reactants, and T_{si} is the self-ignition temperature of the fuel. Fig. 14.7 depicts the relationships among these variables in a WSR. Under the condition of $\Phi=0.8$, $X_{O2}=5\%$, $p=1$ atm, $\tau=1$ s, the self-ignition temperature of a $CH_4/O_2/N_2$ mixture is 1010 K.

Fig. 14.8 shows the normalized $T_{in}-\Delta T$ curves of $CH_4/O_2/N_2$ (*dash lines*) and $CH_4/O_2/CO_2$ (*solid lines*) mixtures for different oxygen molar concentrations and $\Phi=0.8$, together with the combustion conditions proposed by Cavaliere and de Joannon [7], in order to prove that MILD combustion is prone to occur under the same conditions with CO_2 dilution. In Fig. 14.8, it is easy to identify the combustion state for different conditions. When the oxygen concentration of the mixture was low ($X_{O2}=0.05$), only MILD combustion (and quasi-MILD, MILD-like according to Wang et al. [51]) occurred. As X_{O2} increased, feedback combustion and high temperature combustion conditions

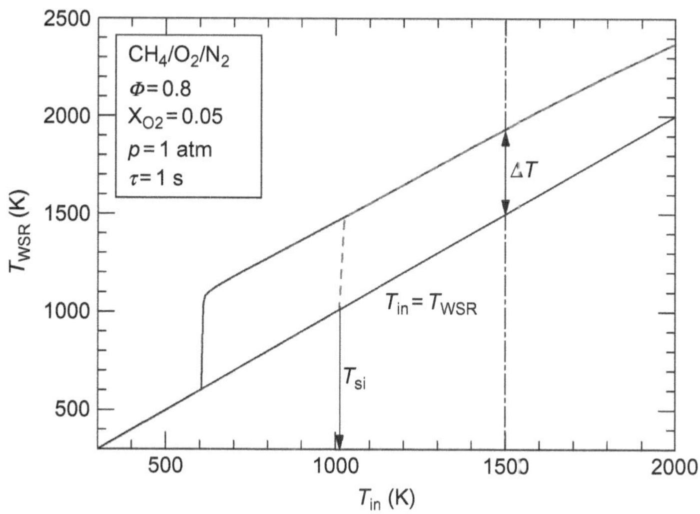

FIG. 14.7 WSR working temperatures (T_{WSR}) versus the inlet mixture temperature (T_{in}) for the $CH_4/O_2/N_2$ mixture at 5% O_2, 2% CH_4, and 93% N_2. *Red line*: Self-ignition reaction temperature; *blue line*: forced-ignition reaction temperature.

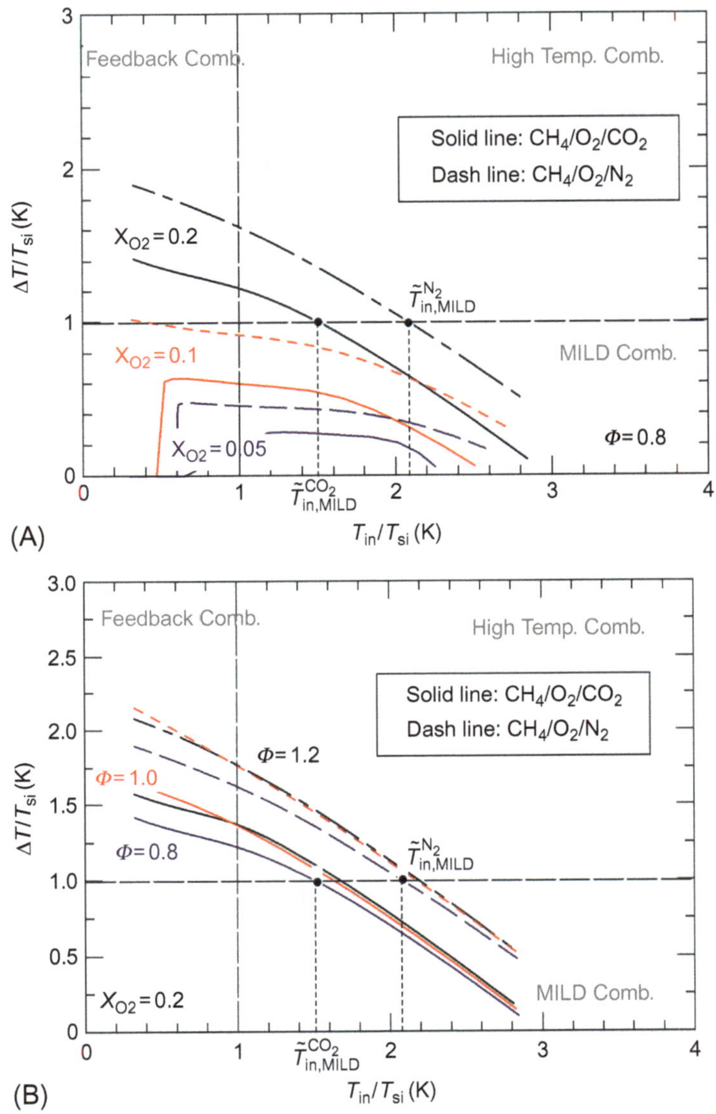

FIG. 14.8 Normalized $T_{in} - \Delta T$ curves of $CH_4/O_2/N_2$, and $CH_4/O_2/CO_2$ mixtures for (A) different oxygen molar concentrations and (B) various equivalence ratios.

also occurred besides the MILD condition. That is to say, MILD conditions are more likely to happen under low oxygen concentration. From the data in Fig. 14.8A, one can obtain

$$\frac{T_{MILD}^{CO_2}}{T_{si}^{CO_2}} = \tilde{T}_{in,MILD}^{CO_2} \le \tilde{T}_{in,MILD}^{N_2} = \frac{T_{MILD}^{N_2}}{T_{si}^{N_2}} \quad (14.2)$$

where the subscript MILD represents the occurrence of the MILD condition. Because for the same X_{O2} and Φ, $T_{si}^{CO_2}$ is always slightly smaller than $T_{si}^{N_2}$ (e.g., see Fig. 14.9A with $\Phi = 0.8$), Eq. (14.2) becomes

$$T_{MILD}^{CO_2} \le T_{MILD}^{N_2} \frac{T_{si}^{CO_2}}{T_{si}^{N_2}} < T_{MILD}^{N_2}. \tag{14.3}$$

Eq. (14.3) proves that for the same X_{O2}, MILD combustion already occurs in the temperature range $\left[T_{MILD}^{CO_2}, T_{MILD}^{N_2}\right]$ with CO_2 dilution, although not with N_2 dilution. For $X_{O2} = 0.2$, the difference between $T_{MILD}^{CO_2}$ and $T_{MILD}^{N_2}$ was about 550 K, which indicated that the preheating temperature could be significantly reduced with CO_2 dilution. Moreover, Fig. 14.8A shows that the temperature increase under CO_2 dilution was less than under N_2 dilution. Thus based on the MILD definition of Cavaliere and de Joannon [7], lower T_{si} together with lower ΔT were beneficial for combustion to "enter" a MILD condition. Because T_{si} is almost the same under N_2 and CO_2 dilutions for the same X_{O2} and Φ, factors that reduce ΔT will help to establish MILD combustion. The normalized $T_{in} - \Delta T$ curves of $CH_4/O_2/N_2$ (*dash lines*) and $CH_4/O_2/CO_2$ (*solid lines*) mixtures for fuel-lean, stoichiometric, and fuel-rich conditions with $X_{O2} = 0.2$ are shown in Fig.8B. Again, ΔT_{CO2} was smaller than ΔT_{N2} under the same conditions for the same T_{in}. Similar to the proof arrived at in Eqs. (14.2), (14.3), it can be demonstrated that MILD combustion is more likely to take place in a CO_2 atmosphere than in a N_2 atmosphere with various equivalence ratios. In summary, based on the MILD combustion definition [7], it is demonstrated in Fig. 14.8 that the MILD condition was easier to establish with CO_2 dilution than with N_2 dilution for the same oxygen molar concentration X_{O2} and equivalence ratio Φ.

A further conclusion about MILD establishment with CO_2/N_2 dilutions can be drawn from the data shown Fig. 14.9. The self-ignition temperatures of CO_2 and N_2 diluted mixtures both decreased as X_{O2} increased from 0.05 to 0.2, and the self-ignition temperatures of N_2 dilution were slightly higher than that of CO_2 dilution, as shown in Fig. 14.9A. The preceding discussion concludes that under the same X_{O2} and Φ, MILD combustion may occur in CO_2 while not in N_2 and as X_{O2} decreases. Therefore, for the normalized $T_{in} - \Delta T$ curve of the $CH_4/O_2/CO_2$ mixture with $X_{O2} = 0.2$, one can find the curve of the $CH_4/O_2/N_2$ mixture with lower X_{O2} ($X_{O2} = 0.124$ when $\Phi = 0.8$) that intersects the former curve at the "MILD combustion occurrence" point, as presented in Fig. 14.9B. Then

$$\begin{aligned} \frac{T_{MILD}^{CO_2}\left(X_{O_2} = 0.2\right)}{T_{si}^{CO_2}\left(X_{O_2} = 0.2\right)} &= \tilde{T}_{in,MILD}^{CO_2}\left(X_{O_2} = 0.2\right) \\ &= \tilde{T}_{in,MILD}^{N_2}\left(X_{O_2} = 0.124\right) = \frac{T_{MILD}^{N_2}\left(X_{O_2} = 0.124\right)}{T_{si}^{N_2}\left(X_{O_2} = 0.124\right)}, \end{aligned} \tag{14.4}$$

and from Fig. 14.6A, we can obtain $T_{si}^{CO_2}\left(X_{O_2} = 0.2\right) < T_{si}^{CO_2}\left(X_{O_2} = 0.124\right)$. Thus,

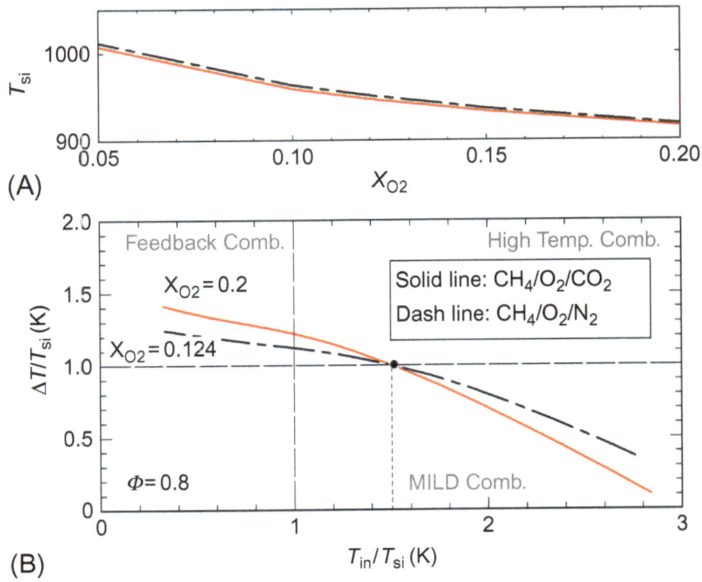

FIG. 14.9 MILD state occurs at higher oxygen concentration in CO_2 dilution than in N_2 dilution: (A) T_{si} versus X_{O2} with $\Phi=0.8$ (*dash line* for N_2 dilution and *solid line* for CO_2 dilution) and (B) normalized $T_{in}-\Delta T$ curves for $CH_4/O_2/N_2$ and $CH_4/O_2/CO_2$ mixtures with $\Phi=0.8$.

$$T_{MILD}^{CO_2}\left(X_{O_2}=0.2\right)\leq T_{MILD}^{N_2}\left(X_{O_2}=0.124\right)$$
$$\frac{T_{si}^{CO_2}\left(X_{O_2}=0.2\right)}{T_{si}^{N_2}\left(X_{O_2}=0.124\right)}<T_{MILD}^{N_2}\left(X_{O_2}=0.124\right). \tag{14.5}$$

Similarly, it can be concluded from Eq. (14.5) that with CO_2 dilution, MILD conditions can be established at a much higher oxygen concentration than with N_2 dilution.

To summarize, based on the definition of MILD combustion [7], in this section, we numerically demonstrated that MILD combustion is more likely to occur with CO_2 dilution than with N_2 dilution under the same conditions (X_{O2} and Φ).

14.4 GENERAL SUGGESTION FOR BURNER DESIGN OF MILD OXY COMBUSTION

Fig. 14.10 shows the configurations of MILD combustion burners used in the previous investigations. The FLOX burner in Fig. 14.10A, a commercial MILD burner by WS GmbH, consists of six air nozzles and a central fuel jet. The six air nozzles supply the preheated air (forward flow), and the other six nozzles exhaust the flue gases (reverse flow) [9]. Fig. 14.10B shows a burner with the configuration of a fuel jet and symmetrical oxidant jets. The MILD oxy

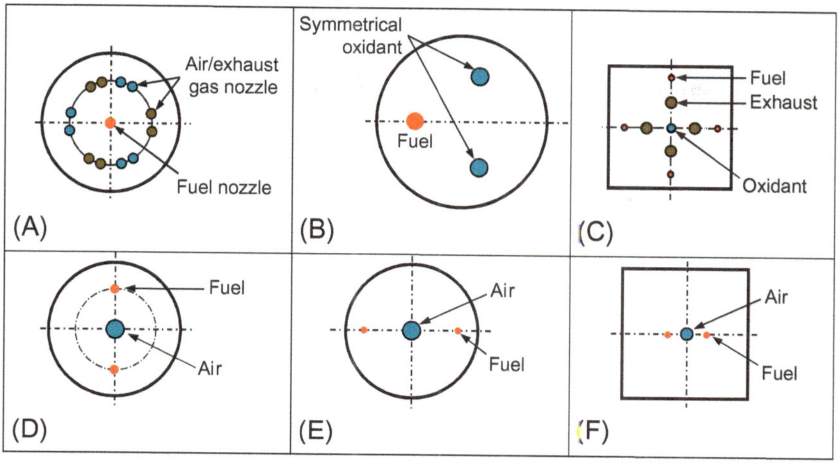

FIG. 14.10 MILD burner configurations: (A) FLOX burner [9], (B) symmetrical jet burner [35], (C) Adelaide combustor [19], (D) RWTH Aachen combustor [25], (E) CORIA-CNRS combustor [56], and (F) IFRF combustor [57].

combustion of propane and pure oxygen was achieved using this configured burner with high momentum oxidant jets [35]. The burner shown in Fig. 14.10C consists of a single central tube located on the axis of the furnace, four side fuel nozzles, and four exhaust gas ports arranged symmetrically in a ring pattern on the same wall [19]. In this arrangement, the incoming air stream was separated from the fuel streams by outgoing exhaust gases, ensuring that reactants were diluted with combustion products before MILD combustion could occur. Other MILD combustion investigators (e.g., Stadler et al. [25] at RWTH Aachen (Fig. 14.10D), CORIA-CNRS (Fig. 14.10E) [56], the IFRF (Fig. 14.10F) [57], Mancini et al. [58], NFK-Japan [8], and Mi et al. [31]) also used similar burner configurations to prompt the occurrence MILD combustion to occur.

From the design experience of the MILD burner of previous investigations, it is found that they all used straight pipe nozzles to discharge reactants into the furnace with high jet momentum. By this approach, both the internal recirculation and mixing are increased, thus MILD combustion is expected to take place. This can be treated as the general design principle for MILD burners.

14.5 MILD OXY COMBUSTION OF PULVERIZED COAL

14.5.1 Experimental Details

The experiments were carried out in a total of 12.5 m a pilot-scale plant with a capacity of 0.3 MW at the Chinese State Key Laboratory of Coal Combustion, Huazhong University of Science and Technology. Fig. 14.11 shows the schematic diagram of the pilot-scale plant and the pipe system. There were four parts to this plant: the combustion furnace, the flue gas purification system, the

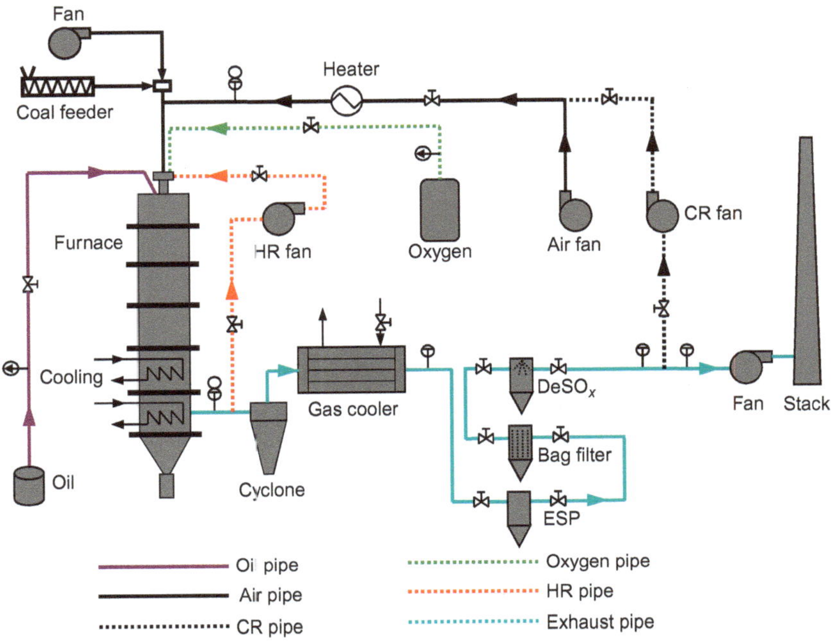

FIG. 14.11 Schematic diagram of the MILD oxy combustion pilot-scale plant and the pipe system. *HR fan*, hot recirculation fan; *CR fan*, coal recirculation fan. *(Reproduced with permission from Li P, Wang F, Tu Y, Mei Z, Zhang J, Zheng Y, et al. MILD oxy-combustion of light fuel oil and pulverized coal in a pilot-scale furnace. Energy Fuel 2014;28(2):1524–35. Copyright 2014 American Chemical Society.)*

supply system, and the instrument control system. The combustion took place in a vertically down-fired furnace (Fig. 14.12). The inner length and diameter of the cylindrical furnace chamber were 6000 and 600 mm, respectively. The furnace had a quartz glass window (Fig. 14.12) through which the reaction zone could be observed. Cooling loops were used to retract heat. A cyclone, a bag filter, an electrostatic precipitator, and an activated CaO desulfurization tower were in the flue gas purification part of the plant. Pure oxygen was stored in a Dewar bottle and was delivered into the furnace by the pressure of the bottle. When air was used as an oxidant, blowers were used to pump the primary air and secondary air into the furnace. Under the oxy-fuel condition, the pulverized coal was delivered into the furnace by the hot exhaust gas recycled without removing H_2O (wet recycle).

For the instrument control system, there was a continuous emission-monitoring system to measure the exhaust gas composition (i.e., O_2, CO, and NO_x) at the furnace exit. The flow rate of the combustion air was controlled by manual valves. S-type thermocouples were used to measure the time-averaged in-furnace and exhaust temperatures. A sampling probe was developed for the in-furnace sampling during combustion. The probe consisted of three stainless

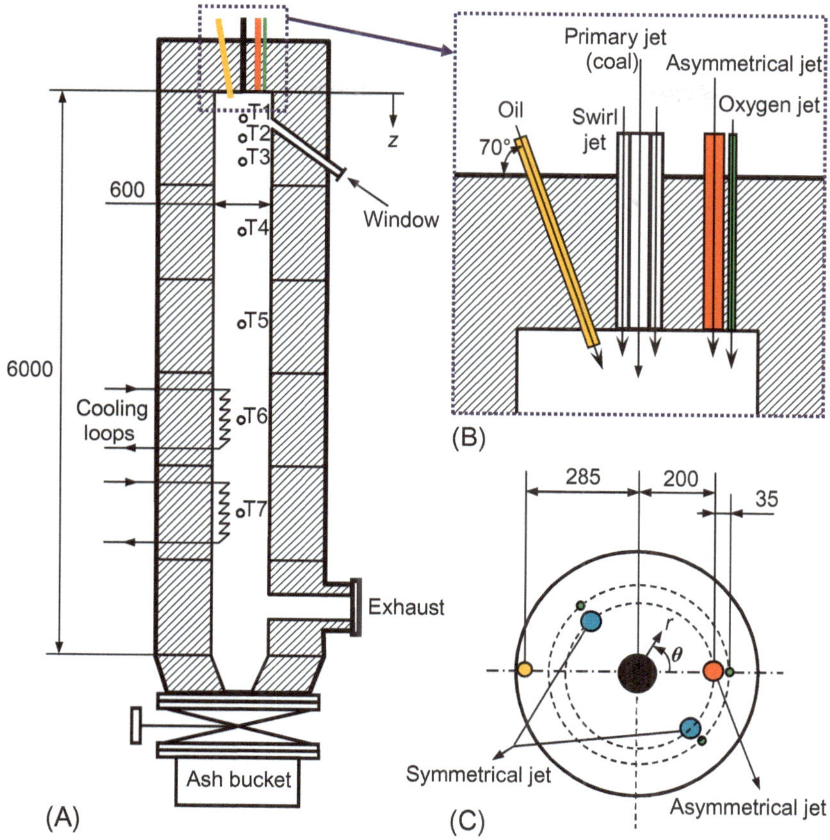

FIG. 14.12 Schematic diagram of the MILD oxy combustion furnace and burner system: (A) the furnace, (B) profile view of the burner, and (C) top view of the furnace top. *(Reproduced with permission from Li P, Wang F, Tu Y, Mei Z, Zhang J, Zheng Y, et al. MILD oxy-combustion of light fuel oil and pulverized coal in a pilot-scale furnace. Energy Fuel 2014;28(2):1524–35. Copyright 2014 American Chemical Society.)*

steel concentric pipes, enabling cold water to circulate in the pipes to rapidly quench hot combustion gases. Fig. 14.12 and Table 14.2 show the measurement locations along the furnace centerline. The concentrations of O_2, CO, and NO_x in the flue gas were measured by a KM950 flue gas analyzer point by point. Although uncertainties of the mean temperatures measured by thermocouples were about ±5 K, the analyzer accuracies were estimated to be $[O_2]=\pm0.2$ vol%, $[CO]=\pm5\%$ of the measured value, and $[NO_x]=\pm5\%$ of the measured value. Total NO_x emission ($NO+NO_2$) concentrations were reported by volume on a dry basis and corrected to 3% O_2 concentration. By correcting to a specific O_2 level, true comparisons of emissions levels could be made because the effect of various degrees of dilutions were removed, while still retaining a familiar mole fraction-like variable.

TABLE 14.2 Measurement Locations Along the Furnace Centerline

Measurement Point	1	2	3	4	5	6	7
z (mm)	180	470	840	1500	2500	3500	4500

According to the discussion on the burner design for MILD combustion in Section 14.4, the direct injection concept was adopted in the present MILD burner design. Two burners of the single asymmetrical jet configuration (MILD Burner I) and the dual symmetrical jets configuration (MILD Burner II) were used as the present MILD burner (Figs. 14.13 and 14.12C). At the same oxidizer flow rate and similar diameters, the single straight pipe could cause higher jet momentum than dual straight pipes, which might result in higher internal recirculation. The MILD oxy combustion characteristics of these two burners were tested and compared. The swirl burner shown in Fig. 14.13C (Swirl Burner III) was originally designed as a conventional oxy-fuel burner and operated under a conventional flame condition as a reference. It had an annular inlet through which the coal and transport stream were injected. The swirl secondary oxidizer stream was injected through the annulus opening surrounding the coal inlet. The swirl number was 0.73, which was defined as a nondimensional number representing the ratio of the angular momentum flux and the axial momentum flux [59,60]. The burner throat opened with a half angle of 15 degrees.

Six cases were tested in the experiment (Table 14.3). The pulverized coal was carried by the primary stream and introduced into the furnace through the central annular pipe (the red tube in Fig. 14.13C). During the coal air combustion (Cases 1–3), the small oxygen nozzles were not used (see Fig. 14.12C and

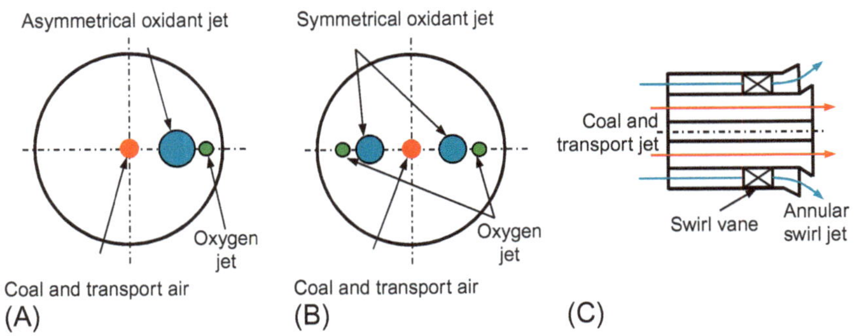

FIG. 14.13 Burner configurations of coal combustion experiments: (A) MILD Burner I: asymmetrical jet configuration, (B) MILD Burner II: symmetrical jet configuration, and (C) reference Swirl Burner III: swirl jet configuration.

TABLE 14.3 Test Conditions of MILD Combustion of Pulverized Coal at 300 kW

Case	Burner Configuration	Oxidant	Capacity (kW)	Condition
1	MILD Burner I	Air	300	MILD
2	MILD Burner II	Air	300	MILD
3	Swirl Burner III	Air	300	Flame
4	MILD Burner I	O_2 + flue gas (oxy-fuel)	300	MILD
5	MILD Burner II	O_2 + flue gas (oxy-fuel)	300	MILD
6	Swirl Burner III	O_2 + flue gas (oxy-fuel)	300	Flame

Fig. 14.13). For Case 6, the oxygen was mixed with the hot exhaust gas recirculated as the oxidant and injected as the swirl jet. For the coal oxy combustion of Cases 4–5, the hot exhaust gas recirculated was injected via the symmetrical or asymmetrical nozzle, while the oxygen was separately supplied through the small oxygen nozzle beside it (Fig. 14.12C and Fig. 14.13).

The primary and secondary oxidizer streams were preheated to 450 and 400 K with electrical heaters, respectively. The diameter of the symmetrical and the asymmetrical nozzle were 26 and 37 mm, respectively (Figs. 14.12C and 14.13). The diameter of the small oxygen nozzle was 13 mm (Fig. 14.13). The velocity of the oxygen jet could be as high as 80 m/s. The mass flow rate of the coal was 49.5 kg/h. The burner operated at the capacity of 300 kW. The primary oxidizer stream accounted for 20% of the total oxidizer volume. For the oxy combustion, the recycle ratio of the hot exhaust gas was ~60%. The recycling ratio was defined as the ratio of the mass flow rate of the exhaust gas being recycled and the total mass flow rate of the exhaust gas.

It is challenging to establish MILD combustion by using pulverized coal with low volatiles because such coal is harder to ignite and burn out than brown coal. The IFRF [61] and Stadler et al. [25] used brown coal with high volatiles (>35%) in their coal MILD experiments. To examine the applicability of MILD combustion with low volatile coals, one type of bituminous coal with high ash and low volatility, obtained from the Qingshan thermal power plant of China, was used in the present experiment. The proximate and ultimate analyses of the bituminous coal are given in Table 14.4.

It is also hard to fire pulverized coal with large particles. In the IFRF experiment of coal MILD combustion [61], the coal was milled to a particle size distribution of 80% < 90 μm. Stadler et al. [25] even milled the coal to a particle size distribution of 90% < 48 μm. To test the pulverized coal being prepared originally for power plants, the coal in the present experiment was obtained directly

TABLE 14.4 Proximate and Ultimate Analysis of Bituminous Coal (Air-Dried)

Proximate Analysis (wt%)				Ultimate Analysis (wt%)					
Moisture	Volatiles	Fixed carbon	Ash	C	H	O	N	S	LCV
1.15	18.74	47.53	32.58	84.60	4.30	8.58	1.45	1.07	21,861 (kJ/kg)

from the Qingshan thermal power plant of China without further milling. The present experiments used the pulverized coal with a mean size of ~110 μm, which was larger than that used in Refs. [25,61].

14.5.2 MILD Burners and Flow Patterns

To investigate the flow pattern and internal gas recirculation caused by the three burner configurations (i.e., Burners I, II, and III), numerical simulations were carried out for nonreacting conditions of the coal air combustion (Cases 1–3). Only the primary and secondary air streams were modeled to obtain the flow field. The Fluent software's code [62] was used for the steady state, implicit, finite volume-based simulation. The pressure velocity coupling was the SIMPLE algorithm. A full hexahedral grid (about 2,000,000 hexahedral cells) was adopted to minimize the grid size, and appropriate refinement of the grid was performed in the regions with higher gradients. The Realizable k-ε model with standard wall functions was used to model turbulence. The Realizable k-ε model has been shown to provide a more accurate prediction of flow evolution and flow spreading of round jets [63]. The boundary conditions were velocity inlet, pressure outlet, and temperature wall boundary. In order to improve the accuracy of the simulations, a second-order scheme was employed for pressure and a second-order upwind scheme for momentum, turbulent kinetic energy, turbulent dissipation, species transport, and energy. The solution convergence was obtained when (a) the residuals were less than 10^{-6} for the energy and 10^{-5} for all the other variables and (b) the variations of the outlet temperature and velocity were allowed to be within 1.0 K and 0.1 m/s, respectively.

Fig. 14.14 shows the contour distributions of the in-furnace nonreacting velocity magnitude and flow pathlines, numerically simulated, for Burners I, II, and III. In the cases of MILD Burners I and II, there was a large-scale recirculation zone downstream from the burner. Also, the recirculation zone for MILD Burner I was significantly larger than that for MILD Burner II. However, for Swirl Burner III, the central primary jet was surrounded by the secondary swirl jet, and the internal recirculation was relatively small. The preceding observations from the numerical simulations suggested that the straightly discharging side jets for the secondary oxidant from MILD Burners I and II had a higher

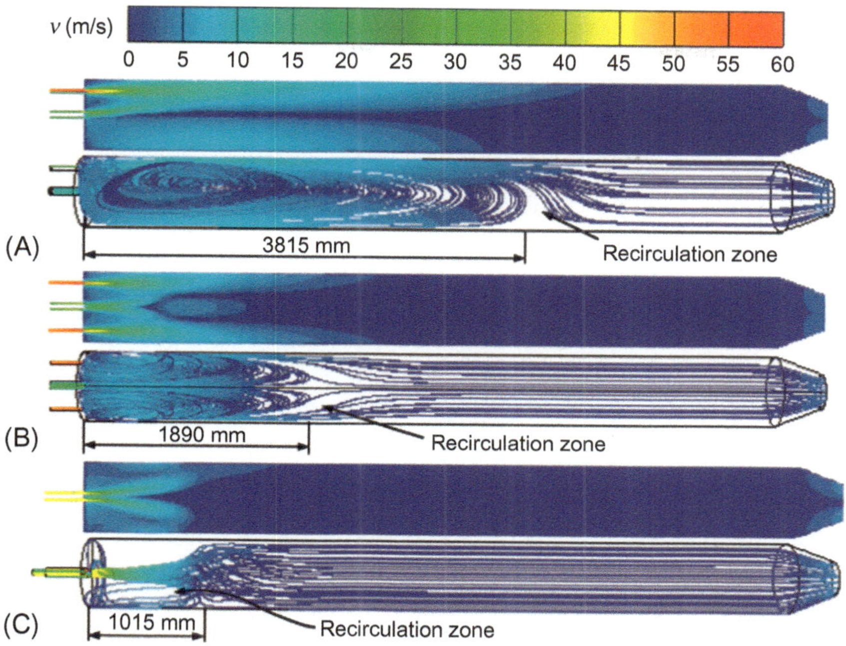

FIG. 14.14 Contour distributions of the nonreacting velocity magnitude and flow pathlines, numerically simulated, for configurations of (A) the asymmetrical jet, MILD Burner I, (B) the symmetrical jet, MILD Burner II, and (C) the swirl jet, Swirl Burner III.

injection axial momentum, thus leading to a stronger recirculation, than the central swirl jet configuration (Swirl Burner III). Therefore the side direct injection burner configuration should be a good choice to achieve MILD oxy combustion of burning solid fuels.

14.5.3 Establishment of the Coal MILD Combustion and the Appearances

To warm up the furnace, Swirl Burner III was adopted with air and light oil, and the furnace operated in conventional mode. It took ~2h to warm the furnace. When the entire furnace temperature exceeded 800K (T7 in Fig. 14.12), Swirl Burner III was switched to MILD Burner I; then MILD combustion of light oil was achieved (Fig. 14.15E) because of the greater recirculation caused by MILD Burner I (Fig. 14.14). The appearance of the combustion changed from a big yellow flame (conventional oil combustion) to no big flame but a dark flamelet (see Fig. 14.15E). Moreover, as the burner switched from Swirl Burner III to MILD Burner I, the furnace temperature and the NO emissions rapidly decreased. When the exhaust gas was externally recirculated, mixed with the oxygen and discharge into the furnace as the oxidant mixture, the MILD oxy-fuel combustion of light oil could be reached (see Fig. 14.15F). For the present

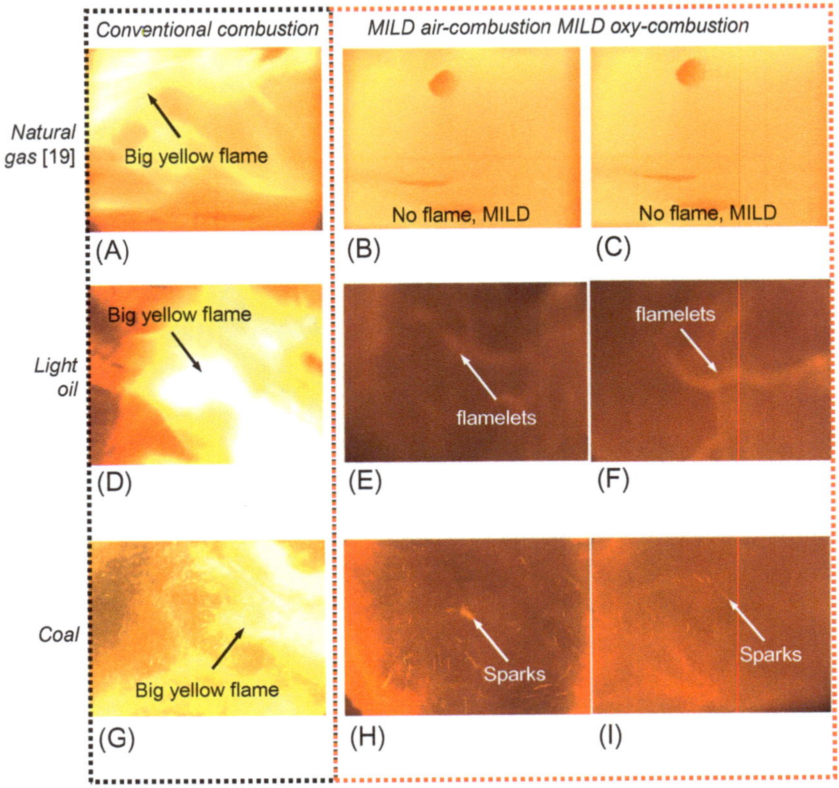

FIG. 14.15 Images showing the in-furnace appearance of the combustion of natural gas [19], light oil, and pulverized coal: (A) conventional flame, (B) MILD air-combustion, (C) MILD oxy-combustion, (D) swirl air jet configuration, Burner III, (E) asymmetrical jet configuration, Burner I, (F) asymmetrical jet configuration, Burner II, (G) swirl jet configuration oxy-fuel flame, Case 3, (H) asymmetrical jet configuration, Case 1, and (I) asymmetrical jet configuration, Case 4.

experiment of burning light oil, the appearance of MILD oxy combustion was similar to that of MILD air combustion (Fig. 14.15E–F); no big flame existed, but a flamelet was observed. For burning coal, the furnace was fully warmed up, and Swirl Burner III was used for conventional air and oxy combustion. The coal MILD air or oxy combustion could be established when the burner was switched from Swirl Burner III to MILD Burner I or II, because of the stronger recirculation caused by MILD Burners I and II (Fig. 14.14). Fig. 14.15G–I shows the photographs of the coal combustion operated at the conventional and MILD conditions. Obviously, in the conventional condition, the visible yellow flame vividly existed. However, under coal MILD combustion, the large yellow flame disappeared, although "ghost" sparks swept rapidly inside the furnace, regardless of air or oxygen with recirculated flue gas as oxidant.

For MILD combustion of light oil and coal in the present experiment, true "flameless" combustion was not observed, and flamelets or sparks always existed, perhaps because dark flamelets are associated with burning droplets and sparks are related to firing char particles. In this case, the big yellow flame appearing in the conventional case became invisible. Therefore the volatile combustible gas burned invisibly while the droplets or char particles fired under the volatile MILD condition and thus remained visible. Likewise, Smart and Riley [64] and Weber et al. [61] found that sparks existed under coal MILD combustion. Dally et al. [65] also observed that glowing red particles fired under sawdust MILD combustion. These observations differ significantly from MILD combustion of gaseous fuel. Fig. 14.15A–C shows the photographs of the conventional and MILD combustion of natural gas obtained from our previous work [19]. No flame was visible under MILD combustion of natural gas. The differences in appearance led to different qualitative criteria for MILD combustion of firing liquid and solid fuels. For gaseous fuels, MILD combustion was qualitatively defined as combustion where no visible flame front occurred [8–10]. For MILD combustion of liquid or solid fuels, the burning of volatiles did not yield a yellow flame or was invisible, while visible dark flamelets or sparks may exist.

Fundamentally, the differences in the appearance of MILD combustion of gaseous, liquid, and solid fuels are caused by their different ignition delay and burnout times. For instance, both the ignition delay time and the burnout time of natural gas are approximately several milliseconds [6]. However, the combustion of pulverized coal is relatively complicated and mainly proceeds in four stages. First, the pulverized coal is preheated. Second, its volatile matter is emitted and ignited. Next, the char burns after reaching its firing temperature. Finally, the char burns out. The heating and devolatilization of the volatiles finish mainly in the first and second stages, which takes several milliseconds. Nevertheless, it takes several seconds to completely burn out the char particles [6]. The local residence time of coal particles in a furnace is typically longer than several milliseconds but less than several seconds. Therefore, for a limited residence time, the relatively long firing time of char particles leads to visible char particles, while the combustion of gaseous volatiles still occurs in flameless conditions without a big yellow flame.

14.5.4 Temperature and Species Distributions

Fig. 14.16 shows the temperature distributions along the furnace centerline for Cases 1–6. Note that the initial oxygen concentration of coal oxy combustion (Cases 4–6) was about 40 vol% and that the hot flue gas was externally recycled without removing H_2O (i.e., wet recycle). It was demonstrated that for all the cases, as the downstream distance (z) from the burner increased, initially the centerline temperature increased before reaching its maximum and then decreased. However, a clear distinction existed between the results measured in conventional and MILD conditions. For instance, the maximal temperature of conventional combustion (Cases 3 and 6) was higher than that

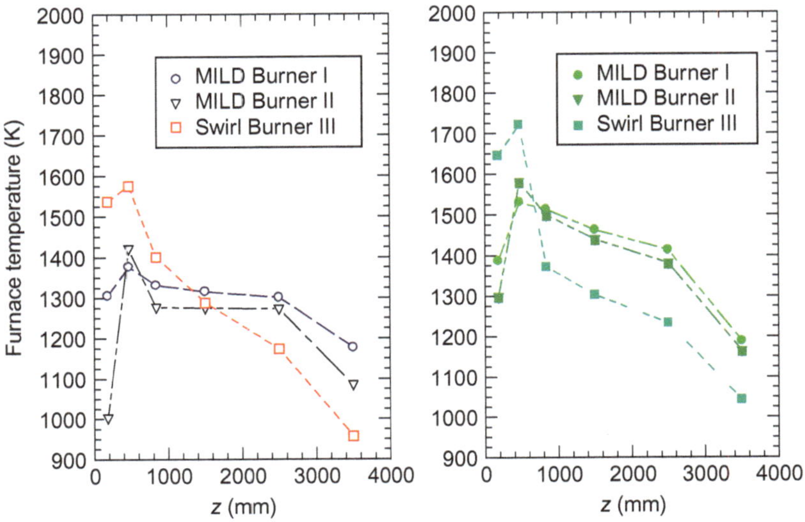

FIG. 14.16 Temperature distribution along the furnace centerline under different conditions of coal combustion (Cases 1–6).

of MILD combustion taken under both air and oxy-fuel conditions. Under air-fuel conditions (Cases 1–3), the maximal temperature for the conventional case was ~1580K, while those for the two MILD cases were close to 1400K; on the other hand, the corresponding maximal temperatures were higher (≈1721K versus 1550K) under oxy-fuel conditions (Cases 4–6). As a result, the temperature drop at $z = 3500$ mm was quite large (>550–675K) for the conventional cases and relatively small (<320–330K) for the MILD cases. Also, at $z > 1500$ mm, the temperature for the MILD cases exceeded that for the conventional cases. It was thus deduced that chemical reactions occurred more uniformly in the furnace in MILD combustion than in conventional combustion. Furthermore, as shown in Fig. 14.16, the maximum temperatures for MILD combustion from MILD Burner I were lower than those from MILD Burner II, because MILD Burner I could produce a higher internal recirculation (Fig. 14.14).

In the present experiment using the identical burner but different oxidant mixtures, the centerline maximum temperature obtained for oxy combustion was ~150K greater than for air combustion (Fig. 14.16). The reason is that the initial oxygen concentration in oxy combustion (40vol%) was much higher than in air combustion (21 vol%). Note that to obtain a nearly identical furnace temperature under both oxy-fuel and air-fuel conditions, the initial oxygen must be higher in the former case than in the latter case because the specific heat of CO_2 is higher than that of N_2. Andersson et al. [66] and Hjärtstam et al. [67] found that when the initial oxygen was 29vol%, the temperature of lignite coal oxy combustion was 100K higher than that of air combustion. Likewise, Croiset et al. [68] used an initial oxygen of 35 vol% in oxy combustion of Canadian western subbituminous coal and found the temperature to be 80 K higher than that in air combustion. Moreover, if H_2O participates in the recycle process, different conditions will result. That is, the furnace temperatures of wet recycle oxy combustion and dry recycle oxy combustion are different. Wall et al. [69] carried out equilibrium calculations and found that in order to match the adiabatic flame temperature of coal air combustion, the theoretical initial oxygen level (vol%) should be about 28% and 35% for wet and dry recycle coal oxy combustion, respectively. The initial oxygen level, coal type, oxidant mixture content (O_2/N_2 or O_2/CO_2), and combustion condition (conventional or MILD) all affect the furnace temperature. Those aspects should be considered in the design of coal MILD oxy combustion systems.

Fig. 14.17 shows the O_2 distribution along the furnace centerline under different coal combustion conditions (Cases 1–6). The centerline oxygen concentrations were relatively higher because of faster mixing between the primary and secondary streams for both conventional combustion cases (Cases 3 and 6) than for the MILD cases. For example, at $z=840$ mm, the oxygen concentration for conventional cases was >10 vol%, while it was <5 vol% for MILD combustion. As a result, chemical reactions under MILD conditions occurred more slowly, thus the in-furnace temperature was lower.

Fig. 14.18 illustrates the NO distribution along the furnace centerline in Cases 1–6. Clearly, higher maximum temperatures (Fig. 14.16) led to higher NO productions. The high temperatures of the conventional combustion in the

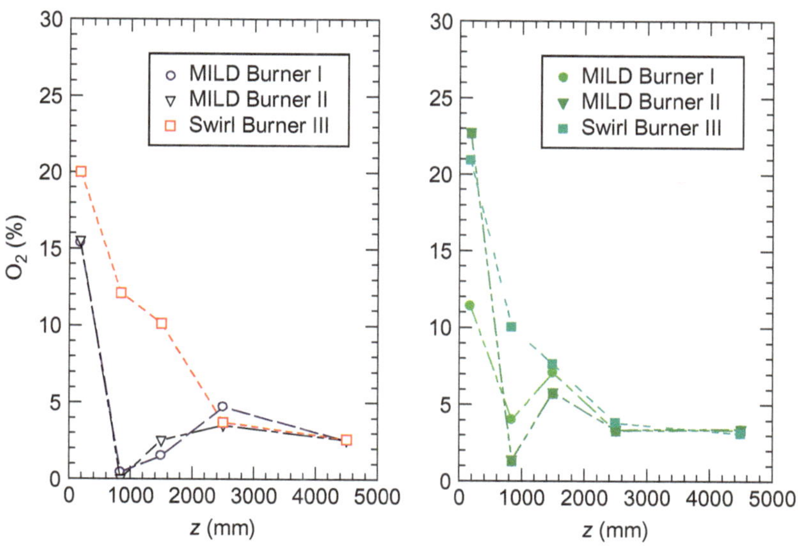

FIG. 14.17 Oxygen distributions along the furnace centerline under different conditions of coal combustion (Cases 1–6).

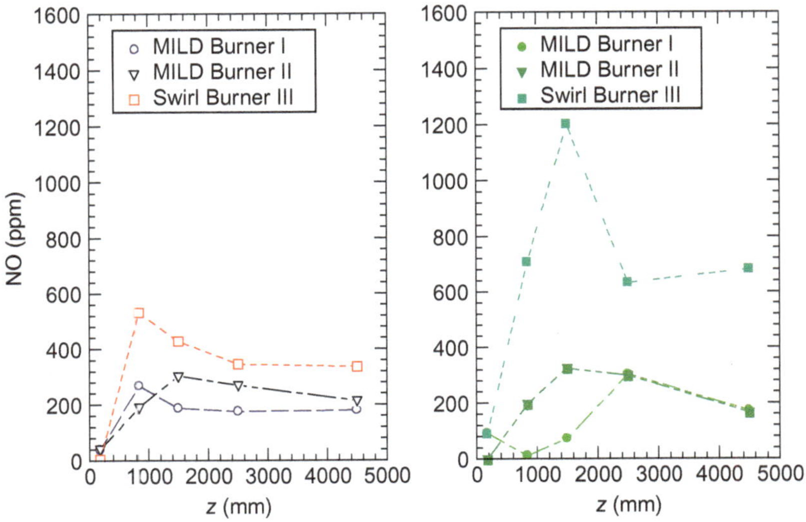

FIG. 14.18 Distributions of NO along the furnace centerline under different conditions of coal combustion (Cases 1–6).

combustion zone produced extremely high NO formations, that is, ~530 ppm for air combustion at $z = 840$ mm and 1200 ppm for oxy combustion at $z = 1500$ mm. In contrast, NO formations from four MILD combustion cases were generally less than 320 ppm. The NO formation was suppressed by the relatively low in-furnace temperature under MILD combustion. Moreover, the centerline maximum temperature occurred at $z = 470$ mm (Fig. 14.16), while the centerline maximum NO production took place at $z = 840$ mm and 1500 mm (Fig. 14.18), respectively, for air combustion and oxy combustion. Namely, the maximum NO formation occurred at a location downstream from that of the maximum temperature because NO always forms at a slow rate.

14.5.5 Exhaust Emissions and Burnout

Fig. 14.19 shows the exhaust NO emissions under different combustion conditions (Cases 1–6). As shown, the conventional oxy combustion (Case 6) yielded the highest NO emission (681 ppm), while both MILD oxy combustion cases (Cases 4 and 5) produced the lowest NO emissions (≈168 ppm). Moreover, the NO emissions from the two MILD oxy combustion cases (Cases 4 and 5) accounted for only 24% of the flame oxy combustion (Case 6), while the NO emissions from the two MILD air combustion cases (Cases 1 and 2) were ~60% of that from the flame air combustion (Case 3). Therefore the NO reduction was relatively stronger in MILD oxy combustion than in MILD air combustion. An analysis and discussion of the NO formation mechanism follows.

For air combustion, it is well known that the fuel-NO route produces more than 80% of the total NO emissions and that the higher reaction temperature

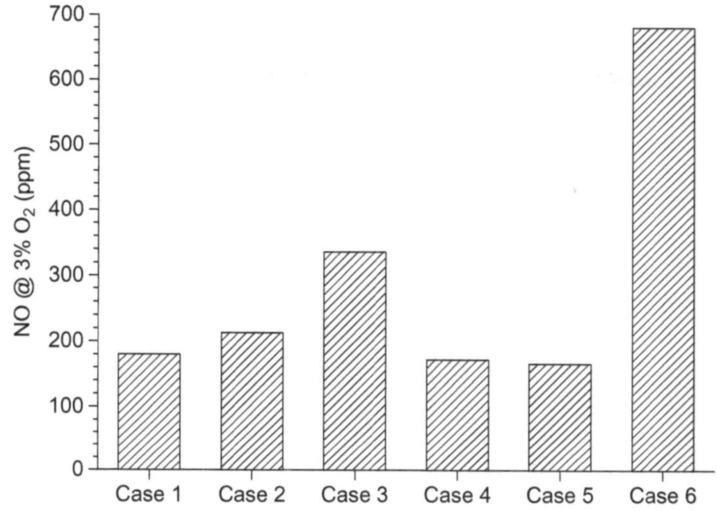

FIG. 14.19 Exhaust NO emissions under different conditions of coal combustion.

results in higher NO production [2,70]. The low temperature of MILD air combustion (Cases 1 and 2) suppressed the fuel-NO and thermal-NO formations; thus the NO emission was ~60% less than in the flame oxy combustion case (Case 3). In addition, the NO emission (213 ppm) of MILD air combustion in the MILD Burner I (Case 1) was 15.5% lower than that (180 ppm) in the MILD Burner II (Case 2). Therefore, for air combustion, the asymmetrical jet burner configuration (MILD Burner I) led to more recirculation (Fig. 14.14) and stronger dilution of reactants by the flue gas, resulting in lower temperatures (Fig. 14.16) and NO emissions, relative to the symmetrical burner configuration (MILD Burner II).

For oxy combustion, the NO emission from MILD oxy combustion (Cases 4 and 5) was only 25% of the NO emission from conventional oxy combustion (Case 6). There were no thermal and prompt NO productions because of the replacement of N_2 with the flue gas mixture (mainly CO_2 and H_2O) [1]. The NO emission was derived mainly from the fuel-NO route. For the present experiment, the relatively low maximum temperature from the two MILD oxy combustion cases (Cases 4 and 5) reduced the fuel-NO production. Moreover, the NO contents from the recycled flue gas could be decreased in the combustion zone by the NO-reburning reaction [1]. Therefore, although the maximal temperature of MILD oxy combustion from MILD Burner II (Case 5) was nearly identical to that of conventional air combustion (Case 3), the NO emission of the former was significantly lower than the latter by 51%.

In addition, for MILD oxy combustion, the NO emission caused by MILD Burner I (Case 4) differed only slightly from that caused by MILD Burner II (Case 5), while the difference increased considerably under MILD air combustion conditions. That is, it is well known that fuel-NO production increases as either the local temperature or the local oxygen concentration in the reaction zone increases. Although Fig. 14.17 suggests that the local oxygen concentration in the reaction zone in MILD Burner I (asymmetrical configuration, Case 4) was higher than that in MILD Burner II (symmetrical configuration, Case 5), the local temperature in MILD Burner I was lower than that in MILD Burner II (see Fig. 14.16). Therefore it does make sense that the NO emission of the former was only slightly higher than the latter.

Fig. 14.20 shows the exhaust coal burnout and CO emissions under different conditions of coal combustion (Cases 1–6). As demonstrated, for the same oxidant (air or O_2/CO_2), the burnout of char was generally poorer for MILD combustion than for conventional combustion. A careful look at the furnace temperature (Fig. 14.16), the oxygen concentration (Fig. 14.17), and the coal burnout (Fig. 14.20) shows that the higher the peak temperature and the oxygen concentration in the main reaction zone ($z < 1500$ mm), the higher the char burnout. A similar observation was made by Dally et al. [65] for sawdust MILD combustion. The short residence time was thought to be the main reason for the low burnout of the char. In their experiment, Stadler et al. [24] improved the char burnout by increasing the residence time for their coal MILD oxy combustion. Moreover, both the volatile content and particle size influenced the char burnout. In the coal MILD air combustion experiment performed by Weber et al. [61], a high volatile coal (Guasare) was used, and the char was nearly completely burned. Su et al. [71] found that the high volatility of coal improved not only the ignition but also the carbon burnout. Du et al. [72] found that the coal burnout increased by ~10% to 15% (depending on the coal type) when the coal particle size was reduced approximately from 150–250 to 75–150 μm.

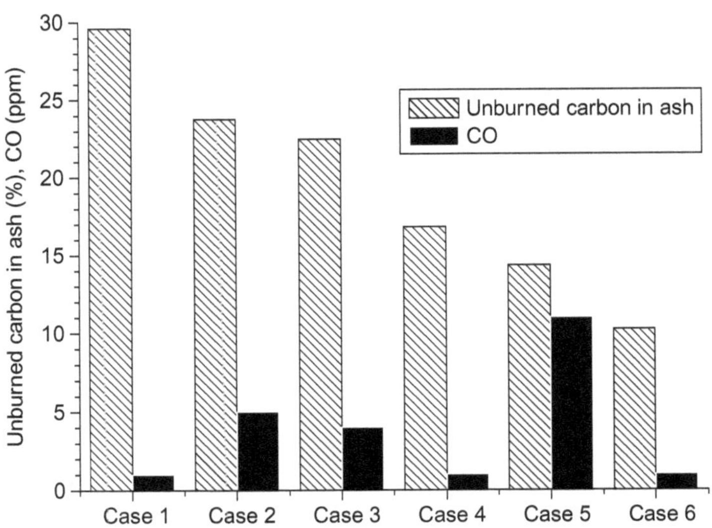

FIG. 14.20 Exhaust coal burnout and CO emissions under different conditions of coal combustion (Cases 1–6).

Therefore the low burnout issue under coal MILD combustion should be resolved by increasing the char particle residence time or by using coal with a higher volatility and smaller particle size.

Fig. 14.20 also demonstrates that exhaust CO emissions from all six cases were extremely low (<11 ppm). The CO was produced and consumed through the reaction of the char and volatile, namely,

$$\text{Char} + O_2 \rightarrow CO, \quad (\text{burnout time : several seconds}), \tag{14.6}$$

$$\text{Char} + H_2O \rightarrow CO + \ldots, \quad (\text{burnout time : several seconds}), \tag{14.7}$$

$$\text{Volatiles} + O_2 \rightarrow CO, \quad (\text{burnout time : several milliseconds}), \tag{14.8}$$

$$CO + O_2 \rightarrow CO_2, \quad (\text{burnout time : several milliseconds}). \tag{14.9}$$

Reactions (14.6), (14.7) took several seconds to burn out, while reactions (14.8), (14.9) could be completed within several milliseconds [6,8]. The extremely low CO emission indicated that volatile gases were fully burned at a limited residence time (more than several milliseconds but less than several seconds), although a high CO_2 concentration for the oxy combustion might have caused reaction (14.9) to occur backward.

14.6 SUMMARY AND CONCLUSIONS

(1) The experimental observations and WSR numerical simulations based on the definition of MILD combustion of Cavaliere and de Joannon [7] confirmed that MILD combustion is more likely to occur under CO_2 than N_2 dilution. In other words, MILD oxy combustion is easier to be established than traditional MILD air combustion under the same initial conditions.

(2) The stability of MILD oxy-combustion is caused by higher values of density, kinematic viscosity, heat capacity, and thermal conductivity of CO_2 than those that occur in a N_2 environment; CO_2 replacement of N_2 also results in a significantly delayed ignition, lower burning rates, and a lower adiabatic equilibrium temperature.

(3) The chemical effect of CO_2 is prominent in ignition delay time, while the discrepancies in adiabatic equilibrium temperature are mainly caused by different physical properties of CO_2 from those of N_2.

(4) Replacing N_2 with CO_2 suppresses Q and the adiabatic equilibrium temperature because of differences in the physical properties of CO_2 and N_2, which promotes the development of MILD oxy combustion.

(5) MILD air and oxy combustion of pulverized coal can be achieved in a pilot-scale furnace without greatly preheating the oxidant. When MILD air combustion or oxy combustion is reached, a quite uniform temperature distribution and very low NO_x and CO emissions are obtained.

(6) The arrangement of straightly ejecting secondary oxidant jets alongside the central primary fuel-oxidant jet leads to an internal large-scale recirculation much greater than does the arrangement of a swirl secondary oxidant jet around the central primary fuel-oxidant jet. Also, the single asymmetrical jet arrangement produces a larger recirculation zone than does the dual symmetrical jet configuration. It is thus understood that burners with an asymmetrical configuration yield lower maximal temperatures in coal MILD combustion than symmetrical arrangements yield. For the present experiment, MILD combustion of burning pulverized coal occurred when a side jet burner was used, while conventional flame combustion took place when a swirl burner was used.

(7) The apparences of firing light oil and pulverized coal under MILD combustion conditions are significantly different from those that occur when burning gaseous fuels. For example, MILD combustion of gaseous fuels occurs invisibly, while sparks may exist when firing light oil or pulverized coal. The tiny sparks result from burning candescent droplets or char particles, which require a relatively long time to burn out. Accordingly, for MILD combustion of solid fuels, the burning of gaseous volatiles, which appears as a large yellow flame in a conventional combustion mode, should be nearly invisible, while sparks from firing char particles are usually visible.

(8) For coal air combustion, both fuel-NO and thermal-NO formations are suppressed in the MILD case. Our experiments found that the NO emission from coal MILD air combustion accounts for about 60% of that from the conventional air combustion. In our experiments on coal oxy combustion, the NO emissions from MILD oxy combustion were only ~25% of those in conventional oxy combustion.

(9) In our experiment, the maximal temperature of MILD oxy combustion from the symmetrical jet burner was nearly identical to that of conventional air combustion from the swirl burner. However, the NO emission of the former was only about half of that in the latter case, because for coal oxy combustion, the NO formation was derived mainly from the fuel-NO route (without thermal and prompt NO). Also, the NO in the recycled flue gas may have been decreased in the combustion zone by the NO-reburning reaction. On the other hand, for air combustion, the NO emission was produced from the fuel, thermal, prompt, and other NO formation routes.

(10) When burning pulverized coal in our experiments, the extremely low CO emission (<11 ppm) indicated that the gaseous volatile was completely burned. However, the burnout of char under MILD combustion was lower than that under conventional combustion. The resolution for this issue is expected to be achieved by increasing the residence time of char particles or by using coal with a higher volatility and smaller particle size.

ACKNOWLEDGMENTS

The authors of this chapter gratefully acknowledge the financial support of grants from the National Natural Science Foundation of China (51406001, 91641122, 51276002, 51396094), the Fundamental Research Funds for the Central Universities, the Specialized Research Fund for the Doctoral Program of Higher Education (No. 20130142130009, 20110001130014). We would also like to thank Professor Bassam Dally for the collaborative research on MILD oxy combustion of gaseous fuels.

REFERENCES

[1] Chen L, Yong SZ, Ghoniem AF. Oxy-fuel combustion of pulverized coal: characterization, fundamentals, stabilization and CFD modeling. Prog Energy Combust Sci 2012;38(2):156–214.
[2] Toftegaard MB, Brix J, Jensen PA, Glarborg P, Jensen AD. Oxy-fuel combustion of solid fuels. Prog Energy Combust Sci 2010;36(5):581–625.
[3] Guo J, Liu Z, Huang X, Zhang T, Luo W, Hu F, et al. Experimental and numerical investigations on oxy-coal combustion in a 35 MW large pilot boiler. Fuel 2017;187:315–27.
[4] Chu S. Carbon capture and sequestration. Science 2009;325(5948):1599.
[5] Wall TF. Combustion processes for carbon capture. Proc Combust Inst 2007;31(1):31–47.
[6] Turns SR. An introduction to combustion: concepts and applications. vol. 287. New York, NY: McGraw-Hill; 1996.
[7] Cavaliere A, de Joannon M. Mild combustion. Prog Energy Combust Sci 2004;30(4):329–66.
[8] Tsuji H, Gupta AK, Hasegawa T, Katsuki M, Kishimoto K, Morita M. High temperature air combustion: from energy conservation to pollution reduction. Boca Paton, FL: CRC Press; 2003.
[9] Wünning JA, Wünning JG. Flameless oxidation to reduce thermal NO-formation. Prog Energy Combust Sci 1997;23(1):81–94.
[10] Katsuki M, Hasegawa T. The science and technology of combustion in highly preheated air. Proc Combust Inst 1998;27(2):3135–46.
[11] Li P, Mi J, Dally BB, Wang F, Wang L, Liu Z, et al. Progress and recent trend in MILD combustion. Sci China Technol Sci 2011;54(2):255–69.
[12] Weber R, Orsino S, Lallemant N, Verlaan AL. Combustion of natural gas with high-temperature air and large quantities of flue gas. Proc Combust Inst 2000;28(1):1315–21.
[13] Weber R, Verlaan AL, Orsino S, Lallemant N. On emerging furnace design methodology that provides substantial energy savings and drastic reductions in CO2, CO and NOx emissions. J Energy Inst 1999;72(492):77–83.
[14] International Flame Research Foundation. http://www.ifrf.net/.
[15] Lallemant N, Breussin F, Weber R, et al. Flame structure, heat transfer and pollutant emissions characteristics of oxy-natural gas flames in the 0.7-1 MW thermal input range. J Inst Energy 2000;73(496):169–82.
[16] Breussin F, Lallemant N, Weber R. Computing of oxy-natural gas flames using both a global combustion scheme and a chemical equilibrium procedure. Combust Sci Technol 2000;160(1):369–97.
[17] Blasiak W, Yang W, Krishnamurthy N, Schéele JV. Flameless oxyfuel combustion for fuel consumption and nitrogen oxides emissions reductions and productivity increase. J Energy Inst 2007;80(1):3–11.
[18] Krishnamurthy N, Paul PJ, Blasiak W. Studies on low-intensity oxy-fuel burner. Proc Combust Inst 2009;32(2):3139–46.

[19] Li P, Dally BB, Mi J, et al. MILD oxy-combustion of gaseous fuels in a laboratory-scale furnace. Combust Flame 2013;160(5):933–46.

[20] Zhang J, Mi J, Li P, Wang F, Dally BB. MILD combustion of methane diluted by CO2 and N2. Energy Fuel 2015;29(7):4576–85.

[21] Tu Y, Su K, Liu H, Chen S, Liu Z, Zheng C. Physical and chemical effects of CO2 addition on CH4/H2 flames on a jet in hot coflow (JHC) burner. Energy Fuel 2016;30:1390–9.

[22] Sabia P, Sorrentino G, Chinnici A, Cavaliere A, Ragucci R. Dynamic behaviors in methane MILD and oxy-fuel combustion. Chemical effect of CO2. Energy Fuel 2015;29(3):1978–86.

[23] Mardani A, Ghomshi AF. Numerical study of oxy-fuel MILD (moderate or intense low-oxygen dilution combustion) combustion for CH4–H2 fuel. Energy 2016;99(15):136–51.

[24] Stadler H, Ristic D, Förster M, Schuster A, Kneer R, Scheffknecht G. NOx-emissions from flameless coal combustion in air, Ar/O2 and CO2/O2. Proc Combust Inst 2009;32(2):3131–8.

[25] Stadler H, Christ D, Habermehl M, Heil P, Kellermann A, Ohliger A, et al. Experimental investigation of NOx emissions in oxycoal combustion. Fuel 2011;90(4):1604–11.

[26] Li P, Wang F, Tu Y, Mei Z, Zhang J, Zheng Y, et al. MILD oxy-combustion of light fuel oil and pulverized coal in a pilot-scale furnace. Energy Fuel 2014;28(2):1524–35.

[27] Tu Y, Liu H, Chen S, Liu Z, Zhao H, Zheng C. Numerical study of combustion characteristics for pulverized coal under oxy-MILD operation. Fuel Process Technol 2015;135:80–90.

[28] Lide D. CRC Handbook of Chemistry and Physics. 90th ed. Boca Raton, FL: CRC Press (Taylor and Francis Group); 2010.

[29] Szegö GG, Dally BB, Nathan GJ. Scaling of NOx emissions from a laboratory-scale mild combustion furnace. Combust Flame 2008;154(1):281–95.

[30] Szegö GG, Dally BB, Nathan GJ. Operational characteristics of a parallel jet MILD combustion burner system. Combust Flame 2009;156(2):429–38.

[31] Mi J, Li P, Dally BB, Craig RA. Importance of initial momentum rate and air-fuel premixing on MILD combustion in a recuperative furnace. Energy Fuel 2009;23(11):5349–56.

[32] Veríssimo AS, Rocha AMA, Costa M. Importance of the inlet air velocity on the establishment of flameless combustion in a laboratory combustor. Exp Thermal Fluid Sci 2013;44:75–81.

[33] Li P, Mi J, Dally BB, Craig RA, Wang F. Premixed moderate or intense low-oxygen dilution (MILD) combustion from a single jet burner in a laboratory-scale furnace. Energy Fuel 2011;25(7):2782–93.

[34] Li P, Mi J. Influence of inlet dilution of reactants on premixed combustion in a recuperative furnace. Flow Turbul Combust 2011;87(4):617–38.

[35] Kumar S, Paul PJ, Mukunda HS. Studies on a new high-intensity low-emission burner. Proc Combust Inst 2002;29(1):1131–7.

[36] Veríssimo AS, Rocha AMA, Costa M. Operational, combustion, and emission characteristics of a small-scale combustor. Energy Fuel 2011;25(6):2469–80.

[37] Curran H, Simmie JM, Dagaut P, Voisin D, Cathonnet M. The ignition and oxidation of allene and propyne: experiments and kinetic modeling. Proc Combust Inst 1996;26(1):613–20.

[38] Donohoe N, Heufer A, Metcalfe WK, Curran HJ, Davis ML, Mathieu O, et al. Ignition delay times, laminar flame speeds, and mechanism validation for natural gas/hydrogen blends at elevated pressures. Combust Flame 2014;161(6):1432–43.

[39] de Joannon M, Cavaliere A, Donnarumma R, Ragucci R. Dependence of autoignition delay on oxygen concentration in mild combustion of high molecular weight paraffin. Proc Combust Inst 2002;29(1):1139–46.

[40] Medwell PR, Kalt PAM, Dally Bassam B. Simultaneous imaging of OH, formaldehyde, and temperature of turbulent nonpremixed jet flames in a heated and diluted coflow. Combust Flame 2007;148(1–2):48–61.

[41] Parente A, Galletti C, Tognotti L. Effect of the combustion model and kinetic mechanism on the MILD combustion in an industrial burner fed with hydrogen enriched fuels. Int J Hydrog Energy 2008;33(24):7553–64.

[42] Petrova MV, Williams FA. A small detailed chemical-kinetic mechanism for hydrocarbon combustion. Combust Flame 2006;144(3):526–44.

[43] Frenklach M, Bornside DE. Shock-initiated ignition in methane-propane mixtures. Combust Flame 1984;56(1):1–27.

[44] Nicolle A, Dagaut P. Occurrence of NO-reburning in MILD combustion evidenced via chemical kinetic modeling. Fuel 2006;85(17–18):2469–78.

[45] van Oijen JA. Direct numerical simulation of autoigniting mixing layers in MILD combustion. Proc Combust Inst 2013;34(1):1163–71.

[46] Guo P, Chen Z. Ignition enhancement of ethylene/air by NOx addition. Chin J Aeronaut 2013;26(4):876–83.

[47] Chen S, Mi J, Liu H, Zheng C. First and second thermodynamic-law analyses of hydrogen-air counter-flow diffusion combustion in various combustion modes. Int J Hydrog Energy 2012;37(6):5234–45.

[48] de Joannon M, Sorrentino G, Cavaliere A. MILD combustion in diffusion-controlled regimes of hot diluted fuel. Combust Flame 2012;159(5):1832–9.

[49] de Joannon M, Sabia P, Sorrentino G, Cavaliere A. Numerical study of mild combustion in hot diluted diffusion ignition (HDDI) regime. Proc Combust Inst 2009;32(2):3147–54.

[50] Li P, Wang F, Mi J, Dally BB, et al. Mechanisms of NO formation in MILD combustion of CH4/H2 fuel blends. Int J Hydrog Energy 2014;39(33):19187–203.

[51] Wang F, Li P, Mei Z, Zhang J, Mi J. Combustion of CH4/O2/N2 in a well stirred reactor. Energy 2014;72(1):242–53.

[52] Wang F, Mi J, Li P. Combustion regimes of a jet diffusion flame in hot coflow. Energy Fuel 2013;27(6):3488–98.

[53] Minamoto Y, Dunstan TD, Swaminathan N, Cant RS. DNS of EGR-type turbulent flame in MILD condition. Proc Combust Inst 2013;34(2):3231–8.

[54] Minamoto Y, Swaminathan N. Scalar gradient behaviour in MILD combustion. Combust Flame 2014;161(4):1063–75.

[55] Li P, Wang F, Mi J, Dally BB, et al. MILD combustion under different premixing patterns and characteristics of the reaction regime. Energy Fuel 2014;28(3):2211–26.

[56] Rottier C, Lacour C, Godard G, Taupin B, Porcheron L, Hauguel R. On the effect of air temperature on mild flameless combustion regime of high temperature furnace. In: Proceedings of the European Combustion meeting, Austria; 2009.

[57] Schaffel N, Mancini M, Schaffel N, et al. Mathematical modeling of MILD combustion of pulverized coal. Combust Flame 2009;156(9):1771–84.

[58] Schaffel-Mancini N, Mancini M, Szlek A, Weber R. Novel conceptual design of a supercritical pulverized coal boiler utilizing high temperature air combustion (HTAC) technology. Energy 2010;35(7):2752–60.

[59] Chigier NA, Beér JM. Velocity and static-pressure distributions in swirling air jets issuing from annular and divergent nozzles. J Basic Eng 1964;86(4):788–96.

[60] Sheen HJ, Chen WJ, Jeng SY, Huang TL. Correlation of swirl number for a radial-type swirl generator. Exp Thermal Fluid Sci 1996;12(4):444–51.

[61] Weber R, Smart JP, vd Kamp W. On the (MILD) combustion of gaseous, liquid, and solid fuels in high temperature preheated air. Proc Combust Inst 2005;30(2):2623–9.

[62] Fluent 6.3. User's Guide. Lebanon: FLUENT Inc; 2006.

[63] Shih T, Liou WW, Shabbir A, Yang Z, Zhu J. A new k-ϵ eddy viscosity model for high reynolds number turbulent flows. Comput Fluids 1995;24(3):227–38.

[64] Smart JP, Riley GS. Combustion of coal in a flameless oxidation environment under oxyfuel firing conditions: the reality. J Energy Inst 2012;85(3):131–4.
[65] Dally BB, Shim SH, Craig RA, Ashman PJ, Szegö GG. On the burning of sawdust in a MILD combustion furnace. Energy Fuel 2010;24(6):3462–70.
[66] Andersson K, Johansson R, Hjärtstam S, Johnsson F, Leckner B. Radiation intensity of lignite-fired oxy-fuel flames. Exp Thermal Fluid Sci 2008;33(1):67–76.
[67] Hjärtstam S, Andersson K, Johnsson F, Leckner B. Combustion characteristics of lignite-fired oxy-fuel flames. Fuel 2009;88(11):2216–24.
[68] Croiset E, Thambimuthu K, Palmer A. Coal combustion in O2/CO2 mixtures compared with air. Can J Chem Eng 2000;78(2):402–7.
[69] Wall T, Liu Y, Spero C, Elliott L, et al. An overview on oxyfuel coal combustion-state of the art research and technology development. Chem Eng Res Des 2009;87(8):1003–16.
[70] Glarborg P, Jensen AD, Johnsson JE. Fuel nitrogen conversion in solid fuel fired systems. Prog Energy Combust Sci 2003;29(2):89–113.
[71] Su S, Pohl JH, Holcombe D, Hart JA. Techniques to determine ignition, flame stability and burnout of blended coals in p.f. power station boilers. Prog Energy Combust Sci 2001;27(1):75–98.
[72] Du S, Chen W, Lucas JA. Pulverized coal burnout in blast furnace simulated by a drop tube furnace. Energy 2010;35(2):576–81.

Chapter 15

Oxy-Steam Combustion

Lei Cai, Chun Zou
Huazhong University of Science and Technology, Wuhan, China

15.1 INTRODUCTION

Oxy-fuel combustion is regarded as a promising technology for CO_2 abatement in coal-fired power generation because it is feasible to produce a high CO_2 concentration in the exhaust gas (>90% by volume), which produces a flue gas that is almost sequestration-ready using existing technologies [1–5]. This factor reduces the risks inherent during the implementation of new technologies. Therefore oxy-fuel combustion has attracted considerable attention in recent years [6–10]. A current example of oxy-fuel combustion is O_2/CO_2 recycled combustion.

Oxy-steam combustion was proposed by Carlos [11] and Seepana and Jayanti [12]. The main difference between this combustion and O_2/CO_2 recycled combustion is that steam, rather than CO_2, is used to moderate the high temperature produced by fuel combustion with oxygen. In oxy-steam combustion, the recycle system is not mandatory. Oxy-steam combustion has the following advantages over O_2/CO_2 recycled combustion: (1) because of the absence of recycled flue gas, the overall system is simple, compact, and easy to start up and shut down; (2) the sizes of the major and auxiliary equipment of the system are smaller than those of the O_2/CO_2 recycled combustion system; (3) the formation of NO_x and SO_x in the boiler can be reduced owing to the introduction of steam; and (4) the pumping costs associated with recycling are relatively low because the transmission medium is water and not flue gas [11,12]. Therefore oxy-steam combustion technology is regarded as the next-generation oxy-fuel combustion technology.

The ignition of pulverized coal is an important combustion characteristic and is crucial to the design of burners and furnaces. For this reason, numerous studies on the ignition of coal in air took place before 1998 [13–18]. The ignition theory points out that three ignition mechanisms occur during the ignition process of coal in air: [13] homogeneous, heterogeneous, and hetero-homogeneous. It is widely accepted that particle diameter, heating rate, oxygen mole fraction, coal type, and ambient gas temperature significantly affect the ignition mechanism [13–15].

Oxy-fuel Combustion. http://dx.doi.org/10.1016/B978-0-12-812145-0.00015-3

Because oxy-fuel combustion has attracted growing interest, many researchers have recently studied the effects of a combustion atmosphere on the ignition of coal. For O_2/CO_2 recycled combustion, because the physical properties, such as volumetric thermal capacity, of CO_2 are different from those of N_2, much of this research has focused on the ignition of coal in O_2/CO_2 mixtures [7–10,19–21]. Most of these studies were performed using a drop-tube furnace (DTF) or an entrained flow reactor (EFR) by pyrometer and/or a high-speed camera. Khatami et al. [7] claimed that in the case of active gas flow conditions, heterogeneous ignition occurred in O_2/CO_2 atmospheres and homogeneous ignition occurred in O_2/N_2 atmospheres. Bejarano and Levendis [8] observed that, in contrast to air combustion, there can be a delay in ignition in oxy-fuel combustion with an O_2 concentration of 21% because the thermal capacity (ρc_p) of CO_2 is greater than that of N_2. Molina and Shaddix [9,10] found that a high concentration of CO_2 had no evident effect on the time for volatile combustion from ignition to burnout under a homogeneous gas-phase ignition regime. However, this high concentration of CO_2 slightly impacted the volatile cloud combustion rate because of the slightly lower diffusivity of volatiles in CO_2 than in N_2. Liu et al. [19] found that the ignition time strongly depended on the furnace gas temperature and coal particle size. Zhang et al. [20] claimed that the volatiles that accumulated on the char surface were mainly converted to CO and coal ignition in ~30%O_2/~70%CO_2, essentially matching that in air. They observed the homogeneous ignition sequence using a high-speed camera. Jovanovic et al. [21] experimentally and numerically studied the ignition of coal and thought that a devolatilization model was crucial for predicting the ignition position of coal in O_2/N_2 and O_2/CO_2 atmospheres.

Because the physical and chemical properties of steam are very different from those of CO_2 or N_2, especially its reactivity, the ignition characteristics of coal in oxy-steam combustion are expected to be different from those in conventional combustion or O_2/CO_2 recycled combustion. Moreover, to fully understand the results of pilot-scale tests of oxy-steam combustion and to accurately predict scale-up performance through CFD modeling, fundamental data concerning coal ignition under oxy-steam conditions are needed. However, few studies have been conducted on the ignition of coal in O_2/H_2O atmospheres.

This chapter presents the ignition behaviors of coal that were studied in O_2/N_2 and O_2/H_2O atmospheres in a DTF by using advanced optical instruments and numerical methods. The ignition characteristics were captured by a high-speed camera, and the particle temperature was measured with a two-color pyrometer. The effects of steam on the ignition process were also analyzed in detail.

15.2 IGNITION OF PULVERIZED COAL IN O_2/H_2O ATMOSPHERES

15.2.1 Coal Sample

All experiments reported in the study used a Chinese bituminous coal, samples of which were air-dried prior to use. The coal sample was ground and sieved

TABLE 15.1 Properties of the Coal Sample

Proximate analysis (%w)	Moisture 6.40	Volatile matter 36.95	Ash 2.90	Fixed carbon 53.75
Ultimate analysis (daf, %w)	Carbon 72.82	Hydrogen 4.88	Nitrogen 0.93	Oxygen + sulfur 21.37

to sizes ranging from 97 to 105 μm. The proximate and ultimate analysis of the coal sample is shown in Table 15.1.

15.2.2 Experimental Equipment

The experiments were conducted by using an electrically heated DTF with a constant wall temperature of 1373 K, which was described in detail in Section 3.4.1. The main schematic structure of the furnace is shown in Fig. 15.1. A FASTCAM SA4 high-speed camera was used to monitor the ignition process through the view ports with a frame speed of 1000 fps. The recordings were made at least in triplicate for each condition. The particle temperature in the reactor was measured via a two-color pyrometer through the view ports, which were positioned along the furnace, as shown in Fig. 15.1. The pyrometer was focused on the centerline of the reactor. The results were exported to a computer by means of

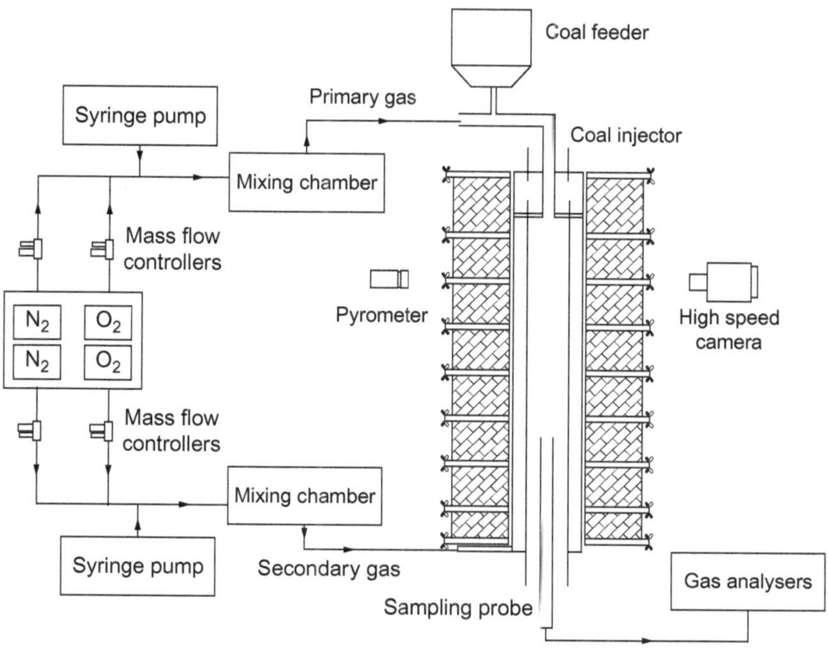

FIG. 15.1 Schematic structure of the DTF.

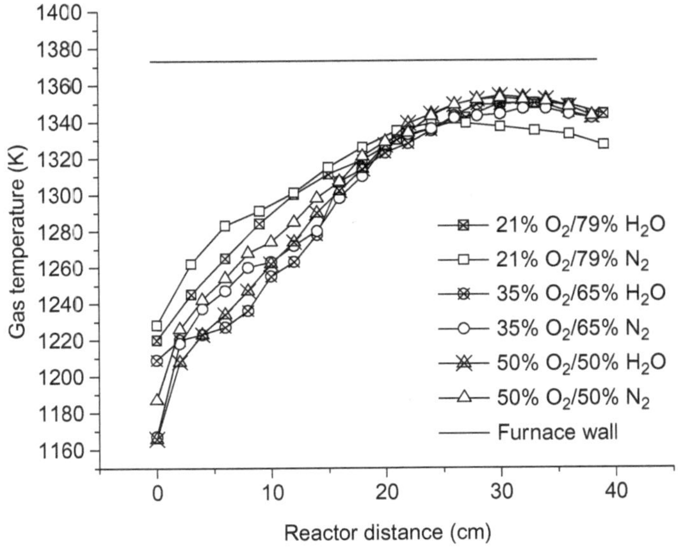

FIG. 15.2 Gas temperature profiles along the furnace centerline.

software. The measurement accuracy of the particle temperature was shown to be about ±30°C [22,23], because some parameters were unavailable for one reason or another. For example, the background temperature was high or the coal particles were smaller than the pyrometer's field of view. Experimental studies were conducted by using several different binary mixtures of O_2/N_2 and O_2/H_2O for each coal sample with different mole fractions of oxygen (21%, 30%, 35%, 40%, and 50%).The composition of the primary gas and the secondary gas was identical in each case.

The gas temperature profiles were measured along the centerline of the furnace at several intervals. As shown in Fig. 15.2, no significant gaps were observed among the gas mixtures.

15.2.3 Experimental Results

Dynamic Behavior of Ignitions

Some typical ignition images involving various oxygen concentrations in the N_2 and H_2O environments were selected from the high-speed film and are shown in Figs. 15.3 and 15.4.

In N_2 background gases, ignition spots were dim at oxygen fractions lower than 35%, whereas the brightness and size of spots increased with increasing oxygen fractions (≥35%). The particles appeared to ignite homogeneously at all oxygen fractions, as depicted in Fig. 15.3. These phenomena agreed with the results of Khatami et al. [7], Molina et al. [9], and Zhang et al. [20].

FIG 15.3 Typical dynamic behavior of ignitions at 1373 K in O_2/N_2 environments at a time interval of 1 ms.

The substitution of H_2O for N_2 in the bulk gas resulted in significantly different ignition processes, as shown in Fig. 15.4. The luminous spots were dim for the lower mole fractions of oxygen of 21% and 30% in H_2O environments, which was similar to those in the N_2 environments. The optical intensities of the luminous spots in the ignition process are shown in Fig. 15.5. It is clear that at the oxygen fraction of 21% in steam, after the onset of ignition, the optical intensities increased more slowly and were lower than those in N_2. At 8 ms after ignition, the optical intensities sharply increased and were slightly higher than those in N_2. It is worth noting that the ignition process in 30% O_2/70% H_2O was similar to that in air.

With increasing O_2 fractions ($O_2 \geq 35\%$), the spots became brighter and larger, the times of ignition sharply reduced, and the particle fragments were more frequently observed in H_2O than in N_2. These phenomena meant that the intensity of the ignition combustion evidently increased. As shown in Fig. 15.4,

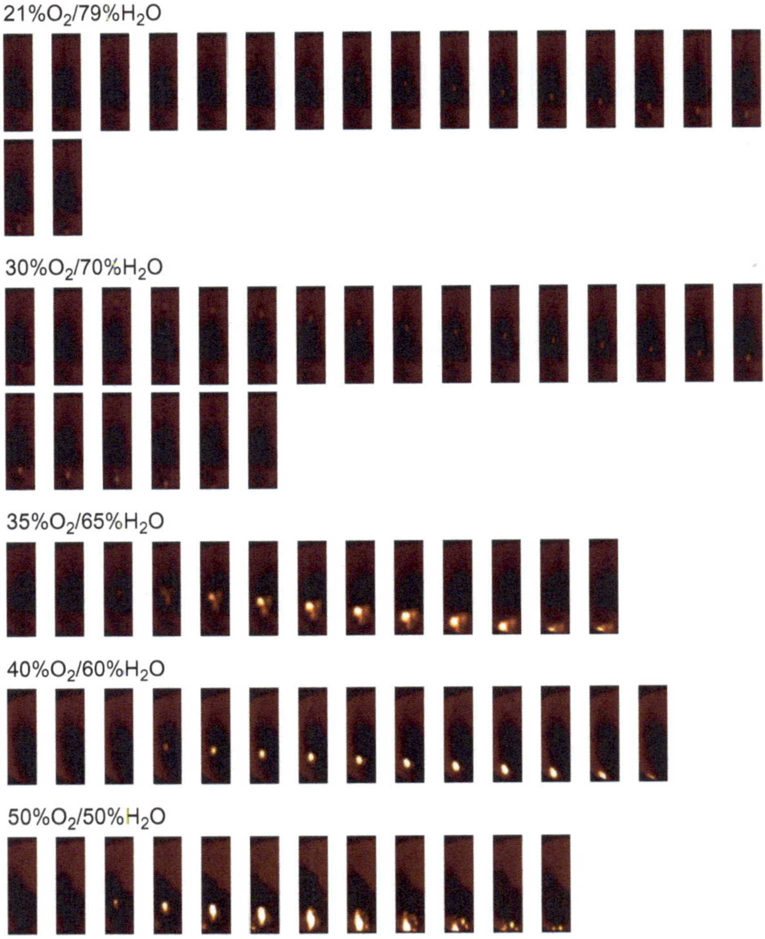

FIG. 15.4 Typical dynamic behavior of ignitions at 1373 K in O_2/H_2O environments at a time interval of 1 ms.

we observed a more drastic increase in the optical intensities in H_2O than in N_2. In other words, the combustion of gaseous mixtures around the coal particle was completed earlier in H_2O than in N_2.

At low oxygen fractions (21% and 30%), steam fractions were high (79% and 70%); thus the thermal capacity of steam had a greater effect, resulting in a dim luminous spot at the ignition of the coal particles. The low temperature of the particles and the high fraction of steam reduced the quantity of volatiles released from the coal. Therefore CO and H_2 took up a greater proportion of the gaseous mixture around the particles than in the corresponding O_2/N_2 atmosphere, which resulted in a slow increase of the optical intensity during the ignition process due to the low calorific value of CO and H_2. It is possible that the onset of the ignition of the particle was mainly caused by the products of

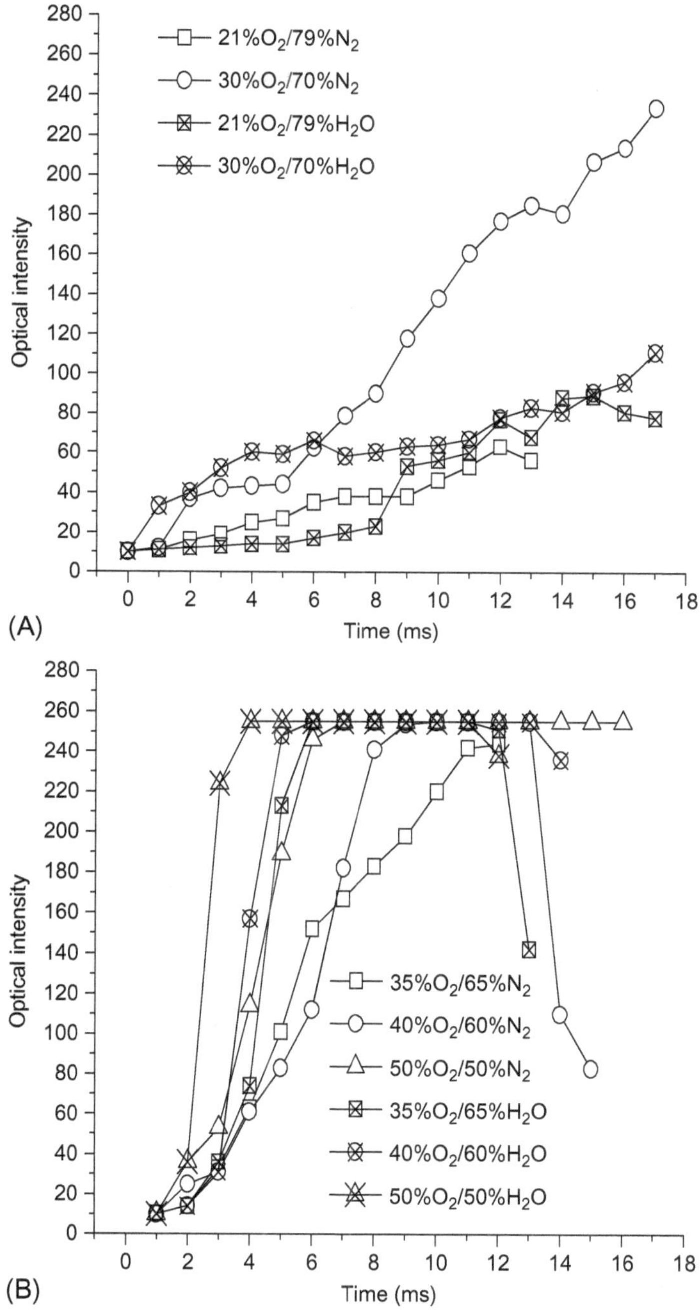

FIG. 15.5 Optical intensities of igniting particle at various O_2 fractions in both bulk gases: (A) 21%O_2 and 30%O_2 and (B) 35%O_2, 40%O_2, and 50%O_2.

the steam gasification reaction and/or the steam shift reaction at a mole fraction of 21% oxygen in steam, after which a significant mass of volatiles was released. This could explain the drastic increase in the optical intensity at 8 ms after the ignition in 21%O_2/79%H_2O, as shown in Fig. 15.5A. At oxygen fractions ≥35%, the evolved volatiles made up most of the gaseous mixture around the particles. The endothermic gasification reaction had little effect on the temperature of the particles because of the feedback of the heat released from the combustion of the evolved volatiles. CO and H_2 enhanced the oxidation rate of the gaseous mixtures, which increased the optical intensity in high oxygen in steam more significantly than in N_2, as shown in Fig. 15.5B.

Particle Temperature

The variation in particle temperatures versus oxygen fractions is shown in Fig. 15.6. It is evident that the particle temperature increased with the increasing oxygen fraction in both bulk gases because the homogeneous and heterogeneous combustion temperatures increased with the increasing oxygen fraction. It is interesting that the particle temperature at the ignition point in H_2O was higher than that in N_2 at an identical oxygen concentration. This phenomenon may be explained in two ways. First, because of the steam gasification and the steam shift reaction, considerable amounts of H_2 and CO were produced in H_2O, and the oxidation of H_2 and CO could have provided extra heat to the particle surface in contrast to that in N_2. Second, the chemical properties of steam may have played a role. The combustion temperature increased when steam was

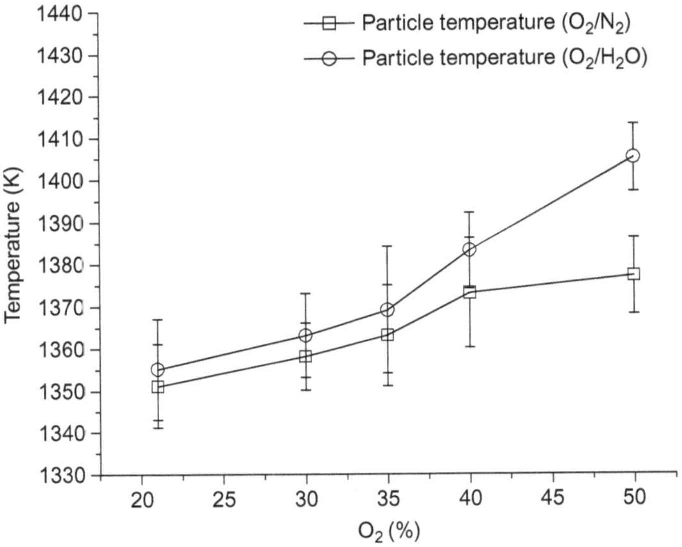

FIG. 15.6 Measured particle temperatures during the ignition process.

added to the gaseous reactant mixtures because of the superequilibrium effect of the steam, resulting in the high temperature of the particles in H_2O [24].

Ignition Distance

The values of ignition distances are shown in Fig. 15.7. Note that the location of the coal ignition in the steam atmosphere was closer to the nozzle of the coal than it was in the N_2 atmosphere in comparable O_2 concentrations. In other words, the ignition of coal occurred sooner in a steam atmosphere than in a N_2 atmosphere at an identical O_2 concentration. In O_2/N_2 atmospheres, the ignition distance dramatically decreased with the increase in the volume fraction of O_2 and the decrease in the volume fraction of N_2, which implied that the oxidation by the high concentration of O_2 played a significant role in the ignition and that the heat absorption of N_2 decreased significantly. However, the ignition distance gradually decreased with the increase and decrease in the mole fraction of O_2 and H_2O, respectively, which was attributed to the fact that the oxidation of H_2O was significant, even though its strength was less than O_2, leading to the conclusion that the oxidation power of the O_2/H_2O atmospheres was stronger than that of O_2/N_2. Therefore the increase in the mole fraction of O_2 in H_2O produced the gradual increase in the oxidation of the atmosphere. As a result, the differences between the locations of ignition points in O_2/H_2O and O_2/N_2 atmospheres decreased from 114 mm to 46 mm when the oxygen volume fraction increased from 21% to 50%.

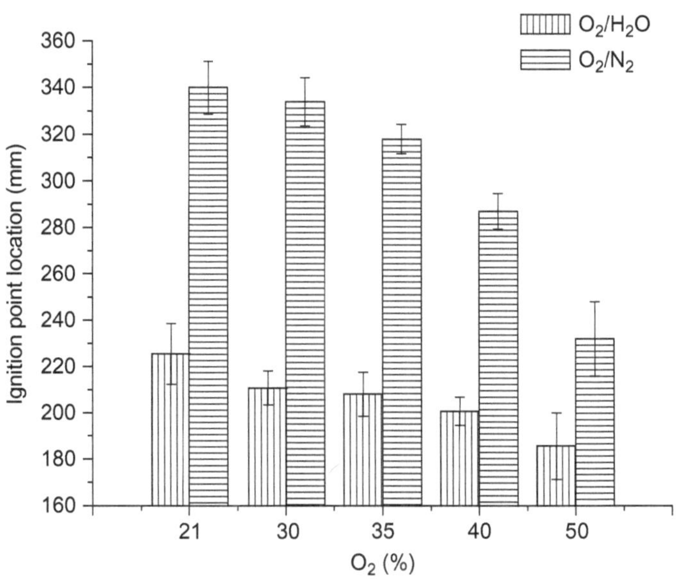

FIG. 15.7 Experimental ignition distances in both O_2/N_2 and O_2/H_2O atmospheres.

15.3 EFFECTS OF H_2O ON THE IGNITION OF PULVERIZED COAL

Steam has three kinds of effects on the ignition of coal in O_2/H_2O atmospheres because of the substitution of H_2O for N_2. First is the effect of thermal capacity. At 1400 K, the mole specific heat of H_2O is 40.9 J/K mol and is 1.3 times higher than that of N_2 (32.3 J/K mol). Similar to CO_2 in oxy-fuel combustion, the higher thermal capacity of steam can reduce the rate of the coal particle's temperature increase, resulting in a low ignition intensity and the delay of ignition.

Second is the effect of the steam gasification reaction:

$$\mathrm{C}+\mathrm{H2O}\rightarrow\mathrm{CO}+\mathrm{H2},\quad \Delta H=130\,\mathrm{kJ/mol} \tag{15.1}$$

This reaction is an endothermic reaction that can reduce the temperature of the coal particle. However, CO and H_2 produced by gasification reaction are beneficial to the ignition and combustion of the homogeneous gaseous mixture (volatiles, CO and H_2). The global effect of the steam gasification reaction on the ignition of the coal particle should be associated with the mole fraction of steam in oxygen.

Third is the chemical effect of the steam shift reaction:

$$\mathrm{CO}+\mathrm{H2O}\rightarrow\mathrm{CO_2}+\mathrm{H2},\quad \Delta H=-40.9\,\mathrm{kJ/mol} \tag{15.2}$$

This reaction is an exothermic reaction, and it produces H_2, which is beneficial to the reduction in the ignition delay time of the gaseous mixture around the pulverized coal.

The coal particle ignited earlier in the case of steam than N_2 at an identical oxygen concentration. Considering that the particle velocities in O_2/H_2O atmospheres are approximately equal to those in O_2/N_2 atmospheres, the ignition delay ratio, which is defined as the ratio of ignition distances in H_2O to those in N_2 for an identical oxygen concentration, is proportional to the ignition distance ratio defined as the H_2O to N_2 ignition distance ratio. The ignition distance ratios for five different oxygen concentrations in steam and N_2 are shown in Fig. 15.8. It is clear that the five values of the ignition location ratio were smaller than 1 and increased with increasing concentrations of oxygen.

According to adiabatic thermal explosion theory [25], the ignition time is given as

$$t_i=\frac{c_\mathrm{v}\left(T_0^2/T_\mathrm{a}\right)}{q_\mathrm{c}Y_{\mathrm{F},0}A\exp\left(-T_\mathrm{a}/T_0\right)} \tag{15.3}$$

where c_v is the specific heat, T_0 is the initial temperature of the local fuel/oxidizer ratio, q_c is the heat of combustion, and $Y_{\mathrm{F},0}$ is the initial mass fraction of fuel. Without considering the steam gasification reaction and the steam shift reaction and the chemical properties of steam, the ignition delay ratios obtained according to Eq. (15.3) for five cases are shown in Fig. 15.8. All ratios from

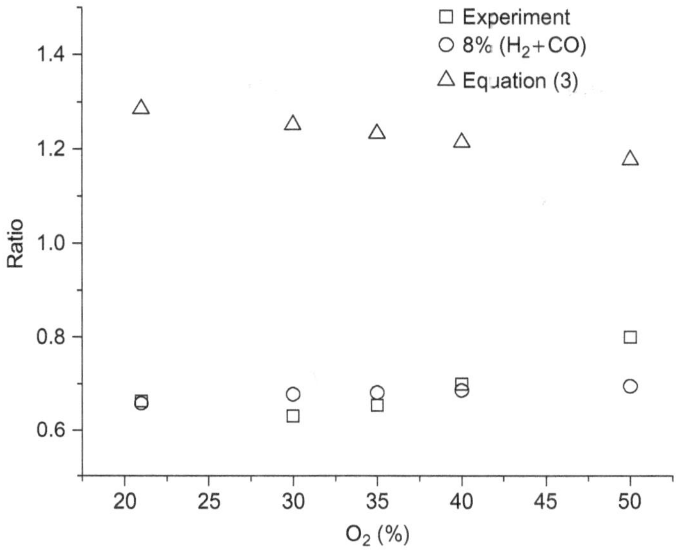

FIG. 15.8 Comparison of the ratio of experimental ignition delay and calculated ignition delay time.

Eq. (15.3) were >1, and the trend was contrary to the experimental results. This meant that the difference in the thermal capacity between steam and N_2 did not cause the reduction in the ignition delay time in the O_2/H_2O atmospheres, in contrast to O_2/N_2 atmospheres. In other words, the chemical effects (the steam gasification reaction, the steam shift reaction, and the chemical properties of steam) should have been the main reasons for this phenomenon. In view of these findings, a CHEMKIN closed homogeneous batch reactor was adopted to calculate the ignition delay times in the cases of the O_2/H_2O and O_2/N_2 atmospheres. For simplicity, CH_4 was used to simulate the coal volatiles' composition. In the case of the O_2/H_2O atmospheres, the products of the steam gasification and/or the steam shift reaction (CO and H_2) mixed with the evolved volatiles, and their total mole fractions in CH_4 were 8%. The calculated ignition delay time ratios are shown in Fig. 15.8. As the figure shows, the calculated results agreed with the experimental results. Therefore the reason for the reduction of the ignition delay time in O_2/H_2O atmospheres, compared with O_2/N_2 atmospheres, was the presence of CO and H_2 produced by the steam gasification reaction and/or the steam shift reaction in the evolved volatiles.

To study the effect of the steam shift reaction on the ignition of coal in oxy-steam, a method similar to the sensitivity analysis for the combustion mechanism was adopted [26,27]. The preexponential factor for the steam shift reaction was changed from 2.74×10^{-1} to 2.74×10^{19}, and the activation energy remained unchanged. The locations of the ignition of coal corresponding to the preexponential factor in 30% O_2/70% H_2O are shown in Fig. 15.9 (the simulations of other O_2 concentrations in O_2/H_2O atmospheres were also conducted, and the

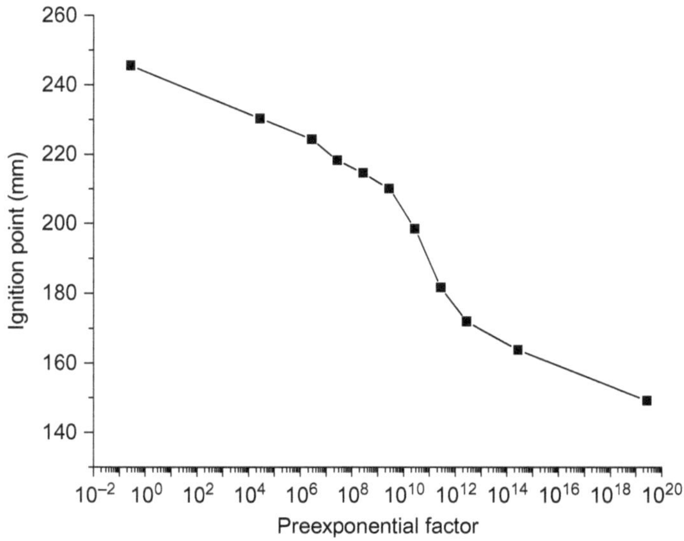

FIG. 15.9 Prediction results of ignition distances by adjusting the preexponential factor of the steam shift reaction.

results were the same, so they are not shown here). The distance in the location of the ignition of coal significantly decreased as the rate of the steam shift reaction increased. That is, the steam shift reaction was distinctly and remarkably beneficial to the ignition of coal in O_2/H_2O atmospheres. The reasons for the shift reaction behavior were twofold. First, more heat could be released because a greater amount of CO was oxidized into CO_2 by the shift reaction at the location of the ignition. Second, additional H_2 produced by the shift reaction was beneficial at the point of ignition. Therefore the steam shift reaction was the main reason that the ignition of coal in O_2/H_2O atmospheres occurred sooner than in O_2/N_2 atmospheres.

A method similar to the study of the role of the shift reaction was applied to investigate the effect of the steam gasification reaction on the ignition process. Fig. 15.10 shows that the predicted ignition locations changed with the different preexponential factors of the steam gasification reaction in a 30% O_2/70% H_2O atmosphere (the simulations of other O_2 concentrations in O_2/H_2O atmospheres were also conducted; the results were the same and are not shown here). As the preexponential factor was varied from 1.125×10^{-9} to 1.125×10^{11}, the ignition distance decreased from 232 to 188 mm, which implied that the steam gasification reaction had positive effects on the ignition of coal in O_2/H_2O atmospheres, even though it was an endothermic reaction. The positive effect of steam gasification can be attributed to the fact that CO produced by the steam gasification was beneficial to the ignition of coal in the reaction (15.3) and to the steam shift reaction in O_2/H_2O atmospheres. However, the steam gasification had much less effect than the steam shift reaction because (1) the steam gasification reaction

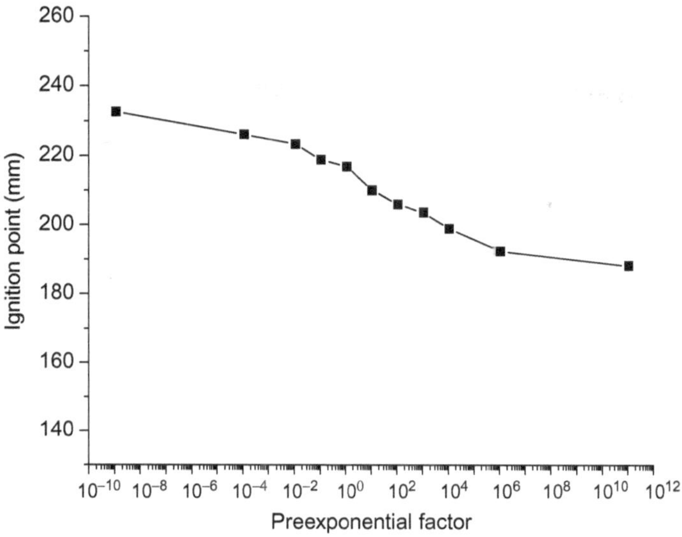

FIG. 15.10 Prediction results of ignition distances by adjusting the preexponential factor of the steam gasification reaction.

was an endothermic reaction, and (2) the reaction rate of the steam gasification was less than that of the steam shift reaction because of the higher activation energy of the former.

15.4 SUMMARY

1. According to the images of the ignition and the numerical results, the pulverized coals ignited homogeneously in all cases. The ignition in the O_2/H_2O atmospheres preceded that in the O_2/N_2 atmospheres at identical mole fractions of oxygen.
2. The steam shift reaction was the primary reason for the advanced ignition of coal in the O_2/H_2O atmospheres in contrast to the ignition in the corresponding O_2/N_2 atmospheres. The steam gasification reaction was a minor reason for the phenomena because it produced CO and H_2, which were beneficial to the ignition.

REFERENCES

[1] Chen L, Yong SZ, Ghoniem AF. Prog Energy Combust Sci 2012;38(2):156–214.
[2] Buhre B, Elliott L, Sheng C, Gupta R, Wall T. Prog Energy Combust Sci 2005;31(4):283–307.
[3] Wall T, Liu Y, Spero C, Elliott L, Khare S, Rathnam R, et al. Chem Eng Res Des 2009;87(8):1003–16.
[4] Wall TF. Proc Combust Inst 2007;31(1):31–47.
[5] Edge P, Gharebaghi M, Irons R, Porter R, Porter R, Pourkashanian M, et al. Chem Eng Res Des 2011;89(9):1470–93.

[6] Rathnam RK, Elliott LK, Wall TF, Liu Y, Moghtaderi B. Fuel Process Technol 2009;90(6): 797–802.

[7] Khatami R, Stivers C, Levendis YA. Combust Flame 2012;159:3554–68.

[8] Bejarano PA, Levendis YA. Combust Flame 2008;153(1):270–87.

[9] Molina A, Shaddix CR. Proc Combust Inst 2007;31(2):1905–12.

[10] Shaddix CR, Molina A. Proc Combust Inst 2009;32(2):2091–8.

[11] Carlos S. International oxy-combustion research network – 2nd workshop, vol. 25, USA. 2007.

[12] Seepana S, Jayanti S. Energy Convers Manag 2010;51(10):1981–8.

[13] Essenhigh RH, Misra MK, Shaw DW. Combust Flame 1989;77:3–30.

[14] Howard J, Essenhigh R. Combust Flame 1965;9(3):337–9.

[15] Annamalai K, Durbetaki P. Combust Flame 1977;29:193–208.

[16] Wall T, Gururajan V. Combust Flame 1986;66(2):151–7.

[17] Zhang D-K, Wall TF. Fuel 1994;73(7):1114–9.

[18] Wall T, Gupta R, Gururajan V, Zhang D-K. Fuel 1991;70(9):1011–6.

[19] Liu Y, Geier M, Molina A, Shaddix CR. Int J Greenhouse Gas Control 2011;5:S36–46.

[20] Zhang L, Binner E, Qiao Y, Li C-Z. Energy Fuels 2009;24(1):29–37.

[21] Jovanovic R, Milewska A, Swiatkowski B, Goanta A, Spliethoff H. Int J Heat Mass Transfer 2011;54(4):921–31.

[22] Binner E, Zhang L, Li C-Z, Bhattacharya S. Proc Combust Inst 2011;33(2):1739–46.

[23] Smart J, Lu G, Yan Y, Riley G. Combust Flame 2010;157(6):1132–9.

[24] Park J, Kim SG, Lee KM, Kim TK. Int J Energy Res 2002;26(13):1141–60.

[25] Law CK. Combustion physics. Cambridge: Cambridge University Press; 2006. p. 309.

[26] Liu N, Ji C, Egolfopoulos FN. Combust Flame 2012;159:465–75.

[27] Liu N, Mani Sarathy S, Westbrook CK, Egolfopoulos FN. Proc Combust Inst 2013;34: 903–10.

Chapter 16

Chemical Looping Combustion

Baowen Wang*, Xin Guo†
*North China University of Water Resources and Electric Power, Zhengzhou, China
†Huazhong University of Science and Technology, Wuhan, China

16.1 INTRODUCTION

To combat global warming and stabilize the atmosphere's CO_2 concentration, immediate measures are needed to reduce CO_2 emissions in fossil fuel combustion [1], especially for coal because of its carbon-intensive property [2]. Moreover, these measures need to be taken without greatly changing the primary structures that deal with the ever-increasing demand for energy. Along with improving the efficiency of energy conversion/utilization and decreasing the use of carbon-intensive fuels, carbon capture and storage is seen as an active and effective measure [3], in which CO_2 capture should be prioritized because of its high cost.

Obtaining a concentrated CO_2 stream is the key challenge to capturing CO_2 in coal combustion and to further sequestration of the captured CO_2 downstream. Among the available technologies [4], chemical looping combustion (CLC) is acknowledged as the new-generation oxy-fuel combustion technology. With this technology, the oxidation of fuel is inititated by the active oxygen carrier (OC), instead of air or enriched gaseous oxygen as is done in the usual combustion mode, thus simultaneously statisfying the increasing energy demand and attaining pure CO_2 capture with a low-energy penalty. Therefore this novel combustion technology has quickly developed and has attracted great interest in academic and industrial communities [5].

Fig. 16.1 shows a mechanistic scheme of the CLC process with coal used as the fuel. The reactor configuration adopted in the CLC system includes fluidized bed prototypes, and the two connected reactors are referred to as air reactor (AR) and fuel reactor (FR). Unlike in the combustion mode in which coal comes in direct contact with air, shown in Eq. (16.1) in Fig. 16.1, in the CLC system, when coal is first introduced to the FR, the pyrolysis (Eq. 16.2a)/gasification (Eq. 16.2b and c)/water gas shift reaction for CO (Eq. 16.2d) and further oxidation with the lattice oxygen [O] involved in the OC(Me_xO_y) (Eq. 16.2e) are inititated. Then the reduced OC(Me_xO_{y-1}) is transferred from the FR to the AR and reoxidized by air to its original state (Me_xO_y) (Eq. 16.2f). In this way, the

Oxy-fuel Combustion. http://dx.doi.org/10.1016/B978-0-12-812145-0.00016-5

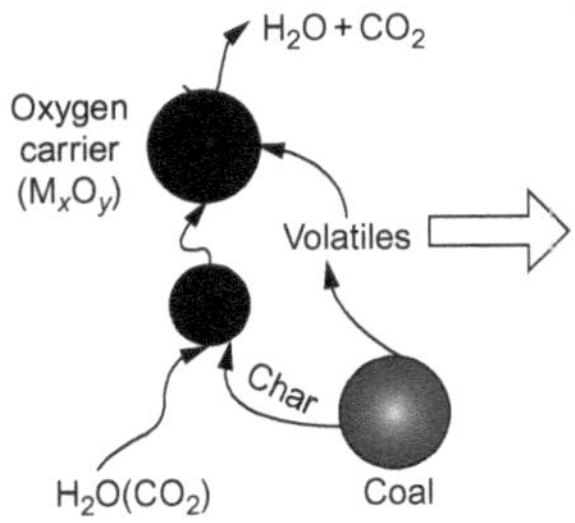

Normal combustion of coal with air (direct contact)

$$\text{Coal} + (n + m/2)(O_2 + 3.76N_2) \rightarrow nCO_2 + mH_2O + 3.76(n + m/2)N_2 \quad \textbf{(16.1)}$$

Chemical looping combustion of coal (unmixed mode)

Step 1: coal oxidation by lattice [O] from oxygen carrier

FR:

$$\text{Coal} \rightarrow \text{volatiles} + \text{char} \quad \textbf{(16.2a)}$$

$$\text{Char(mainly C)} + H_2O \rightarrow H_2 + CO \quad \textbf{(16.2b)}$$

$$\text{Char(mainly C)} + CO_2 \rightarrow 2CO \quad \textbf{(16.2c)}$$

$$CO + H_2O \rightarrow CO_2 + H_2 \quad \textbf{(16.2d)}$$

$$\text{Volatiles}/H_2/CO + Me_xO_y \rightarrow CO + H_2O + Me_xO_{y-1} \quad \textbf{(16.2e)}$$

Step 2: Regeneration of the reduced oxygen carrier

AR:

$$2Me_xO_{y-1} + (O_2 + 3.76N_2) \rightarrow 2Me_xO_y + 3.76N_2 \quad \textbf{(16.2f)}$$

FIG. 16.1 Mechanistic scheme of the CLC by direct use of coal as fuel.

direct contact of air with fuel, as happens in the normal combustion mode, is avoided, although the net heat evolved from (Eq. 16.2a–f) is equal to that of the normal combustion (Eq. 16.1).

The CLC process offers several advantages. One advantage is that, at the full conversion of fuel in the FR, the high concentration of CO_2 can be obtained easily after condensation of the steam in the FR flue gas and made ready for later sequestration downstream with a low energy penalty. Also, the NO_x emission can be effectively inhibited because of the fuel combustion and OC regeneration occurring in the two separate reactors at much lower temperatures ranging from about 800–1000°C. Moreover, thanks to the CLC process's unmixed combustion mode, the irreversibility involved in conventional fuel combustion can be reduced and the associated energy can be utilized in a cascade way. In addition, CLC adapts well with different fuels, such as gaseous/solid/liquid fuels, and has broad potential for application in multiple processes, besides fuel combustion/pure hydrogen or syngas production free of CO_2 emission at the same time.

The CLC concept dates back to 1954 when Lewis and Gilliland invented a new process for pure CO_2 production [6], while the term CLC is credited to Prof. Ishida's group in Japan in the late 1980s; later on, the process was applied to CO_2 capture [7]. Although initially most CLC investigations focused on various gaseous fuels, because of solid fuels' advantages, such as their abundance, easy availability, and low cost, their use in the CLC process became more significant in combating global warming, especially coal because of its higher carbon intensity in contrast to other available fossil fuels.

Utilizing direct use of coal, various CLC systems have operated continuously, from laboratory-scale facilities, to $100\,kW_{th}$-scale facilities [8], to $1\,MW_{th}$ [9] and $3\,MW_{th}$ semiindustrial-scale facilities [10]. The operational practices have not only validated the feasibility of the direct use of coal in a CLC system but also indicated the important role the OC plays in achieving well-performing, economic operations in the system. Moreover, these practices have shown the necessity of fully converting the solid fuel in order to use it efficiently and to achieve CO_2 capture. However, sufficient attention needs to be paid to the detrimental side effects of the evolved sulfur and the transformed minerals that occur during coal's reaction with OC. Therefore, based our group's nearly 10-year progress in this area, we have organized this chapter in four main parts: (1) OC preparation and characterization, (2) reaction characteristics of coal during the CLC process, (3) sulfur's evolution and interaction mechanism with the reduced OC, and (4) the minerals' transformations and their effect on the reduced OC, all of which are meaningful in further development of the direct use of coal in the CLC process.

16.2 OC PREPARATION AND CHARACTERIZATION

The OC is the cornerstone in the use of CLC, and various critera must be met by the OC over the multiple cycles of a fuel's oxidization and of further regeneration of the reduced OC [11] by air, including high reactivity, without great sintering and reactivity decay; structural integrity, without serious attrition and fragmentation; economic competitiveness, without highly expensive ingredients; and environmental benignancy, without toxic wastes discarded into the atmosphere.

The OC generally comprises the active oxide and inert supports, where the active oxides provide the oxygen to adequately oxidize the carbonaceous fuel into CO_2, and the supports are needed to improve the OC's resistance to fragmentation, abrasion, and sintering. On the one hand, as to the active oxides involved in the OC, Ni-, Cu-, Fe-, Mn-, and Co-based active metal oxides have been mainly adopted, among which the former three oxides are more widely applied in the CLC system. However, pros and cons are associated with each metal and the oxide. NiO has high reactivity but thermodynamic limitations, and there are concerns about its toxicity. CuO has exothermic characteristics but a low melting point and possible defluidization, as well as the potential for secondary pollution. Compared to NiO and CuO, Fe_2O_3 has a striking economic strength because of its lower price, an environmental benignancy without potential pollution, and a higher melting point. Therefore Fe_2O_3 is appraised as the active oxide for OC components and is widely applied in the CLC process.

On the other hand, in the selection of an OC selection, in terms of inert supports, a variety of materials have been suggested for different OCs, such as Al_2O_3, bentonite, MgO, SiO_2, ZrO_2, $MgAl_2O_4$, sepiolite, and kaolin. Considering the reactivity, fragmentation, and avoidance in the interactions between Fe_2O_3 and inert supports, Al_2O_3 is recognized as one of the most promising supports [12]. Therefore synthesis of the active Fe_2O_3 combined with the inert Al_2O_3 support are discussed in the following section.

16.2.1 Fe_2O_3/Al_2O_3 Preparation by Sol-Gel Combustion Synthesis

Currently, two main options are available for OC preparation, including "dry chemistry" and "wet chemistry" [13]. For example, in the synthesis of Fe_2O_3/Al_2O_3.OC, the dry chemistry option includes mechanical mixing, freeze granulation, and impregnation; and the wet chemistry option includes co-precipitation, microemulsion, the sol-gel (SG) method, dissolution, and the solution combustion synthesis method. In contrast to the OCs prepared via the dry preparation method, in the wet preparation method, the OCs formed are better able to achieve a uniform mixing condition between the metal oxides and the supports on a molecular scale, have less expensive precursors, and more accurately achieve the desired stoichiometric amount. Among these wet methods, SG is one of the most promising routes because the OCs synthesized by this method are fine and uniform, with strong mechanical strength and high attrition resistance; however, expensive alkoxides must be used. To avoid using expensive alkoxides, the dissolution method was designed in which 2-propanol and metal nitrate were used as substitute metal alkoxides.

Unlike the preceding methods, the solution combustion synthesis method relys on a self-sustained combustion reaction between metal nitrates as the oxidizers and various organic fuels (e.g., citric acid and glycine) to produce nanoscale metal oxide powders. Because of the designated precursors solutions' mixture and the exothermicity of the combustion reaction, the solution combustion synthesis method has more advantages than other wet chemistry methogds, including its rapid and straightforward synthesizing process, its lower energy requirement, and the molecular-level mixing of the precursors. Therefore combining the SG method and the solution combustion synthesis method to produce OC is promising in the sphere of high-performance CLC applications.

Previous studies on the preparation of only Fe_2O_3 and of only Al_2O_3 have reported the use of different methods, including solution combustion synthesis utilizing different organic fuels. However, information on the synthesis of Fe_2O_3/Al_2O_3 composite particles is limited, and the methods of preparing Fe_2O_3/Al_2O_3 particles are confined to impregnation, freeze-granulation, co-precipitation, and SG. Research on novel synthesis methods for Fe_2O_3/Al_2O_3 OC based on a combined sol-gel and solution combustion synthesis (SGCS) method is greatly needed [14].

The materials selected for the preparation of Fe_2O_3/Al_2O_3 OC using the SGCS method were urea as the fuel and complexing agent, because of its ready availability, high exothermicity, and the effective complexation of the amine group (i.e., $-NH_2$) from urea with transition metals; and hydrated metal nitrates as the suitable oxidizers, because of their good water solubility, their low decomposition temperature, and the mild redox reaction between urea and nitrates. Therefore the starting materials were $Fe(NO_3)_3.9H_2O$, $Al(NO_3)_3.9H_2O$, and urea.

Fe_2O_3/Al_2O_3 OC at the mass ratio 6:4 was selected as an example and was designated as Fe6Al4 OC and was prepared by the SGCS method using the following procedures. First, stoichiometric compositions of metal nitrates (including $Fe(NO_3)_3.9H_2O$ and $Al(NO_3)_3.9H_2O$) and urea were calculated at the desired molar ratio on the basis of 0.03 mol of $Fe(NO_3)_3.9H_2O$. The molar ratio of $Al(NO_3)_3.9H_2O$, urea, and deionized (DI) water to $Fe(NO_3)_3.9H_2O$ was correspondingly determined as 1.1, 5.2, and 46, respectively, according to the propellant chemistry theory to ensure that the equivalent ratio of the reducing valence of urea to the total oxidizing valence of nitrates was at unity, and thus the heat released was at its maximum.

Calculated amounts of $Fe(NO_3)_3.9H_2O$, $Al(NO_3)_3.9H_2O$, and urea were weighed and dissolved in DI water in a PYREX beaker. The mixture was placed on a hot plate positioned in a fume hood for good ventilation, stirred while exposed to an air atmosphere, and aged at 75°C until an orange-yellowish-red viscous colloidal suspension (sol) formed. During the formation of the sol, the pH value was periodically tested with a carefully calibrated pH meter. Then the viscous wet sol was dried sequentially at 80°C and 135°C in a temperature-programed desiccator at different durations. Next, the prepared dried gel was transferred to a ceramic dish and ignited in a preheated muffle furnace positioned in the fume hood at 600°C for 15 min. Finally, the burned product was calcined in the same furnace at 950°C for 2 h.

As a new method of preparing OC synthesis, the SGCS method met the basic expectations set for it—rapid and simple synthesizing process, low energy requirements, and good resistance to sintering. In addition, because of the thorough elimination of the detrimental NO_x gases caused by their secondary reaction with the NH_3 that evolved from the urea hydrolysis, the SGCS method did not emit secondary pollutants, thus making it a "green" OC preparation method. Note that the lack of secondary pollutant emissions should be considered as an another serious requirement for OC preparation, especially during the production of large amounts of OC. This method also can perform well in other processes and thus has wide applications, for example, in the preparation of high-temperature CaO-based sorbents, as shown in our previous research [15].

16.2.2 Preliminary Characterization of the SGCS-Prepared Fe_2O_3/Al_2O_3 OC

In the SGCS preparation of Fe6A14 OC at the mass ratio 6:4, after ignition and calcination, the burned and finished Fe6A14 OCs were designated as Fe6A14(ig) and Fe6A14(si), respectively. Their phase identification was first conducted by X-ray diffraction analysis (XRD). After ignition, Fe6A14(ig) consisted of Fe_2O_3 and iron aluminate ($FeAl_2O_4$) instead of Al_2O_3, due to aluminum substitution and the formed complex group $(Fe_{1-y}Al_y)_{1-x}O_{1-3x}$. After 2 h of calcination at 950°C, Fe6A14(si) formed and consisted of separate α-Fe_2O_3 and α-Al_2O_3, which indicated that calcination was the necessary procedure to avoid the formation of a spinel phase between the active oxide and the support. But unlike the synthesis of Fe6A14 OC, the synthesis of the separate phase α-Fe_2O_3 or α-Al_2O_3 could be directly conducted even without calcination. Furthermore, the average crystalline sizes for both the formed Fe6A14 and the two reference oxides were evaluated using the Scherrer formula as $D=0.9\lambda/(\beta \cos\theta)$, where D was the crystallite size in nm, λ was the radiation wavelength (0.15406 nm for Cu $K\alpha$), θ was the diffraction angle, and β was the corrected half width for instrument broadening. It was found that after calcination, the size of all the samples increased more than those after only ignition, particularly the single phase Fe_2O_3 whose crystalline size increased greatly from 27.2 nm at the ignition stage to 107.3 nm after calcination because of the coalescence of neighboring crystallites through calcination. However, the crystalline size of Al_2O_3 increased from 63.9 to only 73.2 nm without much growth in grain size, which proved the good resistance of α-Al_2O_3 to sintering and thus appropriate for use as an inert support for OC synthesis. For Fe6Al4(si) OC, the average crystalline sizes for α-Fe_2O_3 and α-Al_2O_3 were 37.6 and 67.8 nm, respectively, which were lower than their separate counterparts' sizes of 107.3 and 79.1 nm, which may have been due to the aluminum substitution during the formation of sol and the ensuing partial reciprocal solubility between α-Fe_2O_3 and α-Al_2O_3.

As the OC, the Fe_2O_3/Al_2O_3 particle served as the reactant in the CLC, and the associated redox reactions the typical gas-solid reactions instead of a surface reaction like being used as a catalyst in the catalytic process. Therefore, along with surface area, the pore size distribution is important for the reactivity of OC. An N_2 adsorption analyzer was thus used to comprehensively characterize both the formed Fe6Al4 OC and the separate phases of α-Fe_2O_3 and α-Al_2O_3 synthesized using SGCS as references. According to the isotherms and pore size distribution obtained for Fe6Al4(ig) and Fe6Al4(si), the hysteresis loops of the homemade Fe6Al4 OC and separate phases of Fe_2O_3(si) and Al_2O_3(si) were all different from the four typical hysteresis loops provided by the International Union of Pure and Applied Chemistry (IUPAC). Below the relative pressure $P/P_0=0.6$, the adsorption branch increased slowly with the relative pressure and then intersected but did not overlap with the desorption branch, which was far different from the four typical hysteresis loops designated by IUPAC; whereas above $P/P_0=0.6$, the adsorption branch sharply ascended and did not reach the stable status even when $P/P_0=1.0$, much like that part of the H1 hysteresis loop at the higher relative pressure. Therefore the pore types of the homemade Fe6Al4 OC pore types were more complicated and were regarded as being heterogeneous types, which consisted of several typical hysteresis loops [16]. The complexity of the pore shape may have resulted from the preparation process using SGCS, in which a great amount of gas was emitted because of a violent reaction between urea and ammonia nitrate.

Furthermore, in regard to the pore size distribution of the various samples, most of the pores for Fe6Al4(ig) were mesopores within 2–50 nm. When it underwent 2 h of calcination and transformed into Fe6Al4(si), many mesopores in Fe6Al4(ig) increased in size and transformed into macropores (>50 nm) because of the agglomeration and compaction of various neighboring mespores. As references, the pore size distribution of the homemade separate phases, including α-Fe_2O_3(si) and α-Al_2O_3(si), were analyzed. The pores inside the α-Fe_2O_3(si) particle were mainly macropores larger than 50 nm, but for α-Al_2O_3(si), even experiencing serious calcination, most pores were still smaller than 50 nm, with two peaks occurring around 10 and 20 nm, respectively.

Considering the fact that the CLC reduction reaction of the OC with fuel in FR is generally considered as the limiting step relative to the oxidation reaction of the reduced OC in the AR; therefore in order to obtain a preliminary understanding of the reaction characteristics of the SGCS-synthesized Fe6Al4 OC, a reduction reaction with H_2 of 50% volume concentration was performed using thermogravimetric analysis (TGA). The results are shown in Fig. 16.2. The TGA experimental results showed three distinct peaks for the differential weight loss rate (DTG) in the reduction reaction of the burned Fe6Al4(si), which occurred at 384.43°C, 405.26°C, and 813.20°C, accompanied by a left mass on the TG curve reaching 98.19%, 90.92%, and 83.16%, respectively. According to the preceding XRD phase identification, Fe6Al4(ig) consisted of $FeAl_2O_4$ and α-Fe_2O_3. With the reasonable expectation of the complete reaction of both $FeAl_2O_4$ and α-Fe_2O_3 with H_2, the mass fraction of α-Fe_2O_3 present in Fe6Al4(ig) was calculated as 53.92% through the net mass loss for the reduction of α-Fe_2O_3 with H_2 to be divided by the total net loss for the complete reduction of Fe6Al4(ig) with H_2. Furthermore, the following three reactions were inferred. First, ranging from 253°C to 411°C, α-Fe_2O_3 would react with H_2 and form Fe_3O_4 with 98.19% left. Then from 411°C to 611°C, Fe_3O_4 would completely react into Fe with 90.92% left. Finally, with temperatures increasing upward from 611°C, the reduction reaction of $FeAl_2O_4$ would proceed until it terminated at 915°C with 83.20% left. Therefore, in the study, α-Fe_2O_3 present in Fe6Al4(ig) reacting with H_2 experienced two reaction steps, first from

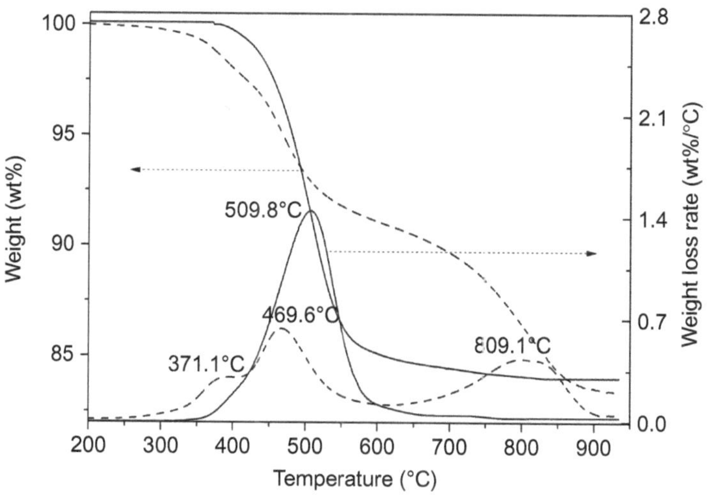

FIG. 16.2 Temperature-programed reduction of the SGCS-made Fe_2O_3/Al_2O_3 OC. *(From Wang BW, Yan R, Lee DH, Zheng Y, Zhao HB, Zheng CG. Characterization and evaluation of Fe_2O_3/ Al_2O_3 oxygen carrier prepared by sol-gel combustion synthesis. J Anal Appl Pyrolysis 2011; 91(1):105–13.)*

α-Fe_2O_3 to Fe_3O_4 and then to Fe. After that, $FeAl_2O_4$ reacted with H_2 and formed Fe and Al_2O_3. Unlike the reaction of Fe6Al4(ig) with H_2, there was only one DTG peak, centered at 506.45°C, for the reduction reaction of Fe6Al4(si) with H_2, which consisted of α-Fe_2O_3 and α-Al_2O_3, as previously mentioned. This sole peak implied that the α-Fe_2O_3 present in Fe6Al4(si) OC directly reacted with H_2 and formed Fe in temperatures ranging from 351°C to 623°C in a single-step reaction.

Furthermore, for both Fe6Al4(ig) and Fe6Al4(si) reacting with H_2, the mass loss in the TG curve actually resulted from the loss of the lattice oxygen present in the oxide carrier. Accordingly, the DTG value implied the transfer rate of lattice oxygen in the oxides to H_2. A higher DTG value meant a higher transfer rate. Therefore the DTG value for Fe6Al4(si) was 1.42%/°C, far more than any of the DTG values of the three sequential reactions involved in Fe6Al4(ig) reacting with H_2, which meant Fe6Al4(si) had a greater oxygen transfer rate and was more suitable for CLC application, because when OC was used in a CLC system, the higher oxygen transfer rate meant less bed inventory in the FR and less recirculation of OC from the AR to the FR [17].

In addition to informative data on the SGCS-made Fe6Al4 OC on the phase identification, structural characteristics, temperature-programed reduction (TPR), good redox properties, and sintering resistance are other concerns related to sustaining good reactivity stability of the OC adopted in CLC processes. In order to evaluate the redox and sintering-resistant characteristics for the synthesized Fe6Al4 OC, four cyclic times of reducing and oxidizing reactions between the sintered Fe6Al4 OC with H_2 in 50 vol % with the balance N_2 and then air were performed using TGA. The results are shown in Fig. 16.3. The results of the mass loss (i.e., TG curve) and those of the corresponding differential mass loss rate (i.e., DTG curve) during the four redox processes showed that although the first redox reaction rates of Fe_2O_3/Al_2O_3 with H_2 and air agreed at 8.2%/min and 41.8%/min, as the reaction continued, the rates related to the reduction by H_2 and oxidization by air increased greatly, stabilizing at about 20%/min and 50%/min, respectively. The lower reduction rate for the first cycle mainly arose from the dense layer of Fe_2O_3 that initially formed on the outer surface and from the uneven pore size distribution for the fresh Fe6Al4 OC during preparation. Both the dense layer formed outside the OC particle and the uneven pore size distribution prevented the H_2 from accessing the active reaction surface. However, after the first redox cycle, abundant pores of different sizes developed over the ensuing redox cycle, with an enlargement in the reduction stage and a contraction in the oxidization stage and with variation in the molar volume of reduction product Fe and oxidization product Fe_2O_3, which was greatly beneficial to an increase in the reaction rates. Furthermore, the stable redox rates that occurred after the first redox cycle and the rich pore distribution also demonstrated that the synthesized Fe6Al4 OC possessed good sintering resistance, which helped sustain the reactivity over many redox cycles in practical CLC systems. Overall, the

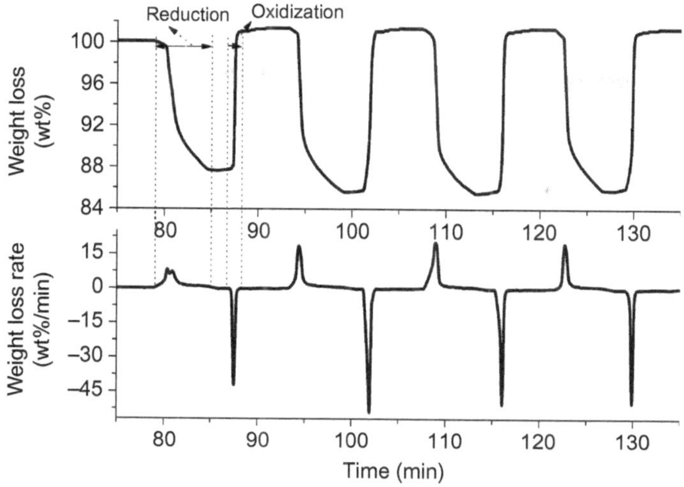

FIG. 16.3 Four redox cycles of the SGCS-made Fe_2O_3/Al_2O_3 OC. *(From Wang BW, Yan R, Lee DH, Zheng Y, Zhao HB, Zheng CG. Characterization and evaluation of Fe_2O_3/Al_2O_3 oxygen carrier prepared by sol-gel combustion synthesis. J Anal Appl Pyrolysis 2011; 91(1):105–13.)*

preliminary characterization for the prepared Fe6Al4 OC using the novel SGCS method developed by our group showed the expected advantages and a great potential for development and use in actual CLC processes.

16.3 CHARACTERISTICS OF COAL-FUELED CLC

16.3.1 Reaction Characteristics of Coal-Fueled CLC

Direct use of coal as fuel for CLC application to simultaneously realize energy provision and CO_2 capture is greatly advantageous because of the wide distribution, easy availability, low cost, and high carbon intensity of coal. Considering the complexity of the pyrolysis/gasification of coal, the oxygen transfer of the OC, and their interactions, using the reaction of typical Chinese coal with Fe_2O_3/Al_2O_3 as an example, we systematically investigated coal reaction characteristics, oxygen transfer and utilization, and their interactions.

First, reactions of three Chinese coals of different ranks with Fe_2O_3 OC under the stoichiometrical condition were performed in an N_2 atmosphere via TGA analysis. The heating rate of 35°C/min was chosen. The temperature was increased to 850°C with an isothermal period of 10 min in order to achieve a sufficient reaction of coal. As a baseline, SGCS-made Fe_2O_3 reactions with the three Chinese coals were conducted without the presence of the inert support Al_2O_3. The experimental results indicated that in temperatures ranging from 200°C to 850°C, the highly reactive Chinese coal of low rank designated as PDS reacted more easily with only Fe_2O_3 at a higher final weight loss (2.5 wt%) than did the other two coals of higher ranks, especially the typical Chinese anthracite with a weight loss of less than 1.0 wt%. However, the final weight losses for the reaction of different coal samples with Fe_2O_3/Al_2O_3 as OC were lower than the corresponding losses for the reaction of coal with Fe_2O_3, which was due to the introduction of the inert support Al_2O_3 into the coal. Also, three reaction stages were observed in the DTG curves of the three coals, which was similar to the finding of Siriwardane et al. [18] Moreover, the corresponding three reactions' characteristic temperatures for the most reactive Chinese coal of the lowest rank were 496°C, 619°C, and 810°C, respectively. The first reaction most likely resulted from the partial pyrolysis of the coal in its mixture with Fe_2O_3, at the mass loss rate of 0.1682 wt%/min. Note that this value (0.1682 wt%/min) was much lower than the pyrolysis rate (5.8 wt%/min) seen for PDS at ~484°C, but if multiplied by 38.2 (i.e., the mass ratio of Fe_2O_3 to the selected coal at the stoichiometric oxygen excess number), the calculated mass loss rate of 6.4 wt%/min was reasonable. The two ensuing reactions happened at 619°C and 810°C, which was attributed to the reaction of the coal to the Fe_2O_3 through the combined primary and secondary pyrolysis of the coal. Such an assumption was also validated by the evolved gas products from the reaction of the coal with Fe_2O_3, which was identified by Fourier transfer infrared analysis at the different reaction stages. The only major gas product was CO_2 at 619°C and 810°C, which came from the reaction of the selected coal with Fe_2O_3 for the later two-stage pyrolysis.

Furthermore, the TGA experimental results showed that all three coals demonstrated three-stage reactions, but as a coal's rank increased, the corresponding characteristic temperature T_m was found to shift to higher values, similar to the trend found for coal pyrolysis alone, which indicated that coal of a higher rank had more difficulty reacting with Fe_2O_3. Thus the application of Fe_2O_3 OC in a coal-based CLC system indicated that low-ranked coal with high volatile matter is preferred.

Finally, it is also necessary to consider the inert support of OC (such as Al_2O_3) in improving resistance to sintering, abrasion, and fragmentation over the cyclic reactions in the CLC system; therefore reactions between the three coals and Fe_2O_3/Al_2O_3 were also investigated. According to the experimental results on the effect of inert support Al_2O_3 on the reaction of coal with Fe_2O_3, similar three-stage reactions occurred, and the corresponding characteristic temperatures also increased with the enhanced coal rank. The characteristic temperature at the first reaction stage was lower than that of Fe_2O_3 without Al_2O_3 because of the improved oxygen transfer aided by the inert support Al_2O_3. However, the later two characteristic temperatures were much higher than those in the case where the coal reacted only with Fe_2O_3 because the effective contact between the Fe_2O_3 with the reactant from coal was weakened as a result of the hindrance effect from the inert Al_2O_3. Only TGA results are presented here, which is a slightly unrealistic situation for the CLC system. Nevertheless, the TG results above provided an insightful understanding on the reaction characteristics of coal with Fe_2O_3 OC, which is advantageous for its application in the realistic CLC process.

16.3.2 Oxygen Ulitilzation During Coal-Fueled Chemcial Looping Combustion

In order to further illuminate the transfer of lattice oxygen involved in the Fe_2O_3/Al_2O_3 OC used and the transformation of coal during its reaction with the OC used, two conversion indexes were developed by our group and presented here [19,20].

First, in the mixture conversion of coal with OC, to reflect the interaction of coal with OC, $X_{\text{coal-OC}}$ was defined as

$$X_{\text{coal-OC}}(t) = \frac{W_{\text{coal-OC}}(t=0) - W_{\text{coal-OC}}(t)}{x_1 \Delta W_{\text{OC}} + x_2 \Delta W_{\text{coal}}} \tag{16.3}$$

$$f = x_1 / x_2 \tag{16.4}$$

where $X_{\text{coal-OC}}$ is the mixture conversion for coal with Fe_2O_3/Al_2O_3 or its reference oxide Fe_2O_3 (%); $W_{\text{coal-OC}}$ ($t=0$) and $W_{\text{coal-OC}}$ (t) are the initial and instantaneous weight loss for the mixture (wt%); x_1 and x_2 are the mass fractions of OC and coal in their mixture (wt%), f is the mass ratio of OC to coal, and ΔW_{OC} and ΔW_{coal} are the maximum weight losses of OC and coal, respectively.

Furthermore, the OC conversion X_{OC} present in the mixture of OC with coal was defined to reveal the real limiting step involved for the reaction of coal with OC. Based on the conservation of the total weight throughout the reaction of OC with coal, the OC conversion can be calculated as

$$W_{\text{coal-OC}}(t) = x_1 W_{\text{OC}}(t) + x_2 W_{\text{coal}}(t) \tag{16.5}$$

$$X_{\text{OC(coal)}} = \frac{W_{\text{coal-OC}}(t) - x_2 W_{\text{coal}}(t)}{x_1 \Delta W_{\text{OC}}} \tag{16.6}$$

where $W_{\text{OC}}(t)$, $W_{\text{coal}}(t)$, and $W_{\text{coal-OC}}$ (t) are the instantaneous weight losses for the coal, the OC, and their mixture at time t. The definitions of the other characters in Eqs. (16.5), (16.6) are the same as those in Eqs. (16.3), (16.4).

To reveal the interaction of coal with OC, based on the previous TG experimental results for pyrolysis of the typical Chinese anthracite and its reaction with Fe_2O_3/Al_2O_3 OC or the reference oxide Fe_2O_3, such two conversion indexes were calculated using Eqs. (16.3), (16.6), respectively, including the OC conversion index X_{OC} for OC alone in its mixture with the lowly reactive Chinese anthracite and the mixture conversion index $X_{\text{coal + OC}}$ for the coal with different OCs were calculated using Eqs. (16.3), (16.6), respectively. According to the calculated results for the two conversion indexes, shown in Fig. 16.4, [20] in the mixture conversion indexes $X_{\text{OC+coal}}$ for Fe_2O_3/Al_2O_3 with the coal, both mixture conversion indexes were below 3% for the coal's reaction with Fe_2O_3/Al_2O_3 or its reference oxide Fe_2O_3. Furthermore, a comparison of the mixture conversion indexes with their relative OC conversion indexes revealed the real rate-limited step for the reaction of the coal with the OCs, and intersections were found between the curves for the mixture conversion index $X_{Fe_2O_3\text{+coal}}$ and the

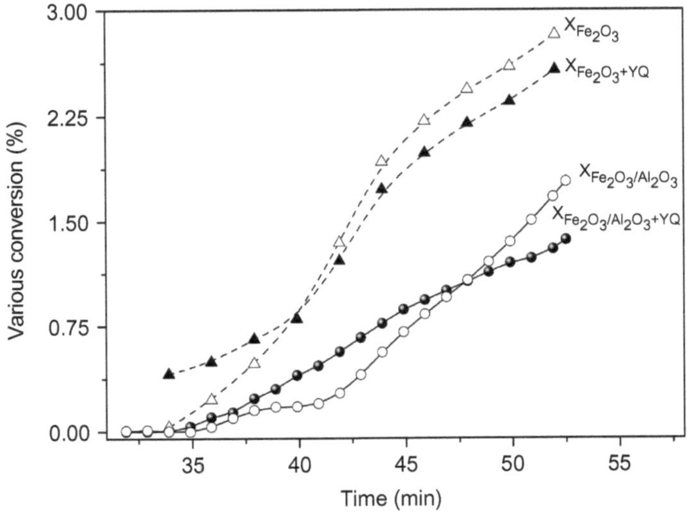

FIG. 16.4 Conversion indexes for YQ reaction with SGCS-made Fe_2O_3/Al_2O_3 OC. *(From Wang BW, Zhao HB, Zheng Y, Liu ZH, Yan, R, Zheng CG. Chemical looping combustion of a Chinese anthracite with Fe_2O_3-based and CuO-based oxygen carriers. Fuel Processing Technol 2012;96(1):104–15.)*

Fe_2O_3 conversion index of $X_{Fe_2O_3}$ or the mixture conversion index $X_{Fe_2O_3/Al_2O_3+coal}$ and Fe_2O_3/Al_2O_3 conversion index $X_{Fe_2O_3/Al_2O_3}$, respectively. Also, the mixture conversion index $X_{Fe_2O_3+coal}$ or $X_{Fe_2O_3/Al_2O_3+coal}$ was higher than that of the OC conversion index $X_{Fe_2O_3}$ or $X_{Fe_2O_3/Al_2O_3}$ before the relative intersection points, which implied that at these reaction stages, the gaseous products emitted from the less reactive coal pyrolysis was sufficient and that the reaction of the coal with Fe_2O_3 was determined by the slow lattice oxygen transfer from Fe_2O_3-based OC to the gaseous products produced from the coal. However, after the intersection points with the temperature increased, the variation trend changed to the contrary, and the mixture conversion index $X_{Fe_2O_3+coal}$ or $X_{Fe_2O_3/Al_2O_3+coal}$ was lower than that of the OC conversion index $X_{Fe_2O_3}$ or $X_{Fe_2O_3/Al_2O_3}$, which meant that the reaction of the coal with Fe_2O_3 or Fe_2O_3/Al_2O_3 was limited by the insufficient supply of gaseous products from pyrolysis of the Chinese anthracite.

These two defined conversion indexes shed light on coal conversion, on oxygen transfer from OC, and on their interactions. Also, the complex reaction processes could be quantitatively evaluated, and the optimized reaction conditions to meet the oxygen transfer and fuel provision could be determined. Thus these two indexes have great potential in the conversion of coal in real CLC systems.

16.3.3 Molecular-Scale Investigation on the Coal-Fueled CLC

Considering that the chemical structure of coal is closely correlated with its reactivity [21], the next step was more in-depth investigation of the variations in the chemical structure of coal during its reaction with OC in order to determine (1) the real rate-limiting factors in the conversion of coal during CLC at the molecular scale and (2) to promote the full conversion of coal.

X-ray photoelectron spectrocopy (XPS) is a powerful nondestructive analysis technique. It is relatively quantitative and surface-sensitive and has been widely applied in research on the elemental composition and chemical state of coal [22]. Therefore, the distribution and evolution of the carbon-related functional groups in the selected Chinese typical raw coal (designated as LZ) and its solid products with Fe_2O_3 were analyzed to gain a deeper insight into the evolution of coal during its reaction with Fe_2O_3. According to our previous research on the distribution of the various C-containing functional groups in the original LZ coal, in a typical Chinese LZ bituminous coal, four such groups were identified as the aromatic and aliphatic carbon (C=C/C–H), carbon bound with one oxygen by a single bond (C–O), carbon bound to oxygen by two bonds (C=O), and carbon bound to oxygen by three bonds ($O{=}C{-}O/COO^-$). Their relative contents in the LZ coal were determined as 71.23%, 11.90%, 11.60%, and 1.41% [23], respectively, by deconvolution and further quantifying the relative area of each curve-fitted XPS spectra for the LZ original coal. Meanwhile, two other carbon components were also found, including carbonates (mainly calcite by our previous analysis of LZ coal) and the π-π*shake-up from the aromatic carbon matrix in coal, but their contents were less than those of the preceding four carbon functional groups. Overall, it was found that the C=C/C–H groups were dominant.

Furthermore, after the reaction of LZ coal with Fe_2O_3, the relative contents of various carbon groups changed greatly. The main C1's envelope shifted to a higher binding energy with more lattice oxygen of Fe_2O_3 incorporated into the carbon matrix of the LZ coal. Thus, among the four main carbon functional groups in LZ coal, the relative fractions of the oxygen-containing functional groups increased. As shown in Table 16.1, compared to the original LZ coal, the relative fractions of C–O, C=O/O–C–O, and COO^- increased to 23.34%, 12.17%, and 5.18%, respectively. The C–O groups' relative fraction increased more than two times because of the propensity for the oxidation of C=C/C–H to form C–O groups. Meanwhile, the relative content of O=C–O after the LZ coal's reaction with Fe_2O_3 at $\Phi=1.0$ increased more than three times from 1.41% in the LZ coal to 5.18% in the solid products of the LZ coal's reaction

TABLE 16.1 Relative Carbon Content of LZ Original Coal and Its Reaction With Fe_2O_3 OC (%)

Sample	C–C/C–H	C–O	C=O/O–C–O	COO–	Carbonates
LZ coal	71.23	11.9	11.6	1.41	3.87
LZ + Fe_2O_3	59.32	23.34	12.17	5.18	N/A

with Fe_2O_3, although the thermal stability of O=C–O was quite low and easily decomposed to form CO_2. Therefore the high increase in the relative content of O=C–O implied that the preferentially formed C–O groups eventually converted into O=C–O groups, whereas C=O/O–C–O groups relative content in the solid product after the reaction of LZ coal with Fe_2O_3 increased slightly from 11.6% in the original LZ coal to 12.17%. The minimal increase implied that, although the formed C–O groups were further converted to C=O/O–C–O groups, these groups acted as the intermediate groups that not only disintegrated to emit CO but also further converted to form O=C–O groups. In addition, as shown in Table 16.1, after the reaction of LZ coal with Fe_2O_3, with an increase in the relative contents of these three oxygen-containing groups, the relative contents of the C=C/C–H groups involved decreased to 59.32%. However, it should be noted that the C=C/C–H groups still were not fully converted and were dominant among all the carbon groups in the LZ coal. Therefore, at the molecular scale, the limiting factor for conversion of LZ coal during its reaction with Fe_2O_3 was the conversion of C=C/C–H groups in the LZ coal.

16.4 SULFUR EVOLUTION AND INTERACTION MECHANISM WITH OC

Sulfur is one of the main elements in coal. In a coal-fueled CLC system, sulfur evolution is a great concern from both an environmental and a CLC operational perspective. On the one hand, different sulfur species will evolve and further interact with the reduced OC in the FR of a CLC system to form different solid sulfur compounds, such as sulfides, sulfites, and sulfates, which make the OC poisonous and partially deactivated. Because of the low melting points of the formed sulfur compounds, potential agglomeration and even defluidization of the bed inventory will occur. The gaseous sulfur species generated in the FR during coal's reaction with the OC greatly harms the compression, transport, and storage of CO_2, and also lowers the CO_2 purity, resulting in difficulty in its separation downstream. On the other hand, in the CLC system's AR, the entrained solid sulfur compounds are further oxidized to form gaseous SO_2, which without further desulfurization will pollute the atmosphere. Therefore, because of the detrimental effect of sulfur and to find ways to effectively reduce its harm throughout the whole CLC process, it is essential to obtain deeper insight into sulfur evolution and its interaction mechanisms with OC.

According to our previous experimental investigations [22,24,25], all these factors (the big differences in the rank and sources of Chinese coal, the great variation in the sulfur forms and content of coals, the complex atmosphere formed in the FR with the enriched steam and CO_2, and so on) will result in the evolution of sulfur and its further reaction with the reduced OC quite complicated. Although extensive experimental research has addressed these issues, a full understanding of the fundamental reaction mechanisms involved in the interactions between the gaseous sulfur species and the reduced OC has not been reached. Therefore, in order to fill in the gaps and further seek explicit and complete clarity on the interaction mechanisms between the reduced OC and the gaseous sulfur species, a theoretical calculation is presented in this chapter. Based on the quantum chemical methods and the widely used density functional theory (DFT), the reaction of NiO with H_2S is exemplified, because NiO OC is more sensitive to sulfur poisoning than Fe_2O_3 and other metal-based OCs, although the Fe_2O_3 OC is focused on in this chapter. On the other hand, H_2S is chosen as the typical gaseous sulfur species because it is always either in the direct use of coal as fuel in the CLC process or in gaseous fuels such as natural gas, refinery gas, and syngas.

To better understand the detrimental effect of the gaseous sulfur species and its harm to the OC, using the Cambridge Sequential Total Energy Package (CASTEP, a commercial software package), our team conducted a DFT-calculated NiO reaction with only CO as a baseline without H_2S added. As to our previous research [26], the gas-solid reaction was dominant in the CLC process as accompanied by the gas adsorption on the solid surface as the primary step prior to its initiation to reaction with the OC, therefore two reference cases were considered for NiO reaction with CO: CO adsorption on the NiO(001) perfect surface and on the defective surface resulting from NiO reduction with oxygen vacancies formed under high heat but low-oxygen reducible atmosphere in the CLC process. The DFT calculation indicated that the CO physically adsorbed on the perfect NiO(001) surface via the Ni top site, while for the defect surface of NiO(001), CO adsorption occurred through the formed oxygen vacancies. Furthermore, far different than in the CLC process, instead of the oxidation of CO by the lattice oxygen in the NiO, the DFT calculation found that CO could be oxidized to form CO_2 through the adsorbed O atom on the NiO. This finding greatly improved our understanding of the reaction of NiO with CO, which has great implications both for improving OC reactivity and for promoting full conversion of the fuel.

Furthermore, in the presence of H_2S, according to our recent research [27], a DFT calculation of the NiO reaction with CO was conducted, and the detrimental effect of H_2S on the CO adsorption was clarified at the atomic scale. It was concluded that H_2S was preferential over the CO for adsorption on the perfect NiO(001) surface, and also altered the exothermic state of only CO molecule adsorption on the NiO(001) surface without H_2S added to the slight enothermic state with the presence of H_2S, and thus further hindered the reaction of CO with NiO. For the defective NiO(001) surfaces that more easily formed under CLC conditions, H_2S sequential dehydrogenation through $H_2S \rightarrow HS + H$ and $HS \rightarrow H + S$ occurred, and the former dehydrogenation pathway was determined as being the rate-limiting step because of its higher dissociation energy value (2.45 eV), while the S was more prone to adsorption on the defective NiO(001) surface with much stronger chemical adsorption energy than both H_2S and HS through the formed O vacancy to form nickel sulfide. Of course, H_2S presence not only hindered the reaction of CO with NiO; it was also found that the presence of H_2S limited the rate of O anion diffusion out of the NiO and negatively influenced the reaction rate of CO with NiO. Therefore fuel desulfurization should be further studied and improved prior to its use in CLC systems.

16.5 MINERALS EVOLUTION DURING COAL-FUELED CLC

Minerals other than sulfur are inherent in coal, and their evolution and interaction with the OC presented many potential problems during the CLC process [28]. These minerals not only evolved but also interacted with the reduced OC to form inert compounds, which deteriorated the reactivity of the OC and also shortened its lifetime. Also, because of the low melting point of the formed inert compounds, a potential for agglomeration of the reduced OC and ash occurred, which could potentially create difficulty in the operation of CLC systems. However, the little research conducted in this area has been confined to laboratory-scale experimental and thermodynamical investigations on the interaction between the OC and ash.

To study the minerals' evolution in coal and further their interaction with the reduced OC, our team selected, as an example, the reaction of a Chinese bituminous coal designated as PDS with Fe_2O_3/Al_2O_3 OC. The morphology and elemental composition of both residues from PDS coal pyrolysis and its reaction with Fe_2O_3 OC and Fe_2O_3/Al_2O_3 (mass ratio 8:2) were characterized using scanning electron microscopy-energy dispersive X-ray spectrometry(SEM-EDX), as shown in Fig. 16.5. The results indicated that the PDS pyrolysis's solid residue mainly consisted of fragmentary white discrete particles and a relatively dark bulk. The EDX analysis showed that the elemental composition of the discrete particles was identified as minerals, mainly consisting of Si, Al, and O. The presence of Al, Si, and O may have resulted from the minerals that naturally occur in the carbon matrix. Furthermore, XRD analysis of the residue from PDS pyrolysis indicated that SiO_2 and some silicates, such as $Fe_2Si_2O_5(OH)_4$, $Al_2Si_2O_5(OH)_4$, and $CaAl_2Si_2O_8{\cdot}4H_2O$, were the main mineralogical compositions. The dark bulk may have been caused by the carbon matrix left after PDS pyrolysis with the atomic fraction of C reached 81.71%, which was much higher than that presented in the discrete particles (32.0% of the atomic fraction). In addition, although most of the residue should be ascribed to the carbon matrix by EDX analysis, it cannot be identified by XRD analysis because of its amorphous characteristics.

In contrast to the solid residue from only pyrolysis of PDS, for the reaction of PDS with Fe_2O_3, the SEM pattern indicated that the solid reaction residue mainly consisted of various discrete white particles of different sizes. Two spots in the SEM pattern were selected and scanned by EDX. It showed that the elemental compositions in the two spots were almost the same, mainly consisting of Fe, O, and C, with their atomic fractions equal to 44%, 50%, and 5.6%, respectively. The valence of the reduced Fe_2O_3 was mainly determined by the Fe_2O_3 to carbonaceous material mass ratio [29,30], and the contained iron oxide could be further inferred as Fe_3O_4 by the Fe to O atomic ratio. When Fe_2O_3 OC was used in a CLC system, the potential defluidization of the FR arose from the deep reduction of Fe_2O_3 further into the lower valence than Fe_3O_4 such as the deficient FeO oxides [31]. Based on our research, if such an undesirable circumstance can be avoided, a high Fe_2O_3 to coal mass ratio should be adopted and applied in real CLC systems.

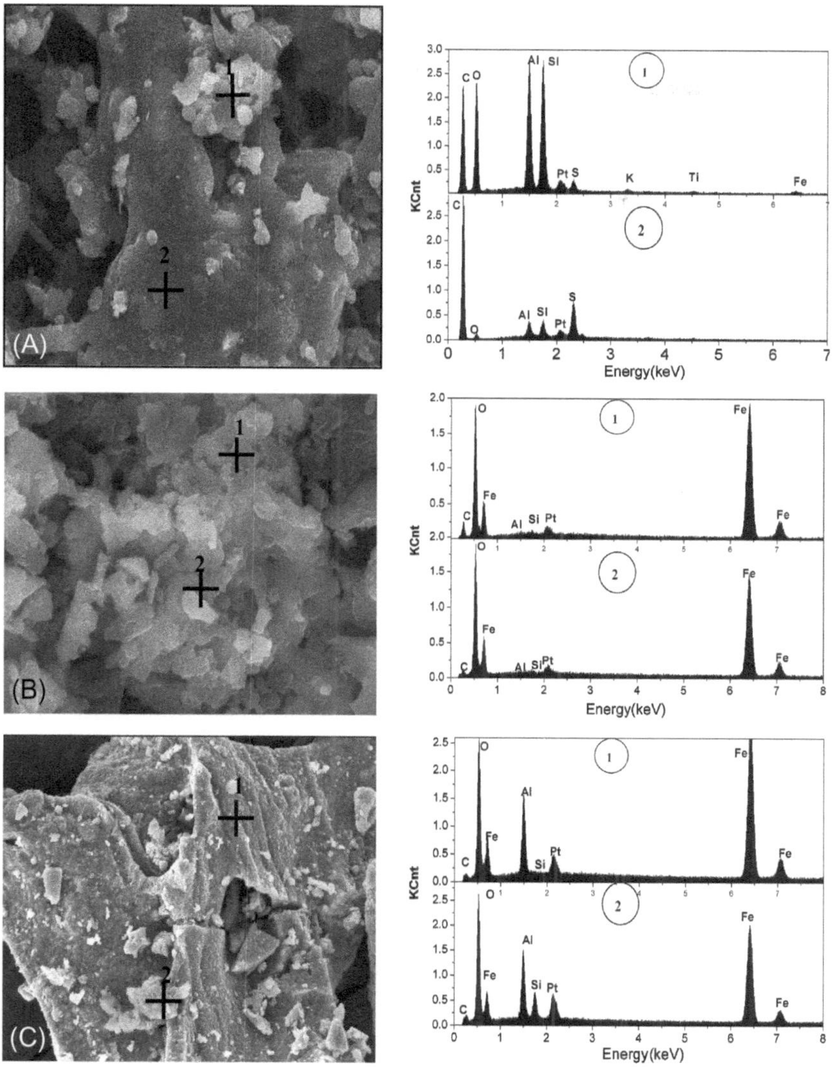

FIG. 16.5 SEM-EDX analysis of the reaction of PDS coal with SGCS-made Fe_2O_3/Al_2O_3 OC: (A) PDS pyrolysis, (B) PDS reaction with Fe_2O_3, and (C) PDS reaction with Fe_2O_3/Al_2O_3 OC. *(From Wang B, Yan R, Zheng Y, Zhao H, Zheng C. Mechanistic investigation of chemical looping combustion of coal with Fe_2O_3 oxygen carrier. Fuel 2011;90(7):2359–66.)*

In order to study the role that inert support played in the preceding research, the solid residues for the reaction of Fe_2O_3/Al_2O_3 with PDS were further investigated with FSEM-EDX. The experimental results showed no sintering due to the high resistance to sintering for Al_2O_3. Also, the atomic fraction of C was around 3%, much less than that of the C left after the preceding reaction of PDS with Fe_2O_3, which proved that the presence of Al_2O_3 improved the reactivity of Fe_2O_3-based OC and promoted the conversion of PDS. Although the Al compositions on the two spots were evenly distributed, with atomic fractions around 15%, the distribution of Si compositions on the two spots differed greatly, and the atomic fraction of Si on one spot was nearly zero, but on the other spot was 4.36%. In order to identify the detailed phases involved after the reaction of PDS with Fe_2O_3 and Fe_2O_3/Al_2O_3, the residues left were identified by XRD. After the reaction of Fe_2O_3 with PDS, the residue mainly consisted of Fe_2O_3, Fe_3O_4, and Fe_2SiO_4. The Fe_2O_3 that was left mainly resulted early from the partial pyrolysis of PDS at the pyrolysis characteristic temperature of 496°C, and the pyrolysis products could not react further with Fe_2O_3 at such a low temperature. Of course, such a circumstance would not occur in real CLC systems. The presence of Fe_2SiO_4 was related closely to the interaction of the reduced iron oxide with SiO_2, which occurs naturally in coal, especially in a local reducing atmosphere. For the reaction of PDS with Fe_2O_3/Al_2O_3 (mass ratio 8:2), the residue after the reaction mainly consisted of Fe_2O_3, Fe_3O_4, and $FeAl_2O_4$. The Fe_2O_3 left was due to the partial pyrolysis of PDS at 487°C prior to the initiation of PDS reacting with Fe_2O_3. The $FeAl_2O_4$ resulted from the interaction of the reduced iron oxide with Al_2O_3, which was included in the Fe_2O_3/Al_2O_3 OC [32]. However, here Fe_2SiO_4 inherent was not identified in the XRD pattern due to the smaller amount of Fe_2SiO_4 formed, likely because the reduced iron oxide reacted more easily with Al_2O_3 than with SiO_2. At a reaction temperature of ~850°C, the reaction enthalpy of the reduced iron oxide with Al_2O_3 was −14kJ, far lower than that of the reduced oxide with SiO_2, which was −4kJ. In addition, the identified phases by XRD for the PDS ash sample mainly contained SiO_2, Fe_2O_3, $Al_4Ti_2SiO_{12}$, $CaAl_2Si_2O_8$, and several silicates, which were far different from the evolved minerals in the reaction between PDS and Fe_2O_3 or Fe_2O_3/Al_2O_3 as previously iterated.

As previously pointed out, inert support was the necessary factor in the OC's ability to sustain good performance over cyclic reactions with coal and air. Al_2O_3 was adopted as a typical inert support. In addition to good resistance to sintering and sustainable structural integrity without fragmentation in the preceding experimiental investigation, its effect on the evolution of minerals and interactions with the reduced OC during the reaction between PDS bituminous coal and Fe_2O_3 was further exemplified using the thermodyanmical simulation method [33]. The simulated results showed that for the PDS reaction with the reference oxide Fe_2O_3 without inert support Al_2O_3 introduced, the main Al species for PDS with Fe_2O_3 was Al_2SiO_5, but when some fraction of Al_2O_3 was introduced into Fe_2O_3, the main Al species was changed to $FeAl_2O_4$, which was completely consistent with the previous FESM-EDX and XRD analyses. The fraction of $FeAl_2O_4$ was found to decrease with temperature because more FeO was produced and further interaction with Al_2O_3 was introduced. Obviously, the interaction of Al_2O_3 with the deeply reduced Fe_2O_3 counterpart than Fe_3O_4 would easily form $FeAl_2O_4$ intermediates and hinder full regeneration of the reduced Fe_2O_3 with air.

Our study showed that sufficient attention should be paid to inert iron compounds such as Fe_2SiO_4 and $FeAl_2O_4$, which were formed from the interactions of the reduced iron oxide with the SiO_2 occurring inherently in coal or with the Al_2O_3 included in the Fe_2O_3 OC. These compounds not only lower the melting point of Fe_2O_3 OC but also deteriorate the reactivity of OC [34]. Thus further research on the mechanisms of forming inert iron compounds are needed. Also studies focusing on effective and practical means for separating coal ash from the reduced OC through density or magnetic differences in CLC systems under long-term, continuous operations are needed.

16.6 SUMMARY

CLC has quickly developed worldwide as a new-generation oxy-fuel combustion technology and has attracted a great deal of attention for industrial-scale application. In this chapter, taking the widely applied Fe_2O_3/Al_2O_3 OC as an example, aspects of great interest about the direct use of coal in CLC processes were discussed, including its preparation by using the novel SG combustion synthesis method and characterization, and direct reaction characteristics with typical Chinese coal were systematically introduced. Also, sulfur evolution and interaction mechanisms with the reduced OC were clarified. In addition, as the essential parts of coal, the dominant minerals' distribution, evolution, and interactions with the OC during its reaction with coal were provided in detail. Our groups' study and the information presented in this chapter provide a comprehensive perspective on CLC of coal.

REFERENCES

[1] CO_2 emission from fuel combustion-highlights. (2015 ed.), International Energy Agency (IEA); Paris, France.

[2] International Energy Outlook 2016 (DOE/EIA-0484), US Energy Information Administration; Washington, DC.

[3] IPCC. Climate change 2014: mitigation of climate change. United Kingdom: Cambridge University Press; 2014.

[4] MacDowell N, Florin N, Buchard A, et al. An overview of CO_2 capture technologies. Energy Environmental Science 2010;3(11):1645–69.

[5] Adanez J, Abad A, Garcia-Labiano F, Gayan P, de Diego LF. Progress in chemical-looping combustion and reforming technologies. Prog Energy Combust Sci 2012;38(2):215–82.

[6] Gilliland RE, Lewis WK. Production of pure carbon dioxide. Patent No: US2665972; 1954.

[7] Ishida M, Jin H. A new advanced power-generated system using chemical looping combustion. Energy 1994;19(4):415–22.

[8] Markström P, Linderholm C, Lyngfelt A. Chemical-looping combustion of solid fuels-Design and operation of a 100 kW unit with bituminous coal. Int J Greenhouse Gas Control 2013;15:150–62.

[9] Ohlemüller P, Busch J-P, Reitz M, Ströhle J, Epple B. Chemical-looping combustion of hard coal: autothermal operation of a $1MW_{th}$ pilot plant. J Energy Resour Technol 2016;138(4), 042203.

[10] Chamberland R, Andrus H, Edberg C. Alstom's chemical looping combustion technology with CO_2 capture for new and existing coal-fired power plants. In: US DOE/NETL CO2 capture technology meeting, Pittsburgh, PA; 2015.

[11] Luo SW, Zeng L, Fan L-S. Chemical looping technology: oxygen carrier characteristics. Annu Rev Chem Biomol Eng 2015;6:53–75.

[12] Johansson M, Mattisson T, Lyngfelt A. Comparison of oxygen carriers for chemical-looping combustion. Therm Sci 2006;10(3):93–107.

[13] Gingaşu D, Mîndru I, Patron L, et al. Copper ferrite obtained by two ferrite obtained by two "soft chemistry" routes. J Alloys Compd 2006;425(1–2):357–61.

[14] Wang BW, Yan R, Lee DH, Zheng Y, Zhao HB, Zheng CG. Characterization and evaluation of Fe_2O_3/Al_2O_3 oxygen carrier prepared by sol-gel combustion synthesis. J Anal Appl Pyrolysis 2011;91:105–13.
[15] Wang BW, Song XY, Wang ZH, Zheng CG. Preparation and application of the sol-gel combustion synthesis-made $CaO/CaZrO_3$ sorbent for cyclic CO_2 capture through the severe calcination condition. Chin J Chem Eng 2014;22(9):991–9.
[16] Hang XX, Jiang XM, Yu LJ, Cui ZG. Change of pore structure of oil shale particles during combustion. Part 1. Evolution mechanism. Energy Fuel 2006;20(6):2408–12.
[17] Lyngfelt A, Leckner B, Mattisson T. A fluidized-bed combustion process with inherent CO_2 separation; application of chemical-looping combustion. Chem Eng Sci 2001;56(10):3101–13.
[18] Siriwardane R, Tian HJ, Richards G, Simonyi T, Poston J. Chemical looping combustion of coal with metal oxide oxygen carrier and coal as fuel. Energy Fuel 2009;23(8):3885–92.
[19] Wang BW, Yan R, Zhao HB, Zheng Y, Zheng CG. Investigation of chemical looping combustion of coal with $CuFe_2O_4$ oxygen carrier. Energy Fuel 2011;25(7):3344–54.
[20] Wang BW, Zhao HB, Zheng Y, Liu ZH, Yan R, Zheng CG. Chemical looping combustion of a Chinese anthracite with Fe_2O_3-based and CuO-based oxygen carriers. Fuel Process Technol 2012;96(1):104–15.
[21] Van Heek KH, Hodek W. Structure and pyrolysis behavior of different coals and relevant model substances. Fuel 1994;73:886–96.
[22] Wang BW, Gao CC, Wang WS, Zhao HB, Zheng CG. Sulfur evolution in chemical looping combustion of coal with $MnFe_2O_4$ oxygen carrier. J Environ Sci 2014;26(5):1–9.
[23] Wang BW, Gao CC, Wang WS, Kong FH, Zheng CG. TGA-FTIR investigation of chemical looping combustion by coal with $CoFe_2O_4$ combined oxygen carrier. J Anal Appl Pyrolysis 2014;105:369–78.
[24] Wang BW, Xiao G, Song XY, Zhao HB, Zheng CG. Chemical looping combustion of high-sulfur coal with $NiFe_2O_4$-combined oxygen carrier. J Therm Anal Calorim 2014;118(3):1593–602.
[25] Wang BW, Cao YM, Li J, Wang WS, Zhao HB, Zheng CG. Migration and redistribution of sulfur species during chemical looping combustion of coal with $CuFe_2O_4$ combined oxygen carrier. Energy Fuel 2016;30(10):8499–510.
[26] Cai X, Wang XH, Guo X, Zheng CG. Mechanism study of reaction between CO and NiO(01) surface during chemical-looping combustion: role of oxygen. Chem Eng J 2014;244:464–72.
[27] Feng YC, Cai X, Guo X, Zheng CG. Influence mechanism of H_2S on the reactivity of Ni-based oxygen carriers for chemical looping combustion. Chem Eng J 2016;295:461–7.
[28] Keller M, Arjmand M, Leion H, Mattisson T. Interaction of mineral matter of coal with oxygen carriers in chemical-looping combustion. Chem Eng Res Des 2014;92(9):1753–70.
[29] Liu G-S, Strezov V, Lucas JA, Wibberley LJ. Thermal investigation of direct iron ore reduction with coal. Thermochim Acta 2004;410(1–2):133–40.
[30] Strezov V. Iron ore reduction using sawdust: experimental analysis and kinetic modeling. Renew Energy 2006;31(12):1892–905.
[31] Cho P, Mattisson T, Lyngfelt A. Defluidization conditions for fluidized-bed of iron, nickel, and manganese oxide-containing oxygen carriers for chemical looping combustion. Ind Eng Chem Res 2006;45(3):968–77.
[32] Abad A, Garcia-Labiano F, de Diego LF, Gayán P, Adánez J. Reduction kinetics of Cu-, Ni-, and Fe-based oxygen carriers using syngas (CO + H2) for chemical looping combustion. Energy Fuel 2007;21(4):1843–53.
[33] Wang BW, Lv HS, Zhao HB, Zheng CG. Experimental and simulated investigation of chemical looping combustion of coal with Fe_2O_3 based oxygen carrier. Procedia Eng 2011;16:390–5.
[34] Shen LH, Wu JH, Gao ZP, Xiao J. Reactivity deterioration of NiO/Al_2O_3 oxygen carrier for chemical looping combustion of coal in a 10 kW_{th} reactor. Combust Flame 2009;156(7):1377–85.

Index

Note: Page numbers followed by *f* indicate figures and *t* indicate tables.

D

E

F

P

R

S

T

U

V

W

FSC
www.fsc.org
MIX
Papier | Fördert
gute Waldnutzung
FSC® C083411

Kolibri 360 GmbH
Ferdinand-Jühlke-Straße 7
99095 Erfurt, Deutschland
produktsicherheit@kolibri360.de